Alphabetum Honorij Thebani, quo magitam f

c d e f g h i k l m n

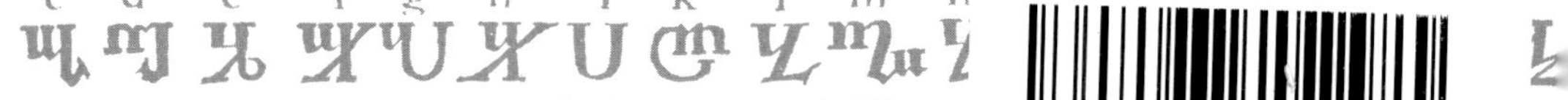

Alphabetum Hichi Fran

c d e f g h i k l m m

Alphabetum secretum Iaimelis Megalopij regis sapientissimi.

c d e f g h i k l m n o p q r s t

Alphabetum Nortmannorum, autore D. Beda Presbytero.

c d e f g h i k l m n o p q r s t

Alphabetum secretum Iohannis Tritenhemij Abbatis Spanheimen

c d e f g h i k l m n o p q r s sch t

Alphabetum Vtopiense, teste doctiss. viro Thoma Mauro D. Erasmi Rot: p̃cip

c d e f g h i k l m n o p q r s

Alphabetum Caroli Magni. Teste Alcunio, Gallo, Philosoph. lib. de v

c d e f g h i k l m n o p q r s t

Alphabetum exploratum Pharamundi regis Francorum per Gallia

c d e f g h i k l m n o p q r s t v

Alphabetum facillimum, secundum anonymum, Trittenhemio notan

c d e f g h i k l m n o p q r s t

Alphabetum cuiusdam summi Alchimistæ, autore Sphanheimens

c d e f g h i k l m n o p q r s t

Alphabetum Dorari q̃ Franci vsi sunt, Hunibaldo Historiographo a

g d c ; f h i k l m m fin. n n fin. o p ph q

betum Vastualdi, qui acta Regum, Ducum & principum per annos Septingento hisce notis primo sermone scripsit autore Hunibaldo.

g d e z e th i k l m n x o p r s t

Alphabetum D. Otfridi Monachi VVissenburgensis. Rabani Fuldensis di

c d e f g h i k l m n o p q r s t

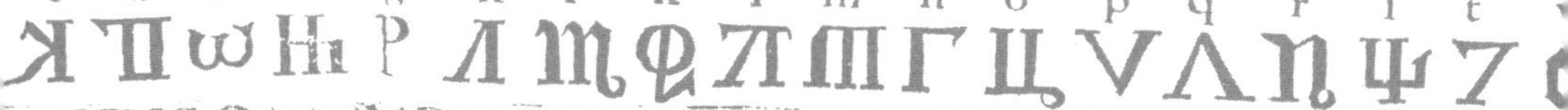

The Enlightenment of Matter

The Definition of Chemistry from Agricola to Lavoisier

To my parents
Anna Maria and Italo

The ENLIGHTENMENT *of* MATTER

The Definition of Chemistry from Agricola to Lavoisier

MARCO BERETTA

SCIENCE HISTORY PUBLICATIONS/USA
1993

ABSTRACT

Marco Beretta, *The Enlightenment of Matter. The Definition of Chemistry from Agricola to Lavoisier.* Doctoral Dissertation at Uppsala University. Institutionen för idé- och lärdomshistoria, Uppsala universitet. xvi + 396 pp. ISBN 0-88135-152-0.

This study examines the historical development of the attempts to reform the nomenclature of early chemistry. It is shown that the historical emergence of chemistry as a scientific discipline during the seventeenth and eighteenth centuries was closely tied to the move to provide a new foundation for its technical language.

Chapter One provides a short overview of the historiography of science and chemistry during the eighteenth century. Chapter Two investigates the relationship between the progress made by some eighteenth-century sciences and the lively discussions on the origin and nature of their language. Chapter Three examines the sources and the development of the lexicons of metallurgy, mineralogy, alchemy, and chemistry during the period 1500 to 1782. Chapter Four analyzes in detail the contents of the *Méthode de nomenclature chimique*, a work published in 1787 in which Lavoisier and his collaborators established the basis of the modern nomenclature of inorganic chemistry. Chapter Five offers an analysis of Lavoisier's masterpiece, the *Traité élémentaire de chimie* (published in 1789), and considers the reception of this work in Europe. A brief appendix reconstructs the development of alchemical and chemical symbolism and discusses their affinities and differences.

Marco Beretta, Uppsala University, Office for History of Science, Box 256, 751 05 Uppsala, Sweden.

This volume is number 15 in the series *Uppsala studies in history of science.*

First published in the United States of America
by Science History Publications/USA
a division of
Watson Publishing International
Post Office Box 493, Canton, MA 02021

Library of Congress Cataloging-in-Publication Data

Beretta, Marco.
The enlightenment of matter : the definition of chemistry from Agricola to Lavoisier / Marco Beretta.
p. cm. — (Uppsala studies in history of science ; 15)
Includes bibliographical references and index.
ISBN 0-88135-152-0
1. Chemistry—Nomenclature—History. 2. Chemistry—Notation—History. 3. Chemistry—Europe—History—18th century. I. Title. II. Series.
QD7.B52 1993
540'.14—dc20 93-8797
CIP

Designed and manufactured in the U.S.A.

Contents

Acknowledgments *vii*

Abbreviations *ix*

Introduction *xi*

Chapter One
The Science of the Past 1
Historiography as a Method of Reasoning 2
Philosophy, Science, and History 5
The History and Definition of Sciences 10
The History of Chemistry in the Eighteenth Century 13

Chapter Two
The Classification of Nature 27
Introductory Note 28
The Catalogue of Nature 30
The Grammar of the Arts 46
"Eadem Natura, Eadem Nomenclatura" 50
The Birth of Medical Philosophy 61

Chapter Three
The Birth of Chemical Language (1500–1782) 73
The Question of the Sources 74
"Die Wissenschaft des Seigerns begriffen" 78
The Characters of Mineralogy 93
"Obscurum per obscurius" 106

The Emergence of Chemical Questions 114
The Phlogistic Nomenclature 124
New Things, New Names 134
The Nomenclature Established 149

Chapter Four
"L'Époque de la Nomenclature Chimique" 159
From One Revolution to Another 160
The Overthrow of Phlogiston 180
The Philosophical Roots of Oxygen 187
The Principles in Action 206
The War between the Lines 214
Immediate Reactions, 1787–1788 221
Immediate Assents 236
The Swansong of Phlogiston 241

Chapter Five
The Chemical Revolution 245
The Permanent Revolution 246
Pedagogy or Science? 258
"Voilà la France, bien déphlogistiquée" 282
"The Philosophers of Laputa" 289
"Die französische Kunstsprache ist nicht für uns" 300
"Si parlerà tra noi lingua bisbetica" 309
The Nomenclature in Spain and Portugal 315
The New Nomenclature in Scandinavia 317

Epilogue 323

Appendix
The Role of Symbolism from Alchemy to Chemistry 329
The Myth of Continuity 330
Symbolism in Alchemy 332
Early Chemical Symbolism 347
The Mathematicization of Chemical Symbols 362

Select Bibliography 369

Index of Names 387

Acknowledgments

This study owes a great debt for the advice, encouragement, and support of three persons. Had I not had the benefit of their generous suggestions, this book would never have appeared.

I began the project in 1983 under the supervision of Professor Gianni Micheli at the University of Milan. Since then Professor Micheli has taken a constant interest in the progress of my research and made many encouraging comments for which I am most grateful.

Professor Ferdinando Abbri has carefully read the whole manuscript, corrected several errors, and offered useful criticism of my interpretative approach. Furthermore, during the past eight years, Professor Abbri has generously shared with me his deep knowledge of eighteenth-century chemistry and provided me with enriching and stimulating insights, both historical and interpretative.

Since my arrival in Sweden in 1987, Professor Tore Frängsmyr has supported my project with great enthusiasm and has generously put at my disposal the facilities of the Office for History of Science. His advice and encouragement have been of the greatest help, particularly in the latter phases of this work.

I am deeply indebted to Dr. Anders Lundgren, who has read the manuscript and made helpful suggestions and criticism.

I have received help and useful comments from Dr. Antonio Clericuzio, Professor Gunnar Eriksson, Dr. Lorella Fontana, Professor Antonello La Vergata, Dr. Daniela Majerna, Dr. Elio Nenci, and Dr. Sven Widmalm.

Special thanks are due to Ulla-Britt Jansson, who has given invaluable help in preparing the manuscript for the publisher.

The normal problems of access to manuscripts and books in various libraries have been considerably eased by the generous collaboration of Dr. Stefano Casati, Professor Virgilio Giormani, Catherine Kounelis, John Neu, Dr. Ezio Vaccari, and Urban Wråkberg.

During my stay in Paris in the academic year 1990–1991, Dr. Michelle Goupil provided me with a great deal of information on Lavoisier's manuscripts. I am particularly grateful to her for allowing me to read and copy her typewritten transcript of Lavoisier's letters of the period 1787 to 1789 and of the correspondence between Guyton de Morveau and Kirwan.

I am grateful to the staff of Uppsala University Library who have made every possible effort to satisfy my requests. This library has probably the best collection of eighteenth-century chemical literature in the world, and the easy access to rare books and manuscripts makes it a unique environment for scholars.

I thank the staff of the *Archives de l'Académie des Sciences* in Paris for facilitating my research in many ways.

I am grateful to Bernard Vowles and Gerald Lombardi for their revision of my English.

I thank the *Society for the History of Alchemy and Chemistry* for permission to reuse my essay "The Historiography of Chemistry in the Eighteenth Century," which appeared in *Ambix*, 39 (1992), and to Professor Renato Mazzolini for permission to publish in the appendix to this work my essay on the evolution of chemical symbolism, which will appear in a forthcoming volume entitled *Non-Verbal Communication in Science Prior to 1900* (edited by Professor Mazzolini).

I am grateful to the *Nobel Committee for Chemistry*, the *Dagmar and Sven Saléns Stiftelse*, and the *Kungl. Humanistiska Vetenskaps-Samfundet* in Uppsala for their financial assistance with publication costs.

The assistance and support I have received from my wife Ilva have been indispensable in improving the quality of both my work and my life.

Uppsala, November 1992

Abbreviations

Arch. Acad.: Archives de l'Académie des Sciences (Paris).

BI: Bibliothèque de l'Institut (Paris).

BIP: Bibliothèque Interuniversitaire de Pharmacie (Paris).

BMHN: Bibliothèque du Muséum d'Histoire Naturelle (Paris).

BN: Bibliothèque Nationale (Paris).

BJHS: British Journal for the History of Science.

BPUG: Bibliothèque Publique et Universitaire de Genève.

DSB: C.C. Gillispie (ed.), *Dictionary of Scientific Biography*, 16 vols. (New York, 1970–1980).

KB: Kungliga Biblioteket (Stockholm).

HARS: *Histoire de l'Académie Royale des Sciences* (Paris).

LC: A.L. Lavoisier, *Correspondance*, 4 vols. (Paris, 1955–1986).

LO: A.L. Lavoisier, *Oeuvres*, 6 vols. (Paris, 1862–1893).

MARS: *Mémoires de l'Académie Royale des Sciences* (Paris).

NTM: Schriftenreihe für Geschichte der Naturwissenschaften Technik und Medizin.

UUB: Uppsala University Library (Uppsala).

Introduction

> "Derjenige, der sich im stillen mit einem würdigen Gegenstande beschäftigt, in allem Ernst ihn zu umfassen bestrebt, macht sich keinen Begriff, dass gleichzeitige Menschen ganz anders zu denken gewohnt sind als er, und es ist sein Glück: denn er würde den Glauben an sich selbst verlieren, wenn er nicht an Teilnahme glauben dürfte. Tritt er aber mit seiner Meinung hervor, so bemerkt er bald, dass verschiedene Vorstellungsarten sich in der Welt bekämpfen und so gut den Gelehrten als Ungelehrten verwirren. Der Tag ist immer in Partein geteilt, die sich selbst so wenig kennen als ihre Antipoden. Jeder wirkt leidenschaftlich was er vermag, und gelangt so weit es gelingen will."
>
> J.W. Goethe, *Naturwissenschaftliche Schriften*, in *Werke*, vol. 8 (Berlin-Darmstadt-Vienna, 1967), 735.

This study aims to show by means of a comparative investigation of different chemical traditions the essential role played by Lavoisier's reform of nomenclature in the history of chemistry. It is in effect striking that despite the fierce criticism and attacks it received, Lavoisier's nomenclature triumphed in an extraordinarily short period of time.

Until the 1780s, in fact, the chemical lexicon remained rooted in the terminology of alchemy, and with the exception of the nomenclature of salts, the names of inorganic and organic substances were almost the same as those used by Paracelsus and his followers. In 1785 Lavoisier was still the only chemist in Europe to use a nomenclature built around the definition of oxygen, whereas less than 20 years later, only Priestley stood out against it. During this period chemistry went through such a lexical transformation that most of its new names were not traceable to any previous nomenclature. In many respects, such as the names of the gases, oxides, bases, etc., the new nomenclature reflected an entirely new hierarchy of matter based on Lavoisier's definitions of an element and of chemical combination. The study of the nomenclature of Lavoisier's chemical system, its sources, and the debates it raised throughout Europe seemed to offer an interesting route to an understanding of the change in the way chemistry was seen at the end of the eighteenth century. The emphasis on the role of language in scientific change is certainly not new, and there are a number of valuable studies that have examined this subject in relation to eighteenth-century chemistry. Hélène Metzger was probably the first fully to appreciate the potential of this approach, and "encouraged by Léon Brunschvicg, she wished to pursue [a] synthetic study of the development of scientific and philosophical thought by a thorough examination of the work of Condillac in relation to that of Lavoisier and the chemists of the end of the eighteenth century."[1] Mindful of the crucial importance of language in science, Metzger wrote articles for a projected *vocabulaire historique* in which she demonstrated with remarkable clarity the impact that change in the meaning of scientific terms had on the development of science.[2] Unfortunately she was not able to complete this project, and her ideas on Lavoisier's nomenclature remained scattered in different essays.[3] A historical dictionary of chemistry is still needed, but the implementation of a project, such as that envisaged by Metzger, would tremendously enrich our understanding of the history of chemistry.

---

[1]Suzanna Delorme, article "Metzger," in *DSB*.

[2]H. Metzger, "Communication pour servir au vocabulaire historique. ALCHIMIE," in *Revue de Synthèse, xvi* (1938), 43–53; *idem*, "Projet d'article pour un vocabulaire historique. ATOME," in *Revue d'histoire des sciences et de leurs applications*, 1 (1947), 51–62.

[3]H. Metzger, "Introduction à l'étude du rôle de Lavoisier dans l'histoire de la chimie," in *Archeion, XVI* (1932), 31–50; *idem*, *La philosophie de la matière chez Lavoisier* (Paris, 1935); *idem*, *La méthode philosophique en histoire des sciences* (Paris, 1987).

More recently, valuable studies by Crosland and Albury[4] have presented new insights and examined many of the historical sources of chemical language. However, a systematic work on the scientific and philosophical sources of the nomenclature of eighteenth-century chemistry remains to be produced.[5] In most cases the role of language in chemistry and more generally in science has been regarded as irrelevant, and unlike historians of literature and philosophy, historians of science have paid little attention to the lexicon of scientific thought. This relative lack of interest[6] is probably due to the fact that science is still regarded as an intellectual endeavor that deals primarily with experiments and ideas rather than with words. Galileo's dislike of the verbosity of the "world of paper" led him to seize on the syllogisms of the scholastic philosophy, and his praise for the language of experiment seems to have affected many historians of science of our time. Indeed, the use of a specific nomenclature during a specific period of time is usually regarded by historians as a subordinate accompaniment to the scientific endeavor. There is no doubt that this assumption is in many historical cases correct, and there are plenty of examples of prominent scientists whose use of the nomenclature of their science has been highly pragmatic, but in a historical perspective, there are a few exceptions that show a completely different attitude to the role of words and their significance. Chemistry provides one of the outstanding demonstrations of the scientific relevance and theoretical implications of linguistic change. The controversy that broke out during the seventeenth century between the followers of Paracelsus, who used a metaphorical and esoteric language, and the metallurgists, who relied on a descriptive nomenclature, tells us more about the development of early chemistry

---

[4]M.P. Crosland, *Historical Studies in the Language of Chemistry* (London, 1962), reprinted 1978; W.R. Albury, *The Logic of Condillac and the Structure of French Chemical and Biological Theory, 1780–1801* (Baltimore, 1972). An excellent and detailed study on the early lexicon of mineralogy and chemistry is Dietlinde Goltz, *Studien zur Geschichte der Mineralnamen in Pharmazie, Chemie und Medizin von den Anfängen bis Paracelsus* (Wiesbaden, 1972).

[5]Crosland himself remarked "a detailed study on the French *philosophy* and a major branch of science such as chemistry would no doubt prove a rewarding enterprise," in M.P. Crosland "The Development of Chemistry in the Eighteenth Century," in *Studies on Voltaire and the Eighteenth Century, XXIV* (1963), 370; see also Crosland's more recent remarks in his "Chemistry and the Chemical Revolution," in G.S. Rousseau and R. Porter (eds.), *The Ferment of Knowledge* (Cambridge, 1980), 389–395.

[6]In addition to Crosland's works, two relevant exceptions to this general indifference are the studies by L. Olschki, *Geschichte der neusprachlichen wissenschaftlichen Literatur*, 2 vols. (Heidelberg-Leipzig, 1919–1922), and M.L. Altieri Biagi, "Lingua della scienza fra Seicento e Settecento," in *Lettere Italiane*, XXVIII (1976), 410–461.

than an investigation of what was actually done in their laboratories. How in fact could the significance of the *fourneau cosmique* invented by the French alchemist Annibal Barlet be compared with that of the assaying furnace described by the German humanist Georg Agricola? What is more interesting is the difference in the meaning attached by these two learned men to the notion and practice of laboratory experience and in the linguistic and epistemological context of their beliefs. Similarly by placing their experimental and theoretical results in their original linguistic context, explaining the origin and use of certain terms, and investigating whether these terms were accepted or resisted, we gain a richer picture of the status of eighteenth-century chemistry and the complex paths of its development.

It has been rightly observed, for instance, that the mutual lack of understanding between Priestley and Lavoisier could be explained by their different philosophies of matter.[7] But the pneumatic experiments performed by the two scientists were not in fact so complicated as to prevent communication and understanding. It was the difference in the terminology that Priestley and Lavoisier employed that made it difficult to translate and reconcile their empirical observations. However, it was not just a matter of the use of the concept of phlogiston, on the one hand, and the introduction of the term oxygen, on the other, that kept Lavoisier and Priestley apart. Behind their distinct lexicons, there were different philosophies of language. Priestley's scepticism concerning any systematic and speculative reform of chemical language illustrates the strength of the British philosophy of language outlined by Bacon and Locke, who warned of the metaphysical dangers inherent in the introduction of neologisms and stressed the primacy of facts and experiments over words. By contrast Lavoisier relied confidently on the use of a few established facts to formulate a general theory and nomenclature, and his idea of subjecting chemical phenomena to the rigor of mathematical expression followed the French philosophical tradition inherited from Descartes through Condillac. The influence of Descartes on the French scientific thought of the eighteenth century is too complex to be summarized in a few lines, but the emphasis laid by Lavoisier, Condillac, Condorcet, d'Alembert, and many others on the importance of the analytical method and on the epistemological role of algebra in all forms of scientific reasoning establishes beyond doubt the filiation from Descartes. As has been pointed out by Aram Vartanian,

[7]See John G. McEvoy, "Continuity and Discontinuity in the Chemical Revolution," in A. Donovan (ed.), *The Chemical Revolution: Essays in Reinterpretation, Osiris*, second series, vol. 4 (1988), 195–213.

"the desirability of a universal physics, as proclaimed by Descartes and his followers, fostered in France a strong faith in the absolute coherence of nature, whose component elements were deemed so completely integrated that no specific phenomena were to be understood except in relation to the whole."[8] From our specific point of view, it is revealing that two balanced scientists, such as Joseph Black and Richard Kirwan, blamed French chemists for their "rage for system," "itch for theory," and Cartesian "*démangeaison*," whereas Lavoisier and Fourcroy often criticized the excessive respect of British scientists for fact gathering and experimentation and argued that a mere collection of experiments hindered the progress of science rather than promoted it.

To illuminate the sources of Lavoisier's chemical nomenclature and the reasons for the strong resistance it encountered, some reference to eighteenth-century philosophy has therefore been necessary. Lavoisier acknowledged his debt to Condillac on several occasions, and many of his scientific memoirs and works show signs of Condillac's influence.[9] Despite these explicit references, however, the extent of this influence has usually been underestimated, and the mention of Condillac has been regarded as a rhetorical device having little real bearing on Lavoisier's chemical theory. In this study, by contrast, I have tried to show that Condillac's philosophy enabled Lavoisier to found his chemical system. Like Hélène Metzger and Léon Brunschvicg, I believe that "it was neither with Bacon nor with Boyle that chem-

---

[8]Aram Vartanian, *Diderot and Descartes: A Study of Scientific Naturalism in the Enlightenment* (Princeton, 1953), 151.

[9]On the influence of Condillac on Lavoisier, see the following studies: Crosland (1978), Albury (1972); H.B. Acton, "The Philosophy of Language in Revolutionary France," in *Proceedings of the British Academy* (1959), 199–219; Francois Dagognet, *Tableaux et langages de la chimie* (Paris, 1969), 15–55; Jurgen Storost, "Zur Herausbildung der Grundsätze der modernen französischen Fachsprache der Chemie im ausgehenden 18. Jahrhundert unter Beachtung des philosophischen Einflusses von Condillac," in *Beiträge zur Romanischen Philologie, XI* (1972), 292–311; L. Rosenfeld, "Condillac's influence on French Scientific Thought," in P. Fritz and D. Williams (eds.), *The Triumph of Culture: 18th Century Perspectives* (Toronto, 1972), 157–168; J. Lambert, "Analyse et chimie condillacienne," in J. Sgard (ed.), *Condillac et le problème du langage* (Geneva, 1982), 369–378; W.C. Anderson, *Between the Library and the Laboratory* (Baltimore and London, 1984), 71–146; Nicolas Rousseau, *Connaissance et langage chez Condillac* (Geneva, 1986), 300–324; W.R. Albury, "The Order of Ideas: Condillac's Method of Analysis as a Political Instrument in the French Revolution," in J.A. Schuster and R.R. Yeo (eds.), *The Politics and Rhetoric of Scientific Method* (Dordrecht-Boston-Lancaster-Tokyo, 1986), 203–225; Trevor H. Levere, "Lavoisier: Language, Instruments, and the Chemical Revolution," in T.H. Levere and W. Shea (eds.), *Nature, Experiment, and the Sciences* (Dordrecht-Boston-London, 1990), 207–223.

istry emerged from alchemy; it was necessary for a mind shaped in the school of Condillac, and therefore practising Cartesian analysis, to precede the experiment, convinced of the necessity of an equation."[10] Algebra as a pattern for chemistry became Lavoisier's general methodological ideal. He was well aware that most chemical phenomena still eluded mathematical quantification, but at the same time, he realized the importance of subjecting chemical reasoning to a rigorous method. The analytical method, so effectively explained by Condillac, enabled Lavoisier to order chemical objects according to an entirely new hierarchy and also ultimately to create his new nomenclature. It is no coincidence that the title of Lavoisier's main work on chemical language was *Méthode de nomenclature chimique,* underlining the theoretical difference between a series of names and a method of naming. A method of nomenclature implied a theoretical framework within which it was possible to name systematically even those substances that had not yet been discovered. This method showed the chemist the path to follow in order to reach the truth. Furthermore, Condillac's analytical method offered Lavoisier a powerful and effective philosophy with which to emancipate chemistry from its primarily practical background and is one of the most revealing examples of the part played by philosophy in a major scientific revolution. In this respect the following words of Ernst Cassirer describe the general aim of this study admirably:

> In antiquity and in the Middle Ages, throughout the Renaissance, and in the great philosophical systems of the seventeenth and eighteenth centuries there is no doubt that philosophy had an independent and distinctive role to play in erecting the structure of scientific knowledge. It was not content with recording the contents of knowledge already acquired or with further working over it systematically. Philosophy then *directed* scientific knowledge toward new goals and opened new paths.[11]

---

[10]Léon Brunschvicg, *De la connaissance de soi* (Paris, 1931), 142.

[11]Ernst Cassirer, *The Problem of Knowledge: Philosophy, Science, and History since Hegel* (New Haven, 1950), 11.

CHAPTER ONE

The Science of the Past

"L'histoire du Monde, sans l'histoire des Savans c'est la statue de Poliphème à qui on a arraché l'oeil."

Denis Diderot, *Encyclopédie*, vol. 1 (Paris, 1751), XLVII

"L'histoire est, en effet, la seule méthode analytique [. . .]. Toutes les fois donc qu'on entreprend d'expliquer une chose humaine, prise à un moment déterminé du temps, [. . .] il faut commencer par remonter jusqu'à sa forme la plus primitive et la plus simple."

Émile Durkheim, *Les formes élémentaires de la vie réligieuse*, 2nd ed. (Paris, 1925), 4.

"La connaissance de soi [. . .] exige que l'*histoire* soit jointe à la *nature*."

Léon Brunschvicg, *De la connaissance de soi* (Paris, 1931), 21.

"L'historicité de l'homme se trouve liée à celle de la Nature [. . .]. Il n'y a pas de véritable histoire de la science sans une histoire parallèle de la *conscience*."

Robert Lenoble, *Histoire de l'idée de nature* (Paris, 1969), 122.

Historiography as a Method of Reasoning

In tracing the development of eighteenth-century science, it would be difficult to overstate the importance of the role played by the idea of history. Bacon's ideal of a systematic historical reconstruction of the useful ideas in science and technology found its most consistent exponents in the French *philosophes*. Condillac, d'Alembert, Turgot, and Condorcet were the leading lights in a change that led the writing of history from a pedantic chronicling of the past to the formulation of a theory of human evolution. Historiography became an effective instrument for expressing an ideological commitment to progress. By showing that all the absurdities, superstitions, and cruelties produced by history were related to specific historical and ideological circumstances, the *philosophes* secularized the chronology of human civilization, denying the role attributed to providence by the theologians, and putting a strong emphasis on the theoretical relevance of the history of science. The history of science provided them with an effective illustration of the fact that the dark ages of superstition could be enlightened by reason through the efforts of men of genius.[1] Furthermore, the historical perspective brought out the contrast between the progressive nature of science, on the one hand, and the immobility of metaphysics and religious belief, on the other. However, to be successful, this revitalized critical historiography needed a philosophical systematization. During the seventeenth century, in fact, historiography had been little more than a recital of events and biographies, and in most cases, the linkage between historical facts was regarded as the natural and necessary effect of divine intervention. This approach was obviously not acceptable to the *philosophes*, who set out to explain the laws of human progress without invoking metaphysical principles.

Etienne Bonnot de Condillac was among the first to perceive the ideological potential of history of science, and he successfully proposed an approach to history that set a standard for almost half a century. This approach incorporated his own empirical philosophy. In his first work, published in 1746,[2] Condillac tried to embrace in a single principle the faculties and activities of human understanding.

[1] "L'histoire des sciences est naturellement liée à celle du petit nombre de grand génies, dont les ouvrages ont contribué à répandre la lumière parmi les hommes," from d'Alembert, *Discours préliminaire de l'Encyclopédie*, E. Picavet (ed.), 2nd ed. (Paris, 1984), 75–76.
[2] Condillac, *Essai sur l'origine des connoissances humaines* (1746), in Condillac, *Oeuvres philosophiques*, vol. 1 (Paris, 1947), 1–118.

Influenced by Locke, Condillac based his epistemology upon human sensations. Sensations and needs were the only sources of a proper understanding of reality. This meant that every abstract notion ought to be reducible to its empirical elements and thus to a series of sensations. When an abstract notion, for example a metaphysical assumption, could not be traced back to any human sensation, it had to be regarded as a product of the imagination and therefore as wrong. The analytical method was the natural and true path to the acquisition of knowledge. Condillac illustrated the nature of this method with an example taken from Descartes' mechanistic philosophy:

> If I wish to know about a machine, I shall decompose it in order to study each part of it separately. When I have an exact idea of each part, and when I can replace them all in the same order as they previously were in, then I shall understand this machine perfectly because I have decomposed it and recomposed it.[3]

However, by comparison with Descartes' epistemology, that of Condillac contained an innovative element. In Descartes' philosophy knowledge came by an exact determination of innate ideas, the truth of which was guaranteed by God; to Condillac knowledge was acquired by a historical process in which God was replaced by nature. The introduction of the historical process into epistemology was justified by Condillac in the following words:

> If we observe the origin and generation of ideas we will see them arise successively, one from another; and if this succession conforms to the manner in which we acquire them we shall have a correct analysis of them. Thus the order of analysis here is precisely the order of the generation of ideas.[4]

The generation of ideas was rooted in the sensations of pain and pleasure. By trying to avoid pain and achieve pleasure, man[5] began to understand and distinguish which things were a threat and which might be useful. Further analysis of external reality made him increasingly

---

[3]Condillac, *La logique* (1780), translated by W.R. Albury with the title *La logique. Logic* (New York, 1980), 79–81. Lavoisier probably had Condillac's example in mind when he wrote: "Une machine n'est que la collection d'un certain nombre d'outils ou d'instruments, réunis pour produire un effet. Toute machine est donc susceptible d'être décomposée, d'être réduite à des éléments simples," in LO, vol. 6, 524.

[4]Condillac, *La logique* (1980), 93.

[5]This conjectural picture of prehistoric man appears frequently in Condillac's discussion of the genealogy of ideas.

capable of determining his needs and resourceful in satisfying them. Knowledge was thus seen as the result of a process in which the combination of primary needs and sensations led man from the simplest ideas on the nature of things to more complex and abstract notions.

These epistemological assumptions had a strong influence on the historiography of the *philosophes.* Condillac claimed that the origin of human understanding was reflected in the history of civilization. In his *Cours d'Etude,* published in 1775,[6] he devoted twelve of the total of sixteen volumes to the history of civilization. In this work, he sought to show that the evolution of humanity parallelled the genealogy of ideas. The primary needs were the driving force behind the first historical progress. The development from the simplest method of satisfying these needs (hunting) to more complex arts and sciences, such as agriculture, gradually raised mankind to successively higher levels of civilization and gave a deeper knowledge of the natural world. Condillac's historiography was therefore based on a study of the collective organization of human needs and sensations: how these needs developed in different climates, how they deviated from the analytical paths, etc. With the application of this philosophy, history became the study of the evolution of man's understanding of the world around man, in other words, of nature. Within this interpretative frame, dominated by the idea of progress, the historical role of the sciences was beneficial and indispensable, and Condillac's periodization of human civilization reflected their development. For example, Condillac saw the beginning of the "true philosophy" as coinciding with the works of Copernicus, Kepler, Brahe, and Galileo,[7] which eventually led to the age of science dominated by Newton.[8] Needless to say, the progress of modern science was associated by Condillac with the fact that scientists reasoned by an analytical method.[9] Furthermore, the history of science provided a wide range of empirical evidence to show that the progress made by human understanding was due to the use of such a method.

---

[6]Condillac, *Cours d'Etude pour l'instruction du Prince de Parme* (Parma, 1775), 16 vols. On Condillac's historical thought, see I. Knight, *The Geometric Spirit: The Abbé Condillac and the French Enlightenment* (New Haven and London, 1968), 223–226; L. Guerci, *Condillac storico* (Turin, 1978).

[7]Condillac, *Cours d'Etude—Histoire Moderne,* in Condillac, *Oeuvres philosophiques,* vol. 2 (Paris, 1948), 191 and ff.

[8]*Ibid.*, 215–220.

[9]*Ibid.*, 222.

Philosophy, Science, and History

Among the works in which Condillac's analogy between epistemology and the history of science was accepted, d'Alembert's *Discours préliminaire* to the *Encyclopédie* is undoubtedly one of the most influential. The first part of the *Discours* deals with the genealogy of ideas, and the main sources that d'Alembert constantly has in mind are Locke's *Essay Concerning Human Understanding*, Condillac's *Essai sur l'origine des connoissances humaines*, and the works of Francis Bacon.

In the *Discours* the empirical origin of knowledge is stated repeatedly. Sensations alone gave rise to the development of understanding, and needs, at first only very basic ones, drove man to look for instruments capable of satisfying them. From this starting point, mankind began gradually to develop primitive mechanical arts and technologies.[10] D'Alembert therefore agreed with Condillac that sensations and needs were the original sources of positive knowledge and that to improve that knowledge it was necessary to employ the analytical method. In the study of nature, in fact:

> The only resource that remains to us in an investigation so difficult, although so necessary and even pleasant, is to collect as many facts as we can, to arrange them in the most natural order, and to relate them to a certain number of principal facts of which the others are only consequences.[11]

The natural order of facts was reproduced in the order in which man acquired his ideas, i.e., from the most simple and empirical notions to more complex and abstract concepts.

If the first part of d'Alembert's *Discours préliminaire* was a philosophical history of man's thinking,[12] the second could be defined as a philosophical history of the natural sciences:

> Our metaphysical analysis of the origin and connection of the sciences has been of great utility in designing the encyclopedic tree; an historical analysis of the order in which our knowledge has devel-

---

[10]D'Alembert, *Discours* (1984), 23–25.

[11]*Ibid.*, 31; English translation by R.N. Schwab, in d'Alembert, *Preliminary Discourse to the Encyclopedia of Diderot* (New York, 1963), 23; hereafter d'Alembert (1963).

[12]The expression "histoire des nos pensées" was used by d'Alembert himself in the first *Éclaircissement sur les élémens de philosophie* (1759) (Paris, 1986), 199.

> oped in successive steps will be no less useful in enlightening us concerning the way we ought to convey this knowledge to our readers.[13]

The chronological starting point chosen by d'Alembert for his historical excursus was the Renaissance.[14] At this stage of history, it was still too early to turn to the study of the natural world; learned men began to cultivate the study of the classics, and "erudition was necessary to lead us to belles-lettres."[15] Significant progress was next achieved with the systematic adoption of the vernacular in place of Latin.[16] The consequent broadening of communication made a substantial contribution to a radical change in traditional views; thanks to Corneille, Racine, Molière, scholars could now turn their eyes towards reality.

With the study of nature, the development of new instruments, and the discovery of new continents, advances became increasingly rapid.

The natural philosophy of Bacon and Descartes and the scientific discoveries of Brahe, Kepler, Galileo, and Newton channelled the spirit of the Renaissance into a logical systematization that formed the foundations of an enlightened period.[17] Even the errors made by these learned men contributed to the emancipation of mankind from superstition. In this respect d'Alembert commented on Descartes' theory of vortices as follows:

> Let us recognize, therefore, that Descartes, who was forced to create a completely new physics, could not have created it better; that it was necessary, so to speak, to pass by the way of the vortices in order to arrive at the true system of the world; and that if he was mistaken concerning the laws of movement, he was the first, at least, to see that they must exist.[18]

Unlike previous historians, d'Alembert recognized that errors constituted necessary and fruitful steps in scientific reasoning and played an extremely important part in science's development. Indeed, the idea of necessity, as is often expressed in the *Discours*, had to be connected with that of historical development. The evolution of scientific knowledge as

---

[13]D'Alembert (1963), 60.

[14]On d'Alembert's preliminary discourse and historiography, see T. Hankins, *Jean d'Alembert: Science and the Enlightenment* (Oxford, 1970), 82–89; J. Shklar, "J. d'Alembert and the Rehabilitation of History," *Journal of the History of Ideas*, XLII (1981), 643–664.

[15]D'Alembert (1963), 64.

[16]*Ibid.*, 65.

[17]*Ibid.*, 87–89.

[18]*Ibid.*, 79.

a whole consisted of a succession of *necessary* stages in which errors gave the impulse for further progress.

The analogy between the genealogy of ideas and the history of science was drawn more explicitly by another prominent *philosophe*. Between 1748 and 1750, Turgot outlined a theory of historical progress in a series of lectures delivered at the Sorbonne.[19] The aim of these papers was to reveal the historical development of mankind down to its biological foundation, to make history a science. To this end Turgot took Condillac's philosophy of history to its logical conclusion. In his *Tableau philosophique des progrès successifs de l'esprit humain* (1750), Turgot claimed that the driving forces of historical evolution were also the driving forces of human understanding:

> The arbitrary signs of speech and writing, by providing men with the means of securing the possession of their ideas and communicating them to others, have made of all the individual stores of knowledge a common treasure-house, which one generation transmits to another, an inheritance which is always being enlarged by the discoveries of each age. *Thus the human race, considered over a period since its origin, appears to the eye of a philosopher as one vast whole, which itself, like each individual, has its infancy and its advancement.*[20] [my italics]

As sensations and needs were the basic sources of human knowledge, human history was the record of the organization of this knowledge in society. It is not surprising, then, that like Condillac and d'Alembert Turgot put a strong emphasis on the history of science. As the senses were the sources of knowledge, the sciences were the sources of historical progress. After all, "sciences, no doubt, [had] their origin in the senses:"[21]

> . . . natural resources and the fertile seeds of the sciences are to be found wherever there are men. The most exalted mental attainments are only and can only be a development or combination of

[19]Anne-Robert-Jacques Turgot, *Plan de deux discours sur l'histoire universelle* (1748), and *Tableau philosophique des progrès successifs de l'esprit humain* (1750); translated by Ronald Meek in his *Turgot on Progress, Sociology and Economics* (Cambridge, 1973); hereafter Turgot (1973). On Turgot's idea of history, see Frank E. Manuel, *The Prophets of Paris* (Cambridge, Mass., 1962), 13–51, and G. Micheli, "Ulteriori sviluppi dell'illuminismo francese: Holbach e Condorcet," in L. Geymonat (ed.), *Storia del pensiero filosofico e scientifico*, vol. 3 (Milan, 1979), 328–332.

[20]Turgot (1973), 41.

[21]*Ibid.*, 96.

> the original ideas based on sensation [. . .]. The same senses, the same organs, and the spectacle of the same universe, have everywhere given men the same ideas, just as the same needs and inclinations have everywhere taught them the same arts.[22]

If the sensations were always the same, however, it was difficult to see how any form of change and progress could occur in history. To this problem Turgot answered that the immense variety of permutations of sensations and external objects created the conditions for historical development:

> It is from this interlinking [between sensation and external objects], and from the order of the laws which all these ideas follow in their continual variations, that man acquires the consciousness of reality. Through the relation of all his different sensations he becomes aware of the existence of external objects. *A similar relation in the succession of his ideas reveals the past to him.*[23] [my italics]

The complex process of linking the right sensations to the corresponding objects improved as man accumulated enough experience to correct his own mistakes. As Frank Manuel has pointed out, in Turgot's view "constant inconstancy, eternal change and progress, were the true distinctions of mankind."[24]

In the narrative part of his reconstruction, Turgot underlined the crucial role played by sciences in general, and mathematics in particular, in the "successive advances of the human race." Natural sciences, being derived from sensations, were actually the most important means of improving human knowledge. In Turgot's view, the history of science showed that naturalists were the first to apply the analytical method to abstract notions and to eliminate the deceptive effects of imagination from human reasoning. Significantly, Bacon, Galileo, Descartes, Kepler, Newton, Locke, and Condillac were the heroes of Turgot's history of mankind:

> We owe to these great men the pattern and the laws of analysis, the lack of which had for such a long time retarded the progress of metaphysics, and even that of physics.[25]

---

[22] *Ibid.*, 42.
[23] *Ibid.*, 63.
[24] Manuel, *Prophets of Paris* (1962), 21.
[25] Turgot (1973), 96.

Undoubtedly, the most complete and thorough application of Condillac's philosophy of history came with the publication in 1795 of Condorcet's *Esquisse d'un tableau historique des progrès de l'esprit humain.* This seminal work begins with the claim that "man is born with the ability to receive sensations," to combine them, to compare them, to recognize them, and to remember them.[26] The sources of knowledge thus lay in the functioning and organization of intellectual faculties. A study of the functioning of these faculties in social groups led to an understanding of the history of mankind:

> If one confines oneself to the study and observation of the general facts and laws about the development of those faculties, considering only what is common to all human beings, this science is called metaphysics. But if one studies this development as it manifests itself in the inhabitants of a certain area at a certain period of time and then traces it on from generation to generation, one has the picture of the progress of the human mind. *This progress is subject to the same general laws that can be observed in the development of the faculties of the individual, and it is indeed no more than the sum of that development realized in a large number of individuals joined together in society.*[27] [my italics]

To Condorcet, too, most progress and achievements in the history of mankind derived from the development of arts and sciences, whose origins he also linked to the empirical sources of human understanding.[28]

The emphasis placed by the *philosophes* on the importance of the history of science was not a mere effort to enlarge the sources of universal history: the history of science showed that the investigation of nature gave man the means to satisfy his needs and to improve the conditions of life of an increasing number of individuals. For centuries theology and superstition kept man in a state of ignorance and fear; science, by contrast, emancipated man and provided him with the means to dominate nature. The democratic value of science was inherent in the fact that scientific reasoning had its roots in human sensations and therefore belonged to all human beings without exception.

---

[26]Condorcet, *Sketch for a Historical Picture of the Progress of the Human Mind* (1795) English translation (London, 1955).

[27]*Ibid.*, 3–4.

[28]On Condorcet's historiography, see Manuel, *Prophets of Paris* (1962), 62–81; Alexandre Koyré, *Études d'histoire de la pensée philosophique*, 2nd ed. (Paris, 1971), 103–111; Keith Michael Baker, *Condorcet: From Natural Philosophy to Social Mathematics* (Chicago, 1975), 343–371.

The History and Definition of Sciences

The success of the philosophy of history propounded by the *philosophes* is attested by the increasing number of histories of sciences published in the second half of the eighteenth century throughout Europe. It is true that the practice of reconstructing the past of science was common among naturalists long before the 1750s, and that there were in existence many works dealing with the history of medicine, botany, mathematics, astronomy, and physics,[29] but these works were descriptive bibliographies and chronicles of scientific events rather than real attempts to penetrate the past of the particular science concerned. In most of them, in fact, the authors treated the sources to which they referred as though they were contemporary and paid little attention to the different background and theoretical contexts to which the science of the past belonged.

The publication of the works of the *philosophes* in the second half of the century brought about a radical change in this situation. Scientists began to realize that the history of their own science was of much more than mere academic value,[30] and that an appropriate treatment of it could have important philosophical and political implications. A study of history enabled naturalists to demonstrate the progressive character of science by comparison with other forms of knowledge, such as theology, which seemed to be static; they could point to the benefit accruing to mankind from the application of scientific discoveries to technology; naturalists could therefore claim greater institutional and political consideration. At the same time, the historical perspective gave the naturalists of the eighteenth century greater confidence in their capacity and

---

[29]Among the most representative historical works of this period, the following must be mentioned: D. Leclerc, *Histoire de la Medicine, où l'on voit l'origine et les progrès de cet Art, de siècle en siècle, les sectes qui s'y sont formées, les noms des Médicins, leurs découvertes, leurs opinions, et les circonstances le plus remarquables de leur vie* (1695) (Amsterdam, 1723); J.C. Heilbronner, *Historia Mathesos universae a Mundo condito ad saeculum P.C.N.XVI., praecipuorum Mathematicorum vitas, dogmata, scripta et manuscripta complexa* (Leipzig, 1742); J.F. Weidler, *Historia Astronomiae* (Wittenberg, 1741); Carl Linnaeus, *Bibliotheca Botanica* (Amsterdam, 1736); *idem.*, *"Incrementa botanices,"* in *Amoenitates Academicae*, vol. 3 (Stockholm, 1756), 377–393; N. Regnault, *L'origine ancienne de la physique moderne*, 3 vols. (Amsterdam, 1735).

[30]The change in this perspective has been illustrated in detail in the valuable study by Dietrich von Engelhardt, *Historisches Bewusstsein in der Naturwissenshaft von der Aufklärung bis zum Postivismus* (Freiburg-Munich, 1979); see also Vincenzo Ferrone, "Clio e Prometeo: La storia della scienza tra illuministi e positivisti," in *Studi Storici* XXX (1989), 339–357.

their opportunities. The scientific achievements of the past were still regarded with great respect, but the theoretical subservience of the Renaissance to classical sources was itself a thing of the past. Whereas the discovery of a Greek scientific manuscript in the sixteenth and seventeenth centuries would be treated as a scientific discovery, such an event would have been regarded in the eighteenth century as no more than an archaeological curiosity. To the naturalists of the Enlightenment, in fact, the contemplation of the past of their science served as an illustration of the progress of scientific reasoning rather than as a useful source for their own investigations.

Inspired by the works published by the *philosophes*, an increasing number of scholars began to reconstruct the past of science for the ideological purpose of demonstrating the superiority of scientific reasoning over other forms of knowledge. The increasing popularity of the historical perspective is illustrated not only by the hundreds of publications devoted to the history of science but also by the institutional recognition that such works now received. In Sweden, for instance, the Royal Academy of Sciences instituted a series of regular publications and lectures entirely devoted to the history of sciences as early as 1746.[31] In Paris a chair of history of medicine was established in 1795.[32] By the end of the eighteenth century, the Göttingen theologian Johann Gottfried Eichorn was editing the vast and ambitious project of a *Geschichte der Künste und Wissenschaften*, which included, among other things, the history of medicine, chemistry, mathematics, technology, and physics, and which involved many prominent scientists of the University of Göttingen.[33] The popularity and the spread of scientific historiography in Germany was underlined by the physicist

[31]The first concrete evidence of this early institutional attention to the history of science was the publication of a booklet on the history of mathematics by Pehr Elvius, *Historien om mathematiska vetenskaper* (Stockholm, 1746). On this important chapter of Swedish scientific historiography, see S. Lindroth, *Kungl. Svenska Vetenskapsakademiens Historia 1739–1818* (Stockholm, 1967), vol. 1, 460–464.

[32]The professorship was instituted at Paris's *École de medicine* in favor of Jean Goulin, author of the *Mémoires littéraires, critiques, philologiques, biographiques et bibliographiques pour servir à l'histoire ancienne et moderne de la medicine*, 2 vols. (Paris, 1775–1776).

[33]*Geschichte der Künste und Wissenschaften seit der Wiederherstellung derselben bis an das Ende des achtzehnten Jahrhunderts* (Göttingen, 1796). The history of each discipline was compiled by a specialist in the field: Kästner (mathematics), Gmelin (chemistry), Poppe (technology), Buhle (philosophy), etc.

On the genesis of Eichorn's project, see Luigi Marino, *I maestri della Germania. Göttingen 1770–1820* (Turin, 1975), 184–187, and Giovanni Santinello (ed.), *Storia delle storie generali della filosofia*, vol. 3:2 (Padua, 1988), 696–697.

and philosopher Georg Christoph Lichtenberg, who criticized it as follows:

> I think that nowadays the history of sciences is practiced in too much detail, to the great disadvantage of science itself.[34]

The effectiveness of historical reconstructions was also emphasized by the success and popularity of the historical method proposed by the *philosophes*. One significant example of the new historiography was Bailly's *Histoire de l'astronomie ancienne*.[35] In the introduction to this important work, the French astronomer set out the main principles of scientific historiography endorsed by the French Enlightenment. The passage of time, the penetration of the secrets of nature, and the spread of human understanding were interwoven in a single thread that gave coherence and an ontological foundation to the study of history. The belief was illustrated by Bailly in a beautiful passage of his *Discours préliminaire, sur la manière d'écrire l'Histoire de l'Astronomie, & d'esposer les progrès de cette Science*:

> While the great men keep the sciences on the move, adding new discoveries to the number of truths, history spreads these truths, history passes down the knowledge, like the waters from the hills, which the slopes channel out over the plains. This blessing from the heights belongs to the country; the highest knowledge belongs equally to all men. We have attained it by degrees; the means of investigation have been drawn from nature.[36]

One of the main characteristics of the history of science was therefore that of showing the non-specialist the paths followed by the great men of science in discovering the truth. Bailly took this concept even further and declared:

> The main use of the history of a science is to meet the needs of people of learning, who wish to learn more; it must make the science known to those who have no idea of it. Men have shared the toil & the labour; each has his district, his duty & his glory: history is a means of

[34]C. Lichtenberg, *Vermischte Schriften*, vol. 1 (Göttingen, 1867), 296.

[35]J.S. Bailly, *Histoire de l'astronomie ancienne depuis son origine jusqu'à l'établissement de l'école d'Alexandrie* (Paris, 1775). Bailly also published an *Histoire de l'astronomie moderne depuis la fondation de l'école d'Alexandrie jusqu'à l'époque de MDCCXXX*, 3 vols. (Paris, 1779–1783), and the *Lettres sur l'origine des sciences et sur celle des peuples de l'Asie* (Paris, 1777). On Bailly's historiography, see I. Bernard Cohen, *Revolution in Science* (Cambridge, Mass., 1985), 221–223 and 511–512.

[36]Bailly (1775), V.

> communication between different classes; it is an account to the human race of the labours of a few individuals.[37]

From a reading of this passage, it is no exaggeration to say that Bailly regarded the history of science as a means of intellectual democratization. Science was by definition a progressive kind of knowledge; its history gave everybody access to it. The logical universality of scientific thought was guaranteed by its history, which illustrated the paths followed by reason in revealing nature. The analogy and relation between science and its history were formulated by Bailly as follows:

> Science is only the product, the succession of the operations of genius; & its history is the history of men & of their thoughts.[38]

The history of science thus dealt with the course of human thinking by which mankind proceeded from ignorance to knowledge. Indeed, while outlining the origin of astronomy, Bailly claimed that like all the other sciences, astronomy originated in human needs.

The approach of the *philosophes* was successfully applied by many other scientists of the period, who argued the importance of the history of various sciences along similar lines.[39]

The History of Chemistry in the Eighteenth Century[40]

The extraordinary increase in the number of chemical and mineralogical observations and discoveries that occurred in the late seventeenth and early eighteenth centuries forced chemists and naturalists to consider new theories of the ultimate chemical constitution of matter. Stahl's phlogiston theory, the diffusion of Newtonian ideas in the works of such chemists as Keill and Freind, and the progress made by the mechanistic view of chemical phenomena taught by Lemery and Boerhaave were steps along the road to modern chemistry.

---

[37] *Ibid.*, VII.

[38] "La science n'est que le produit, la succession des opérations du génie; & son histoire est l'histoire des hommes & de leurs pensées," *ibid.*, XVI; by the term "*génie*," Bailly meant the creativity of human understanding.

[39] In their histories of several sciences, Montucla, Saverien, Goguet, Macquer, Sprengel, Kästner, etc., all supported the analogy between the history of science and the functioning of human understanding, and they traced the origins of sciences back to man's needs and sensations.

[40] This subchapter is drawn from my essay, "The Historiography of Chemistry in the Eighteenth Century: a Preliminary Survey and Bibliography," *Ambix*, 39 (1992), 1–10.

As a result of the debates provoked by these problems, eighteenth-century chemists began to reflect on the methodological and theoretical structure of their science, and to formulate a theoretical definition of chemistry, capable of symbolizing its new experimental status.[41] From this initial need, there developed an interest in a historical reconstruction of chemistry that could define its main characteristics while outlining its advances. By the second half of the seventeenth century, chemical historiography had laid the philological foundations for the later authors; much of this effect was achieved by the lively dispute between the Danish chemist Oluf Borch and the German physician and erudite Hermann Conring. Borch asserted the antediluvian origins of chemistry,[42] whereas Conring was much more wary of establishing too rigid a periodization without previous verification of the reliability of the sources.[43] However, the works of both were biographical and bibliographical dictionaries rather than genuine histories of chemistry.[44]

---

[41]The first evidence of this search for a scientific identity was the introduction of chemistry into university curricula and scientific academies in the first half of the eighteenth century. See Christoph Meinel, "Artibus Academicis Inserenda: Chemistry's Place in the Eighteenth-Century and Early Nineteenth-Century Universities," in *History of Universities*, 7 (1988), 89–115.

[42]"Chemiae incunabula in antiquissima prospiciunt tempora. Natam ante diluvium ex Tubalcaini (qui aliis Nationibus Vulcanus est) historia sagaciores colligunt, eo inducti, quod ferri, aerisque metalla, illa Tubalcaini Magisteria inveniri, fingi, formarique, nequeant, ni ratio prius cognoscatur, minerarum naturas investigandi, coquendi, purgandi, segregandi, quae universa reperire non nisi Divini ingenii est, reperta autem prosequi, fabri cuiusvis proletari," in O. Borch, "De ortu et progressu chemiae dissertatio" (1668), in J.J. Manget, *Bibliotheca Chemica Curiosa* (Geneva, 1702), vol. 1, 1. In this work Borch defended his belief in the historical existence of Hermes Trismegistus. The Danish chemist published two works devoted to the history of ancient chemistry: the *Conspectus scriptorum illustrorum* (Copenhagen, 1697), which is a biographical dictionary of the main alchemists, and the *Hermetis Aegyptiorum et chemicorum sapientia ab Hermani Conringii animadversionibus vindicata* (Copenhagen, 1674), in which he answered Conring's scepticism concerning the existence of Hermes and of chemical knowledge among the Egyptians.

[43]H. Conring, *De Hermetica medicina, libri duo quorum primus agit de medicina, pariterque de omni sapientia veterum Aegyptiorum, alterum nontantum Paracelsi, sed etiam chemicorum, Paracelsi laudatorum aliorumque, potissimum quidem medicina omnis, simul vero et reliqua universa doctrina examinatur* (1648), 2nd edition (Helmstadt, 1669). It was in this second enlarged edition that Conring challenged Borch's assertion of the antediluvian origins of chemistry.

[44]Despite these intrinsic limitations, the work of Conring and Borch formed rich sources for all eighteenth-century studies. On the controversy between Conring and Borch and more generally on the definition of chemistry in the seventeenth century, see J. Weyer, *Chemiegeschichtsschreibung von Wiegleb (1790) bis Partington (1970)* (Hildesheim, 1974), 17–25; and R. Halleux, "La controverse sur les origines de la chimie de Paracelse à Borrichius," in *Acta conventus neo-latini Touronensis*, vol. 2 (Paris, 1980), 807–819.

The works that appeared at the beginning of the eighteenth century had a completely different impact from that of earlier ones; this was due both to the complexity and richness of the contents and to the variety of the approaches. Their number, too, is astonishing: more than 50 studies dealing with the history of chemistry were published, mostly in France, Germany, Sweden, Britain, and Italy.[45]

Eighteenth-century chemical historiographers, like the *philosophes*, were inspired by a rational and cumulative conception of the evolution of science, containing a more or less prominent theme of progress. These general characteristics, however, did not prevent a surprisingly rich plurality of approaches; the historical writings of Senac, Bergman, Macquer, Gmelin, and Fourcroy, for instance, reflected their different definitions of chemistry asserted in their own science.

Many other factors helped to increase and enrich this differentiation. In Germany, Sweden, Britain, and Italy, for example, the emphasis on the utility of chemistry in the narration of its history was essentially intended to win scientific respectability for a discipline that still suffered from a marked institutional discrimination. In France, where chemistry enjoyed great prestige and recognition thanks to the works of Lemery, Geoffroy, Homberg, Bourdelin, etc.,[46] the reconstruction of its history helped to renew the epistemological foundation. Another motive of eighteenth-century chemists in dealing with the history of their own discipline was to show the difference between the new experimental spirit and the "raving darkness" of the alchemists. For this purpose history became the most effective and successful tool. Here it must be mentioned that despite the new empiricism, most eighteenth-century chemists still used the ambiguous nomenclature of the hermetic and alchemical tradition.[47] In the century in which physics, astronomy, and even natural history had finally established their own technical nomenclatures, chemistry still had a lexicon rooted in alchemical symbolism. Chemists were both well aware of and dissatisfied with this situation, but to solve the problem by creating an alternative lexicon was a daunting task. Therefore, to show their independence from alchemy, eighteenth-century chemists found it more effective to reconstruct the history of the tradition to which they claimed to belong. To this end, early histories of chemistry emphasized the utility of pure experimentation and its appli-

[45]See the bibliography in my "Historiography of Chemistry" (1992).

[46]On the early development of French chemistry and its impact, see Frederic Lawrence Holmes, *Eighteenth-Century Chemistry as an Investigative Enterprise* (Berkeley, 1989), 34–82.

[47]On these difficulties, see M.P. Crosland, *Historical Studies in the Language of Chemistry* (1962), 2nd ed. (New York, 1978), 65–130, and Chapter 3.

cability to the arts, as opposed to the abstract and metaphysical speculations of alchemical thought. The arts of metalworking, mineralogy, and pharmacology practiced by the most ancient civilizations were cited as examples of chemical science in its modern and "positive" sense. A connection was thus established between the origin of chemistry and the experimental techniques of the ancients. To assert the complete independence of chemistry from any metaphysical evanescence, it became usual to draw a sharp but sometimes forced distinction between the development of alchemy and that of chemistry. This anti-alchemical stance was shared to differing degrees by almost all eighteenth-century historians of chemistry, although the unanimity of the chorus against alchemy did not deprive chemical historiography of a rich variety.

One of the first to deal with the historical evolution of chemistry was Senac. In his *Discours historique sur l'origine de la chymie*, published in 1723 as an introduction to his well-known course of chemistry,[48] Senac criticized Borch's erudite approach and claimed that the problem of the origins of chemistry was much simpler:

> At the beginning chemistry was the art of working metals for the satisfaction of human needs [. . .]. A useless science does not deserve to be studied; chemistry, on the contrary, combines utility with delight and we can see it in an infinite number of arts, that would have never reached their present perfection without its help.[49]

For these practical and utilitarian reasons, the goals of chemistry were very distinct from those of alchemy, where all the laboratory experiments were subordinated to the search for the philosopher's stone, or *elixir.*

Despite this preliminary distinction, Senac did not hesitate to include alchemists such as R. Bacon, Albertus Magnus, and A. DaVillanova, among the founders of chemistry.[50] Moreover, being a physician himself, Senac saw the historical role of medicine as playing an essential part in the reform of the Aristotelian doctrine of the four elements. Paracelsus, Tachenius, Van Helmont, Stahl, Geoffroy, Homberg, and Boerhaave were the major scientists who understood the

---

[48]J.B. Senac, *Nouveau cours de chymie suivant les principes de Newton et de Sthall* (1723) (Paris, 1737), anonymously published. The authorship of this work has long been questioned. However, in a letter to T. Bergman, Macquer wrote, "L'auteur du nouveau cours de chymie . . . est Mr. Senac premier medicin du Roy," in *T. Bergman's Foreign Correspondence* (Stockholm, 1965), 230. Also, Fourcroy ascribed the *Cours* to Senac (*Encyclopédie Méthodique: Chimie, pharmacie et métallurgie*, vol. 3 (Paris, 1796), 741.

[49]J.B. Senac, *Nouveau Cours* (1737), XVII and LXX.

[50]*Ibid.*, LXXXV.

importance of applying chemical knowledge to medicine and pharmacology.

With the work of Venel, we come to a more interesting historiographical attempt. Venel, the author of the article *Chymie* in the *Encyclopédie*,[51] intended to provide chemistry with a more accurate definition. The polemical targets of his article were the mechanistic interpretation of matter and the inclination of some chemists to try to reduce chemistry to physical and quantitative principles. Venel deplored these attempts to establish a theory of the internal constitution of matter without any empirical basis. What was written "on this subject by Boyle, Newton, Keill, Freind, Boerhaave [was] clearly characterized by a lack of experience."[52] A correct view of chemistry could not be obtained from the abstract formulations of these authors; instead, it was necessary to turn to the ancients, as the only ones whose works were "rich in experience and authentic chemical knowledge."[53] Matter was too complex for its mysteries to be penetrated by vague hypotheses or mathematical formulas. Chemistry was a qualitative science that dealt with the separations and reunions of the constituent principles (*principes*) of bodies.[54] Venel backed up his definition of chemistry with an interesting reconstruction of its history. Although drawing a sharp distinction between chemistry and physics, Venel did not assert a continuity between alchemy and modern science. In this connection Venel passed sarcastic judgments on *antiquaires chimistes*, such as Borch, with their belief in chemistry's antediluvian origins.[55] Venel argued that early chemistry originated in attempts to satisfy developing human needs. For people living in northern Europe, for instance, it was natural for the opportunities offered by mining to arouse their interest and for them to develop a rudimentary knowledge of the nature of metals and minerals.[56] Likewise, the search for new pharmacological remedies extracted from plants and minerals led the ancient physicians to further investigation of the chemical manipulation of substances. Venel devoted much of his historical reconstruction to the

---

[51]Gabriel François Venel, "*Chymie,*" in *Encyclopédie ou Dictionnaire des sciences des arts et métiers*, vol. 3 (Paris, 1753), 420–437. I have used here the reprinted version included in the *Encyclopédie Méthodique*, vol. 3, 262–303.

[52]*Ibid.*, 263.

[53]*Ibid.*

[54]"La chimie est une science qui s'occupe des séparations et des unions des principes constituants des corps, soit opérées par la nature, soit opérées par l'art dans la vue de découvrir les qualités de ces corps, ou de les rendre propres à diverses usages," *ibid.*, 274.

[55]*Ibid.*, 280.

[56]*Ibid.*, 285.

development of early modern chemistry, outlining the positive contribution made to chemical investigation by physicians, such as Paracelsus, Van Helmont, Becher, and Stahl. In keeping with his general definition of chemistry as a science completely distinct from physics, Venel assigned the contributions of Boyle and the Newtonians to the physical rather than the chemical domain. Not surprisingly, the heroes of Venel's history were those chemists, namely, Stahl and Rouelle, who supported a qualitative vision of matter.

As Fourcroy pointed out,[57] Venel's article made an impact on eighteenth-century chemists primarily because it showed chemistry to be an autonomous discipline whose history revealed that many arts benefitted from its achievements and that its doctrines had to be clearly distinguished from those of the alchemists.

Venel's ideas were particularly well received in France. His qualitative philosophy of matter and his aim of penetrating its structure through experiment rather than through reasoning were taken as a model by Diderot, who, in the *Prospectus* to the *Encyclopédie*, ranked chemistry among the most developed branches of knowledge:

> From the experimental knowledge or from the history, perceived by the senses, of *exterior qualities* which are *sensible*, *apparent* etc., of *natural bodies*, reflection has led us to the artificial investigation of their interior and hidden properties; and this art is called *chemistry*. *Chemistry* is the imitator and rival of nature; its object is almost as extensive as that of nature itself. . . .[58]

This claim was repeated by Diderot in his *Introduction contenant l'histoire des progrès de cet art*, which was intended as an introduction to his notes on the chemistry lectures delivered by Rouelle in 1756.[59] Like Venel, Diderot saw matter in qualitative terms and reconstructed the history of chemistry accordingly. Stahl's chemical philosophy was in fact compared in importance by Diderot to Newton's celestial mechanics.[60]

---

[57]"L'article de Venel, a eu une grande réputation parmi les chimistes, parce qu'il tendoit manifestement à disculper la *chimie* des reproches qu'on lui avoit faits, & à faire prendre à cette science le rang qui lui étoit dû dans l'ordre des connoissances humaines," *ibid.*, 303.

[58]D. Diderot, "Prospectus" to the *Encyclopédie*, vol. 1 (1751), L; English translation in d'Alembert (1963), 155.

[59]Published in D. Diderot, *Oeuvres complètes*, vol. 9 (Paris, 1981), 179–208.

[60]"Ce n'est pas assez de réveler; il faut encore que la révélation soit entière et claire. Il est une sorte d'obscurité que l'on pourrait définir, *l'affectation des grands maîtres*. C'est un voile qu'ils se plaisent à tirer entre le peuple et la nature. Sans le respect qu'on doit aux noms célèbres, je dirais que telle est l'obscurité qui règne dans quelques ouvrages de Stahl et dans les Principes mathématiques de Neuton," in Diderot, *Pensées sur l'interpretation de la Nature*, *ibid.*, 68–69.

Venel's and Diderot's historical reconstructions were challenged in 1766 by Macquer, one of the most influential chemists in Europe. In his short *Discours préliminaire sur l'origine et les progrès de la chymie*, which serves as an introduction to his celebrated *Dictionnaire de Chymie*,[61] Macquer was careful to provide a general concept of the history of the science:

> As the History of any Science ought to relate the labours, the discoveries, and the errors of the Cultivators of that Science; and to shew the obstacles which they have been obliged to surmount, and the mistaken paths into which they have sometimes been misled; it cannot therefore fail of being very useful to persons engaged in the same pursuits [. . .].
>
> We shall here treat no further concerning the particular History of Chemists, than as it tends to illustrate the general History of Chemistry. Our intention is, to shew the several states in which this Science has appeared; the revolutions it has undergone, and the circumstances which have favoured or retarded its progress.[62]

The emphasis on the doctrines rather than the lives of chemists was typical of the priorities observed by the *philosophes* in their historical writings. Indeed, Macquer's history of chemistry owed a considerable debt to the approach of the *philosophes*. In this respect it is interesting to consider the argument used by Macquer in denying the antediluvian origin of chemistry:

> What we call Science is the study and knowledge of the relations which subsist betwixt certain facts, which necessarily must be supposed to have been previously discovered. But the discovery of these facts must have been effected merely by the desire of gratifying the senses. For the most active and penetrating mind could have done little, in comparison with the internal perception of a craving want, which must be obeyed [. . .].
>
> The first men therefore being necessitous, must for that reason have been the first artists. They seized on the beginnings of the arts by a natural effort, very different from that improved reasoning, which alone can give birth to Science, and which requires the space of several succeeding ages for its formation and perfection.[63]

In this important passage, Macquer denied the ancient origin of chemistry on the basis not of a prejudice against alchemy, but of an

---

[61]P.J. Macquer, *Dictionnaire de chymie* (Paris, 1766), vol. 1, V–XXVI.
[62]*Ibid.*; English translation, *A Dictionary of Chemistry* (London, 1771), I; hereafter Macquer (1771).
[63]*Ibid.*, II.

original philosophy of history. The definition of modern science as being the result of the reduction of remote techniques of manipulation of chemical processes to the patterns of present-day science was not legitimate. The awareness of the historical perspective was underlined by Macquer as follows:

> If *Stahl* had lived before the deluge, all the efforts of that genius which unfolded the mysteries of nature by means of the most sublime chemistry, would probably have been reduced to invent a method of forging an ax; and also the great *Newton*, who could measure the universe, and calculate the infinities, might perhaps have exhausted the whole force of his mind in reckoning so far as the number ten, if he had been born among those American nations whose best arithmeticians cannot exceed three.[64]

Thus according to Macquer, chemistry was an achievement of later times when historical and intellectual conditions were conducive to its progress. Early chemistry was therefore to be regarded as the product of the combined efforts of those metallurgists, physicians, pharmacists, and naturalists who during the sixteenth and seventeenth centuries provided it with its experimental sources. This new periodization of the history of chemistry was consistent with the general periodization of the history of science by the *philosophes* and with the belief that science was a modern achievement that originated during the Renaissance.

Although his survey was brief, Macquer's approach to the history of chemistry also met with wide approval, especially in France. Whereas in Germany, Italy, and Sweden, the historiography of chemistry primarily served practical and utilitarian purposes,[65] in France the reconstruction of the past of their science gave chemists a clearer picture of its present theoretical state and offered pointers to the future. For this reason historical research by French chemists had as its focus the development of chemistry from the seventeenth century onwards, rather than such matters as the origin of its name and its prehistory. The following two examples will better clarify this characteristic.

In the early 1780s, Louis Bernard Guyton de Morveau was commissioned to write the first volumes on chemistry of the monumental *Encyclopédie Méthodique*. In the second *Avertissement* to the first volume,[66]

---

[64]*Ibid.*, III.

[65]For a brief survey of other national approaches to the history of chemistry, see Weyer, *Chemiegeschichtsschreibung* (1974), and my "History of Chemistry" (1992).

[66]L.B. Guyton de Morveau, "Second avertissement, pour servir d'introduction aux articles qui suivent, et pour indiquer quelques correction à faire à ceux précèdent," in *Encyclopédie Méthodique* (1789), vol. 1, 625–635.

Guyton presented a short history of pneumatic chemistry from the discovery of the active role of gases up to 1789. In a few concise pages, he dealt with the theme of error and truth in science. The choice of theme was inspired by the attacks of certain French and foreign chemists on Lavoisier's oxygen theory. Lavoisier was castigated for lack of originality, since his pneumatic hypothesis concerning the calcination of metals had already been proposed by the French physician Jean Rey, who had published a work in 1630 in which he observed that air caused the gain in weight in calcined lead and tin. To defend Lavoisier from the accusation of plagiarism, Guyton put forward the subtle argument that in the history of science, as in the history of human understanding, there exist both sterile truths and fertile errors. It was not the first time in history that a scientific discovery had been neglected when it was announced in a period not ready for it. On the other hand, many incorrect scientific interpretations and theories had been the subject of debate for decades before ultimately leading to an original and correct solution of the problem concerned. This was illustrated by the historical example of Rey's discovery:

> He enjoyed the best reputation he could obtain from a century unprepared for his discovery; being himself unaware of it, he seemed *to guess the cause of a specific fact rather than discover a principle whose consequences must form a theory and enlighten a part of the operations of nature.*[67] [my italics]

These requirements were fully satisfied by the interpretation of Lavoisier, who not only observed the active role of air in the calcination of metals, but also placed it in a wider theoretical frame that eventually led him to construct a theory of pneumatic phenomena. As early as 1778, Guyton had remarked:

> It is with justice that we say *the theory of M. Lavoisier*; we do not conceal that Jean Rey had the idea of the absorption of air by the calcined tin, but one idea is not a theory and the inventor is he who makes us to possess it. The work by M. Lavoisier made a demonstrated truth of an opinion.[68]

Thus the historical perspective helped Guyton to define the differences between the chemical investigations of the seventeenth century

---

[67] *Ibid.*, 627.

[68] Guyton de Morveau, *et al.*, *Élémens de chimie théorique et pratique*, vol. 3 (Dijon, 1778), 177.

and those performed by his contemporaries. This reasoning, which may sound self-evident to us, with our more sophisticated sense of history, was in the eighteenth century extremely innovative and helped to give scientists greater confidence in their intellectual abilities and their social function.

In his own contribution to the *Encyclopédie Méthodique*, Antoine François Fourcroy developed the interpretations of Macquer and Guyton into a full history of chemistry. Fourcroy's article "Chimie," written in 1796 for the third volume,[69] represents the final flowering of eighteenth-century chemical historiography.

Fourcroy divided his work into eight parts in order to present a "comparative picture of the great discoveries of different eras and their incorporation in a single body of doctrines by the French chemists."[70]

"The history of chemistry," he wrote, "may be divided into three periods: the mythological or legendary one, the obscure or inexact one, and the certain one."[71] By certain (*certe*), Fourcroy meant the period of unequivocal certainty regarding sources, which began only with the establishment of the first scientific academies in the second half of the seventeenth century. It is not, therefore, surprising that only five pages of the entire article were devoted to ancient and medieval chemistry and alchemy,[72] whereas more than 400 pages of the remaining text were concerned with the modern period. By his indifference to remote antiquity, Fourcroy intended to show that the great advances achieved by eighteenth-century chemistry owed little to early chemical arts. Only in the sixteenth and seventeenth centuries, coinciding with the first discoveries of and observations on the active role of gases by Van Helmont, Mayow, Boyle, and Rey, and the further investigations by Becher, Boerhaave, Hales, and Stahl, did chemists begin to realize that new horizons were opening. At first these experiences, whose innovative significance people often failed to grasp,[73] formed only random pieces of an unfinished mosaic.[74] Between 1760 and 1772, and in particular with the contributions of Black, Caven-

[69]A.F. Fourcroy, *Encyclopédie Méthodique*, vol. 3 (Paris, 1796), 303–781.

[70]*Ibid.*, 303.

[71]*Ibid.*, 313.

[72]*Ibid.*, 313–319. In compiling this section, Fourcroy confined himself to an abstract of Bergman's works on the history of medieval chemistry, *Dissertatio de primordii chemiae* and the *Dissertatio inauguralis sistens chemiae progressus a medio saec. VII ad medium saec. XVII*, published in Uppsala in 1779 and 1782.

[73]On the historical insignificance of the early discovery made by Rey, Fourcroy's opinion was not different from that of Guyton.

[74]Fourcroy, *Encyclopédie* (1796), 342.

dish, and Priestley, the range of pneumatic chemistry broadened enormously, and gases became the characteristic object of the science. But "the peculiar versatility of opinion" of Priestley[75] and, more generally, the distaste of British chemists for general and abstract theories prevented them from ordering their fundamental experiences in a coherent theory of pneumatic phenomena.

Allowing for Fourcroy's "scientific nationalism," his criticism of the British chemists was to some extent justified. The history of chemistry did not lack empirical observations and brilliant experiments, of which the British were probably the best producers; what was missing was a clear understanding of the meaning of these experiments, a theory forming a coherent connection between the various elements. In brief, what was needed was a definition of chemistry in its general and basic aspects.[76] This gap was not filled until Lavoisier appeared on the scene. In this respect Fourcroy observed that though

> . . . not all discoveries are due to his works [Lavoisier was for chemistry what] Kepler, Newton, Locke and Euler had been for mathematics and geometry. Lavoisier established a new pathway [. . .] he successfully changed both the experimental art and the way of reasoning in chemistry [. . .]. In short he has been one of those philosophers who change the state of knowledge, giving it a completely new movement and direction.[77]

Fourcroy, by emphasizing Lavoisier's role in the theoretical evolution of chemistry rather than his experimental contributions, pointed out Lavoisier's new epistemological significance and his achievement in synthesizing a huge collection of scattered facts in a single theory. With Lavoisier chemistry became "a completely new science, very different from what it used to be before this memorable revolution."[78]

According to Fourcroy, the main roots of the chemical revolution were threefold: 1.) the systematic experimentation that led to the discovery of gases; 2.) the technological progress that led to the construction of sophisticated and accurate instruments, such as the eudiometer, the calorimeter, the gasometer, the balances, etc., capable of isolating and quantifying the evanescent pneumatic substances;[79] 3.) the theo-

---

[75] *Ibid.*, 380.

[76] Fourcroy had named this foundation of all chemical operations and experiences "general or philosophical chemistry," *ibid.*, 306–312.

[77] *Ibid.*, 415–416.

[78] *Ibid.*, 715.

[79] The third part of the article (pp. 319–414) was devoted to the history of chemical instruments.

retical interpretation of pneumatic phenomena that led Lavoisier and his associates to revolutionize the view of chemistry with regard to its experimental structure and theoretical principles.

These three causes were examined by Fourcroy in detail, and many of his interpretative insights and sources still provide a valuable and original approach to the history of chemistry today.

Another important contribution by Fourcroy to this field was in establishing the position of the history of chemistry in relation to science. Interestingly Fourcroy regarded the history of chemistry as an essential part of theoretical chemistry[80] and intended to distinguish it more sharply from the old scholastic historiography:

> The HISTORY of chemistry, constituting in my plan, the second part of those extensive departments of information which comprehend the whole of the science, is not what is usually understood by this name in the greater number of chemical books. The first pages of these treatises, the introductions, the preliminary discourses ordinarily present an imperfect sketch of the principal epochs of the science; a sketch, decorated to no purpose with the title of the History of chemistry, with regard to the manner in which I apply these terms in my work. This part, which I regard as a complete and very essential member of the science, taken in its utmost extent, requires details and developments of which are only very incomplete sketches in the works we possess. It ought not only to consider and fix all the epochs of chemistry, to distinguish them carefully from each other, but even to examine all the discoveries made at different periods, to shew their connection with each other, their succession, the progression they have followed, the manner in which each truth has been discovered. . . .[81]

The emphasis on the route to discovery of the truth is not accidental. The analogy between the early progress of chemistry and chemical reasoning itself played an important role in Fourcroy's chemical philosophy. The teaching of chemistry had to show beginners that science, as its history exemplified, followed a chain of ideas that moved from the simple up to the increasingly abstract and complicated. The change in the disposition of chemical matter that occurred in most French chemistry textbooks in the second half of the eighteenth century shows the importance of this belief. The treatment of the mineral

---

[80]"The first is the Theoretic part, or the THEORY of the science. The second comprehends its HISTORY," from A.F. Fourcroy, *A General System of Chemical Knowledge*, vol. 1 (London, 1804), 34.

[81]*Ibid.*, 35–36.

kingdom, the nature of acids and salts, which were indeed the most simple and known parts of chemistry, were moved to the beginning in these textbooks whereas the investigation of vegetable chemistry was placed at the end. This reversal of the traditional order was often justified with explicit references to the analytical method and to its usefulness in scientific reasoning. The history of science implemented this methodological change by revealing to students that the attainment of scientific truths throughout history followed the analysis of ideas rather than the dreams of the imagination.

CHAPTER TWO

The Classification of Nature

"Dich verwirret, Geliebte, die tausendfältige Mischung
Diese Blumen über dem Garten umher;
Viele Namen hörest du an und immer verdränget,
Mit barbarischen Klang, einer den andern im Ohr.
Alle Gestalten sind ähnlich, und keine gleicht der andern;
Und so deutet das Chor auf ein geheimes Gesetz,
Auf ein heiliges Rätsel, O, könnt' ich dir, liebliche Freundin,
Überliefern sogleich glücklich das lösende Wort!"

Johann Wolfgang Goethe, *Die Schriften zur Naturwissenschaft*, vol. 9 (erste Abteilung) (Weimar, 1954), 67.

"La nature ne fait rien d'incorrect. Toute forme, belle ou laide, a sa cause; et, de tous les êtres qui existent, il n'y en a pas un qui ne soit comme il doit être."

Denis Diderot, *Essai sur la peinture*, in Id., *Oeuvres complètes* (Paris, 1875), vol. 10, 465.

"Il n'y a point [. . .] d'erreurs dans la nature; tous les vices qu'on impute au naturel sont l'effet des mauvaises formes qu'il a recues."

J.J. Rousseau, *Julie ou la Nouvelle Héloise* (1758), 5me partie Lettre 3 à Milord Edouard (Paris, 1967), 426.

> "L'homme est l'ouvrage de la nature, il existe dans la nature, il est soumis à ses loix, il ne peut s'en affranchir, il ne peut même par la pensée en sortir."
>
> D'Holbach, *Système de la nature* (1770) (Paris, 1990), vol. 1, 37.

Introductory Note

As was shown in the first chapter, the redefinition of science that took place in the eighteenth century was considerably influenced by a new interest in approaching science through its history. However, the process also necessitated a systematic reform of technical language and nomenclature. The success of experimental methods of investigation in the eighteenth century, the emphasis on empiricism, and the skepticism towards metaphysical explanation combined to invest the actual language of scientific literature with a new importance. The weight now attached to rigorous description of phenomena put a premium on terminological accuracy, particularly in those sciences that lacked an institutional tradition. A new definition of an emerging science could only be provided after a methodical reform of the naming and classification of its objects, an unequivocal correspondence between name and object being considered a prerequisite for successful experimental investigation. This need was particularly urgent in those sciences, such as chemistry, whose traditional nomenclature was ambiguous or inadequate.

The obscurities of alchemical symbolism had already been criticized by Vannoccio Biringuccio and Georgius Agricola during the Renaissance, but few attempts had been made to create a durable alternative. Despite such dissatisfaction, the language of chemistry remained encumbered with alchemical terminology until the second half of the eighteenth century.

However, in order to place in its context the theoretical significance and conceptual structure of the philosophy[1] that led to the reform of chemical nomenclature, it is necessary to give a brief account of the problem of language and classification in the arts and natural sciences of the eighteenth century. Despite the heterogeneity of the

[1]When discussing the scientists, I use the term *philosophy* to denote the method and the theoretical principles to which they referred.

fields concerned, we can find several relevant characteristics that were shared by those sciences that neither had an academic tradition nor relied on mathematical patterns. The scientific principles and methods of eighteenth century botany, zoology, mineralogy, chemistry, and medicine[2] were based on empirical observation, analytical description of phenomena, systematic experimentation, and qualitative interpretation. Naturally, this general empiricism varied from author to author and from science to science, but the increased number of new discoveries and "objects" was a characteristic shared by all Baconian sciences. The increase in the number of known plants, chemical substances, animal species, and diseases was only clearly recognized as a distinctive characteristic of science during the eighteenth century. This proliferation of objects, together with the empirical turn of mind of the scientists of the Enlightenment, stimulated the search for new and more satisfactory schemes of classification.

In making this effort, they wanted both to construct new technical languages and to assert the autonomy and epistemological identity of their sciences, seeing the one as inseparably linked to the other. The scientists of the Enlightenment were aware that the progress achieved by astronomy, optics, mechanics, and mathematics was due both to their solid and continuous tradition and to their precise and rigorous language. By means of quantification, these sciences were able to classify many phenomena that did not at first appear to be connected.[3] After Galileo, Kepler, Descartes, and Newton, mathematics became the *medium* bringing together the mind and the physical world by closer and closer approximations. Moreover, the Galilean mathematicization of nature and its success as an explanatory tool revealed the inadequacy of the qualitative terminology used in scholastic natural philosophy and the scientific importance of choosing appropriate means of expression.

In their admiration for the Copernican revolution, the *philosophes* wished to extend the scientific method proposed by Descartes and Galileo to all natural sciences, and there is no doubt that the reform of the nomenclatures of the Baconian sciences was partly inspired by the enormous success of the mathematicization of celestial mechanics.

It is natural to wonder why the question of a reform of scientific language had not arisen before the eighteenth century. A number of

---

[2]A discussion of the distinction between exact and Baconian sciences can be found in the essay by Thomas S. Kuhn, "Mathematical versus Experimental Traditions in the Development of Physical Science," in *The Essential Tension* (Chicago, 1977), 31–65.

[3]On the influence exerted by the concept of quantification during the eighteenth century, see T. Frängsmyr, John Heilbron, and Robin Rider (eds.), *The Quantifying Spirit in the 18th Century* (Berkeley, 1990).

factors help to explain this. First, the Baconian sciences—medicine is a good example—still had a qualitative structure, based on a well-established terminology inherited from the classical sources. Second, sciences like botany, zoology, mineralogy, and chemistry did not finally establish their institutional status within the universities and the scientific academies until the first half of the eighteenth century. Third, it was only between about 1650 and 1750 that the collection of new observations and the discovery of new objects reached such proportions as to raise the question of classification and a new terminology. And last but not least, only with the philosophical ferment of the Enlightenment were conditions conducive to the discussion of scientific language created. This fourth factor was probably the most influential. The philosophical writings of Condillac, Voltaire, Maupertuis, Rousseau, Herder, Monboddo, and many others on the origin and structure of language were widely quoted by most French naturalists when discussing a reform of the nomenclature of their science. The concern with language was also intimately connected with another philosophical theme of the Enlightenment: naturalism. The new conception of nature established by the *philosophes* generated a growing interest in the systematic classification of phenomena. Eighteenth-century scientists were commonly of the opinion that in observing nature it was more important to follow its unfolding process than to try to find a deductive or hypothetical explanation for it. This naturalistic attitude was born of opposition to mechanism, whose application to Baconian sciences turned out to be impossible. In the subsequent move towards naturalism in the Baconian sciences, the *explanation* of natural laws gave way to the *description* of their visible effects. This attempt faithfully to reproduce the extraordinary mutability of nature emphasized the importance of the tools used. It therefore became essential to have at one's disposal a precise method of classification and an effective procedure for naming things and phenomena. Language increasingly became the most important vehicle for the acquisition of scientific knowledge. The connection between naturalism, epistemology, and language was a credo of the Enlightenment, and its influence on the progress of natural history and chemistry shows that the link consisted of more than a mere occasional interaction.

The Catalogue of Nature

A glance at the secondary literature on the Enlightenment is enough to show the fundamental role played by the theme of nature. The central position of nature pervaded all intellectual activity from literature[4] to philosophy, but it was over natural sciences that it exercised complete hegemony.[5]

The idea of nature had always contained something both general and vague. In its general sense, nature represented the total sum of all things in time and space and in their essential structure. In another sense it stood for the principle of development of all things both inanimate and animate. In other contexts it indicated more specific individual phenomena.[6] Use in the first sense implied a mechanistic outlook on the world and in the second a vitalistic one. According to Lenoble,[7] alternation in the supremacy of these views of nature marks the epochs in the history of Western thought. The Renaissance philosophers and scientists, for instance, took a decidedly vitalistic view of nature, whereas that of the seventeenth century, especially after Descartes, was mechanistic. Inevitably, therefore, the image of nature established a century later both challenged and at the same time reaffirmed views held in the past.

The eighteenth-century *philosophes* had to resolve the difficulty of forming a new conception of nature capable of reflecting their development of Descartes' philosophy, without reverting to Renaissance mysticism and vitalism. This process was also complicated by the fact that at the beginning of the eighteenth century, science was essentially based on a mechanistic method whose universal and objective validity, according to Descartes, rested on a metaphysical foundation and was guaranteed by the goodness of God. By contrast, in the first half of the eighteenth century, belief in the kingdom of God and his supremacy in the universe became more and more limited, and science came to

---

[4]On the influence of the idea of nature on literature and *belles lettres*, see the monumental study by Jean Ehrard, *L'idée de la Nature en France dans la premiere moitié du XVIIIe siècle*, 2 vols. (Paris, 1963), although the observations on natural sciences are often unconvincing and superficial.

[5]On the influence of the idea of nature on the Enlightenment, see Cassirer, *The Philosophy of the Enlightenment* (Princeton, 1953), 37–92; Lovejoy, *The Great Chain of Being* (1960), 183–207; Paul Hazard, *European Thought in the Eighteenth Century* (1946) (Harmondsworth, 1965), 307–349; Vartanian, *Diderot and Descartes* (1953); Jacques Roger, *Les sciences de la vie dans la pensée française du XVIIIe siècle* (Paris, 1963), 206–212, 468–474, 527–542, 676–682; Robert Lenoble, *Esquisse d'une histoire de l'idée de nature* (Paris, 1969), 339–384; Georges Gusdorf, *Dieu, la Nature, l'homme au siècle des Lumières* (Paris, 1972), 299–354; Gianni Micheli, article "Natura," in the *Enciclopedia Einaudi*, vol. 9 (Turin, 1980), 715–756. On the relation between the idea of nature during the eighteenth century and the emerging of a new philosophy of language, see L.J. Jordanova (ed.), *Languages of Nature: Critical Essays on Sciences and Literature* (London, 1986), 15–47 and 86–116.

[6]Micheli (1980), 715.

[7]Lenoble (1969).

be regarded as the most effective weapon against the abuses of metaphysics. With the acceptance of Newton's natural philosophy, it became clear in France that God could no longer serve as the foundation of science. The superiority of the Newtonian method over that of Descartes was based on his renunciation of any research into primary causes.[8] This distinction between description and explanation, however, does not mean that science renounced its universality and objectivity: on the contrary, the new idea of nature conceived by the French *philosophes* helped to lay a new philosophical foundation for science, which saved the Cartesian method from the dangers of both relativism and dogmatic metaphysics.

Schematically, the concept of nature that emerged from the philosophy of the Enlightenment embraced five main assumptions that formed a starting point for methodological argumentation on science. The first of these was that nature was benevolent[9] and that mankind had to be in harmony with it in order to understand the true structure of things. Nature was conceived as a permanent substratum whose constancy enabled its mysteries gradually to be revealed to man's accurate observations.

Tied in with this view was the assumption of the unity of nature: "the claim that the conception of nature as *un tout* was the necessary condition and starting point of natural science, this being, of course, essentially the converse of Descartes' doctrine of unity of science."[10] This axiom led to a belief in the interdependence of all natural sciences.

Nature was, moreover, uniform in all its parts, and this particular aspect was seen as a guarantee of the possibility of formulating general and universal laws on the nature of phenomena. The uniformity of nature inspired the *philosophes* to make systematic use of *analogy*, to establish a link between different phenomena. Fourth, faith in the constancy of the laws of the universe gave rise to a diffuse optimism concerning the possibility of improving society on a natural basis and

---

[8]In the following famous passage, Voltaire explained the reason for the superiority of Newton's philosophy: "Ce grandhomme 'Newton' pendant soixante ans de recherches, de calculs et d'expériences, a été obligé de *se contenter du simple fait qu'il a découvert.* Jamais il n'a fait d'hypothèse pour expliquer la cause d'attraction des planètes et celle de la lumière; il a démontré que cette gravitation existe. . . . Qu'on s'en tienne là, et qu'on n'imagine pas pouvoir faire par un roman, ce que Newton n'a pu faire par ses mathématiques" [my italics], Voltaire, *Défense du Newtonianisme* (1739), in *Oeuvres Complètes* (Paris, 1784), vol. 31, 254.

[9]On the goodness of nature, see Hazard, *European Thought* (1965), 332–349.

[10]Vartanian, *Diderot and Descartes* (1953), 154–155.

of bringing major changes in the political organization of the state and administration.

Finally, nature was identified with reason. "Nature and Reason were united by an indissoluble tie. That much was certain. Nothing was clearer, nothing was more firmly established, or more often inculcated by the sages. Nature was rational, Reason was natural."[11] This analogy was probably the most characteristic feature of the outlook of the Enlightenment. The order of nature reflected the order of our understanding and vice versa.

Given this conceptual framework, it is clear that the foundations of science and its objectivity and universality were determined by its accordance with nature. The same idea sustained the attempt to base literature, morals, and laws on naturalistic principles.

This wish to trace all scientific, philosophical, political, and economic phenomena back to their natural origin explains the popularity during the eighteenth century of Lucretius' *De rerum natura*.[12] The *philosophes* saw in this work a mirror of their efforts to strike a balance between a mechanistic and a vitalistic view of nature. Moreover, they found in it many naturalistic arguments tending to eradicate teleologism from science and philosophy. The tension in the poem between a continuous evolutionary process and the primitive movements of the atoms showed the *philosophes* that a rational explanation of natural phenomena could be provided without either a theological foundation or the use of too rigidly mechanical models. Another theme of the poem that struck a responsive chord in the eighteenth century was the linking of science to political and ethical concerns, such as the fight against religious and superstitious beliefs and the search for a naturalistic and biological foundation for human ethics.

Lucretius was especially influential in France,[13] where his ideas and the doctrines of Epicurus that he expounded had already been spread by Gassendi, the libertines, Molière, and, later, Pierre Bayle.[14]

In 1770 Paul Henry Thiry d'Holbach, who promoted the best

---

[11]Hazard (1965), 307.

[12]According to Cosmo Alexander Gordon, *A Bibliography of Lucretius* (London, 1962), 58 editions of *De rerum natura* were published during the eighteenth century.

[13]"France was the most favorably minded toward Lucretius and the Epicurean philosophy than any other country in Europe," George Depue Hadzsits, *Lucretius and His Influence* (London-Calcutta-Sydney, 1935), 322.

[14]In his *Dictionnaire historique et critique*, 2nd ed. (Rotterdam, 1702), vol. 2, 1919–1928, Bayle devoted a long article to Lucretius, praising his philosophy of nature and saying: "jamais homme ne nia plus hardiment que ce poëte la providence divine," 1920.

French version of the poem (translated by La Grange and possibly revised by Naigeon and Diderot),[15] published his *Système de la Nature,*[16] which was basically a paraphrase of the doctrines of Epicurus and Lucretius. According to d'Holbach, all phenomena, their internal and external properties and their combinations, contributed to the essence of nature. Nature encompassed the whole universe, and therefore the only way to know her was to concentrate on her manifestations. The means of contact between man and nature was sensation. Thus, what in the philosophical system of the seventeenth century was under the sole jurisdiction of God could now be explored by the epistemological power of *sensation,* which, residing in a stimulation of the human body, represented a link between the derivation of knowledge and the properties of matter.

However, this mixture of psychological and scientific materialism was not enough to constitute a scientific philosophy capable of meeting the demands for a unifying and universal theory of knowledge made by eighteenth-century scientists. Sensation was in fact an invisible and undetectable point of correspondence between man and nature.[17] Only at the moment of its provocation could sensation be recognized as such and the contact between man and nature established. Besides, an intrinsic property of sensation was that of belonging exclusively to individuals. It was under the pressure of this crucial philosophical problem that the study of language developed. It was assumed that the external world corresponded to our system of sensations, which found its objective expression in language. Words were the medium linking individual sensations and the objects to which they related; and at the same time, words were the only means of collective communication. Thus, if a method of coining names of universal validity were devised, the problem of knowledge would be solved without physical or metaphysical intervention. Language was regarded as the product of the biological conformation of the human organ to external circumstances and consequently a natural means of expression.[18]

---

[15]La Grange was the tutor of d'Holbach's children. The supposition that Diderot and Naigeon revised this edition is made by Naville (1967) but not confirmed by Gordon (1962), who believes that it was only Diderot who recommended La Grange to undertake the translation of the poem, whereas he found no evidence of any active collaboration with d'Holbach and Naigeon.

[16]D'Holbach, *Système de la Nature, ou des lois du monde physique et monde morale* (London, 1770), 2 vols., which had eight editions in the eighteenth century.

[17]This property of sensations could have logically led to Berkeley's idealism, had not the possibility of relating them to the external world not previously been demonstrated.

[18]See Chapter IV.

With the adoption of this belief, nature suddenly became cognizable without external intervention. The explanation of phenomena by reduction to a primitive cause was replaced by empirical observation and description of the natural world. The study of external reality could no longer be bound to the imaginary world of the *systèmes* that explained it rather than tried to understand it. It was now necessary to observe the development of nature and all its effects without preconceptions.

From these theoretical premises, it might seem that the only possible epistemological conclusion was a reassertion of Baconian empiricism. However, the spirit of the Enlightenment demanded not merely the reduction of knowledge to sensibility but the understanding of reality according to rational principles. The fight of the *philosophes* against the *Systèmes* did not imply the rejection of reason and its methods, but the creation of a new kind of reason, more faithful to the structure of nature.

D'Alembert, who was one of the most coherent interpreters of this position, declared in a famous passage of the *Discours préliminaire*:

> It is not at all by vague and arbitrary hypotheses that we can hope to know nature; it is by thoughtful study of phenomena, by the comparisons we make among them, by *the art of reducing, as much as that may be possible, a large number of phenomena to a single one that can be regarded as their principle.* Indeed, the more one reduces the number of principles of a science, the more one gives them scope, and since the object of a science is necessarily fixed, the principles applied to that object will be so much more fertile as they are fewer in number. This reduction which, moreover, makes them easier to understand, constitutes the true "systematic spirit."[19] [my italics]

This famous quotation reflects an attitude very common among French philosophers and scientists. Schematically it expresses the intention to classify and study nature according to the order of phenomena in their natural development rather than with subjective representations. It is for this reason that the act of description assumed a prominent role in the process of achieving any kind of knowledge.

The description, however, could not be totally congruent with the object described. This was impossible for two reasons. First, the ordinary language could not correspond fully to reality; if the converse was assumed, the *philosophes* would have had to accept the linguistic realism of Linnaeus, which, on the contrary, they strongly criticized. Sec-

[19]D'Alembert (1963), 22.

ond, it was impossible to give a satisfactory account of the infinite modifications of nature within a static categorization. For this purpose it was necessary to regard words and descriptions as conventions capable of ordering phenomena in classes. On the other hand, a conventional use of words deprived scientific investigation of its objective and universal foundation.

Therefore the question the scientists of the Enlightenment had to face concerned how to order nature in the most *natural* way without falling into a static or, worse, metaphysical realism. The idea of comparing phenomena in their visible effects helped the *philosophes* to find a solution. The investigation of remote causes having been rejected, "all we can do is to perceive some particular effects, to compare them, to combine them and finally to recognize among them an order relative to our own nature rather than appropriate to the existence of things we are considering."[20] At the same time, the author of this passage, the French naturalist Buffon, was equally critical of absence of method and the danger involved in a purely empirical concentration on minute details and particular phenomena. Rejecting the artificial methods conceived by men meant, to Buffon, adopting the method taught by nature itself.

According to the *philosophes*, this tension between nature and system was an intrinsic characteristic of the epistemological relationship between man and nature. But, they claimed, the proportion of the *esprit de système* could be reduced if not eliminated.

So, if the Enlightenment thinkers denounced the intolerable supremacy of imagination in the philosophical systems of the seventeenth century, they did not renounce a systematic way of reasoning. Condillac, who was the main inspiration in the philosophical struggle against the *Systèmes*, was aware of the structural insolubility of the conflict between reason and nature, and between system and analysis. He asserted that even if the problem as such was impossible to solve, it could be presented and discussed very differently. Although in the philosophical systems of Descartes, Spinoza, and Leibniz imagination and hypothesis prevailed over analysis and experience, it was now possible, with the advances in knowledge that had been achieved, to reverse matters by basing natural philosophy on observation rather than ideas. "We conclude," said Condillac, "that we can create true systems only when we make enough observations to grasp the chain of phenomena. Now we

---

[20]Buffon (1954), 10.

have seen that we can neither observe the elements of things nor the forces that move living bodies; we can only see their remote effects. Therefore the best principles we can use in physics are phenomena able to explain other phenomena."[21] The task of the philosopher was to control, but by no means to destroy, the *système* by submitting our imagination to the tribunal of experience.

The long chains of phenomena and the need to provide exact descriptions of them created an awareness of the importance of a precise scientific language. This is the reason for the spread throughout the eighteenth century of classificatory systems and methods, nomenclatures, taxonomies, etc. described as *naturelles*. In order to demonstrate the superiority of the new *esprit systématique* over the old systems, the *philosophes* aimed to show the intimate connection between their ideas and the structure of nature. The task was to shift philosophy from a subjective to an objective foundation. Naturalism, in all its possible varieties, was considered the most effective doctrine to supersede the subjectivity of the philosophy of Descartes and the *a priori* dogma of religion.

The positions declared by d'Holbach, d'Alembert, Buffon, and Condillac were shared by most of the French scientists of their generation. The immense quantity of data and observations collected in the first half of the eighteenth century needed rational inventorying rather than fanciful interpretation. A mechanistic view of nature could no longer explain the complexity of reality. In chemistry, natural history, medicine, and the other applied sciences, the corpuscular hypotheses and the mechanical models did not satisfy the emerging need to order the scattered universe of the Enlightenment. The naturalistic approach, on the other hand, was a solution that met this need for unification and synthesis. Even the most conventional of scientific problems was solved by reference to nature. Everything that was true and simple was so because it was natural. By contrast, what was complicated or wrong was regarded as a deviation from the pathways of nature. This belief is well expressed in the following claim by Buffon:

> Human understanding [*esprit*] cannot create anything, and it produces only after being fertilized by experience and meditation; its knowledge is the germ of its creations; but *if it imitates Nature in its process and in its working, if it ascends by contemplation to the most sublime truths, if it unites and connects them, if from them it forms a whole [. . .], it*

[21]Condillac, *Traité des systêmes*, in *Oeuvres*, vol. I, 207.

> *will establish immortal monuments on indestructible foundations.*[22] [my italics]

Another eloquent example of this attitude to nature was the establishment in 1789–91 of the Commission of Weights and Measures in Paris, which was entrusted with the reform of the whole system of measures of the *ancien régime.*[23] In principle, this might be expected to have been a routine matter. But the exigencies of practical economics led the scientists to create a more rational and convenient system.

There is no doubt that social and political circumstances played an important part in the decision to undertake this task. In eighteenth-century France, the number of different systems of weights and measures was considerable, and the revolutionary enthusiasm for bringing unity to the kingdom implied the standardization of weights and measures and also of the monetary system. The reform of taxation and of agriculture, in which Lavoisier was actively involved in the early 1790s, required a more rational system of measurements. Nevertheless, this rationalization was designed to construct not merely a more effective set of social conventions but, as Crosland pointed out, "a system of measurement which would be in conformity with nature."[24]

For this purpose various commissions were set up by the Assemblée National between 1789 and 1791. The reform engaged the most prominent French scientists, such as Borda, Lavoisier, Lagrange, La place, Haüy, Bailly, Monge, Delambre, Condorcet, and Tillet, as well as the best technicians and instrument makers. In the first *Rapport* to the Academy of Sciences, presented in 1791, Borda, Lagrange, Méchain, Laplace, Monge, and Condorcet declared:

---

[22]"L'esprit humain ne peut rien créer, il ne produira qu'après avoir été fecondé par l'expérience et la méditation; ses connoissances sont les germes des ses productions: mais s'il imite la Nature dans sa marche et dans son travail, s'il s'élève par la contemplation aux vérités les plus sublimes, s'il les réunit, s'il les enchaine, s'il en forme un tout, un système par la réflexion, il établira, sur des fondemens inébranlables, des monumens immortels," Buffon, "Discours sur le style," in *Oeuvres complètes de Buffon* (Paris, 1822), vol. 1, 5.

[23]The best study on the Commission and its purposes is the essay by Maurice P. Crosland, "Nature and Measurement in Eighteenth-century France," in *Studies on Voltaire and the Eighteenth Century*, LXXXVII (1972), 277–309. More focussed on the social and economical context is the reconstruction by Witold Kula, *Measures and Men* (Princeton, 1986), 228–264. See also G. Bigourdan, *Le système métrique des poids et mesures. Son établissement et sa propagation graduelle* (Paris, 1901); J. Fayet, *La Révolution française et la science: 1789–1795* (Paris, 1960), 442–467; and Frängsmyr, Heilbron, and Rider (eds.) (1990), 207–242.

[24]Crosland, "Nature" (1972), 277.

> The idea of relating all measurements to a unit of length taken from nature occurred to the mathematicians as soon as they realized the existence of such a unit and the possibility of determining it. They saw that this was the only way to exclude all arbitrary elements from the system of measurements and to be sure of preserving it unchanged. . . . They felt that such a system did not belong exclusively to any nation. . . .[25]

The dimension of the earth was the universal and natural standard of reference. One-fourth of the earth's meridian would be the real (*réelle*) unit of measure and a ten-millionth part would be the usual (*usuelle*) unit. These being the sources of the system, no nation would accuse the French scientists of nationalism.

To underline the universality of the choice of the natural unit, Lavoisier wrote a paper that remained unpublished on the history of measurements among ancient civilizations. He emphasized the fact that very disparate civilizations incorporated the dimension of the earth in their systems of measurement:

> This uniformity in measures among people of different customs, different governments, thousands of leagues apart, cannot be the effect of chance, and recently the suspicion has begun to form that all the measurements of the ancients were originally subject to a general system. There has been more and more confirmation of the assumptions first made by Bailly and by Paucton as more is discovered of the history of antiquity.
>
> From their research it emerges that, long before anything that ancient history has passed down to us, there existed a people in remote antiquity who devised and produced a metrical system deduced from the dimensions of the earth.[26]

Thus, the very first civilization conceived a metric system based on the same natural pattern as the one proposed by the French scientists in 1791. The fact that a primitive civilization could have achieved such a result was seen by Lavoisier as the evidence of its natural origin. Despite all the linguistic, political, geographical, and historical differ-

---

[25]"Rapport fait à l'Académie des Sciences, sur le choix d'une unité de mesure. Par MM. Borda, Lagrange, Laplace, Monge et Condorcet," in *Histoire de l'Académie Royale des Sciences* [1783] (Paris, 1791), 7.

[26]Lavoisier, "Éclaircissements historiques sur les mesures des anciens," manuscript of 1790 or 1791, published in LO, vol. 4, 699. In his *Histoire de l'astronomie moderne* (1779–1783), vol. 1, 156, Bailly stated: "Les anciens ont eu l'idée de rendre leurs mesures invariables *en les prénant dans la nature*, et cette idée, encore sans exécution chez nous, semble avoir été rempli par eux" [my italics].

ences, when it was guided by nature, mankind arrived at the same conclusion. But what was it, then, that made the metric system of the French a substantial advance on the ancient ones? Once again the answer was to be found in its closer conformity to nature. One of the essential advantages of the simple circumstance that the meter was a ten-millionth part of one-fourth of the earth's meridian lay in the decimalization of the units. The multiples and sub-multiples of the unit of measurement coincided with the numerical relationships of the decimal scale, which could be used to express any measurement. The simplicity and the efficiency of the decimal scale were due to the natural fact that the hands of a man have ten digits:

> We generally observe that decimal arithmetic has been used by all nations which history has reported. It is probable that the conformation of the hand by the fact that each man has ten fingers had determined a sort of predilection in favour of the number ten. . . . It is in the way that this number became *naturally* the basis of the arithmetical scale.[27] [my italics]

The dimensions of the earth and the decimal scale were both models derived from nature; the dimension of the earth was derived from an imaginary "real," rather than arbitrary, unit, and the decimal system from the biological conformation of the human body. The construction of a natural system of weights and measures encountered all sorts of technical complications from the beginning, and many scientists gradually realized the impossibility of keeping their initial philosophical ambitions intact. After 1796 the Convention ceased to regard the platinum prototype of the "universal" meter as natural, and some years later, when tracing the history of the metric system, Delambre was more cautious in claiming an exact basis in nature.[28] With the significant progress accomplished in geodetic research, many technical and physical problems arose, and it became clear that measurement of the dimension of the earth relied on the conventional exactness of numbers rather than on the real structure of nature.

However, what is interesting in this episode from the history of science is the fact that nature was the main inspiration in this first

---

[27] *Instruction sur les mesures déduites de la grandeur de la terre* . . . (Paris, 1794), quoted by Crosland (1972), 298–299; on the decimal scale, see also the "Rapport fait à l'Académie des Sciences par MM. Borda, Lagrange, Lavoisier, Tillet et Condorcet," in *Histoire de l'Académie Royale des Sciences* [1788] (Paris, 1791), 6. See also the preliminary discourse by Méchain and Delambre of their *Base du système metrique décimal* (Paris, 1806), vol. 1.

[28] Jean-Baptiste Delambre, *Rapport historique sur les progrès des sciences mathématiques depuis 1789 et sur leur état actuel* (Paris, 1808), 4–10 and 62–73.

attempt to provide a general and universal metric system. By using nature as a scientific model, eighteenth-century scientists were able to give their investigations a philosophical foundation and to connect their thoughts and theories with the intrinsic structure of phenomena.

The establishment of the new system involved the creation of a new nomenclature for weights and measures consistent with the philosophy behind it. Prieur de la Côte d'Or, a pupil of the chemist Guyton de Morveau, who, as we shall see in the next chapter, was one of the leading protagonists in the reform of chemical nomenclature, was commissioned to create a new terminology. Prieur's proposals were derived from the Greek language and provided names composed of generic and specific elements. To the generic element *déca-* (ten), *hecto-* (hundred), or *kilo-* (thousand) was attached the specific coordinate of measurement (length, weight, capacity, etc.) giving, for example, the name *hectometre* to denote a length of 100 meters, *hectogramme* (a weight of 100 grams), and *hectolitre* (a capacity of 100 liters) (see Fig. 1).[29] All the names reflected the decimalization of the system conceived in 1790. The reform of the nomenclature of weights and measures was adopted in 1795, and after initial resistance and hostility, it became the most successful and universal system of measurements ever created. For the first time in history, a national system of weights and measures extended beyond the geographical and political boundaries of one nation, being adopted during the nineteenth century by Spain, Italy, and other European countries. Though its evolution was slower than expected by Condorcet and his colleagues, in the twentieth century, the decimalization of nature became by far the most practical framework for conceiving physical measurements.

Less successful, but no less significant, was the attempt to reform the measurement of time. In September 1792, a new calendar was created. By a decree approved retrospectively by the Convention in October 1793, the revolutionary day was divided into ten parts, each composed of 100 minutes or 10,000 seconds.[30] The immediate aim of

[29]C.A. Prieur, *Rapport fait au nom du Comité d'Instruction publique, sur la necessité et les moyens d'introduire dans toute la République les nouveaux poids et mesures précédemment décretés* (Paris, 1794); *idem.*, *Note instructive sur les poids et mesures* (Paris, 1794).

[30]"Le jour, de minuit à minuit, est divisé en dix parties; chaque partie en dix autres, ainsi de suite jusqu'à la plus petite portion commensurable de la durée" (Article XI of the "Décret de la Convention National concernant l'ère des Français," reprinted in Bernard Deloche and Jean-Michel Leniaud (eds.), *La culture des sans-culottes* (Paris, 1989), 76.). It is interesting to note in passing that Lavoisier was one of the protagonists of the decimalization of time; see J. Guillaume, "Lavoisier Anti-clérical et révolutionnaire," in *Révolution française 26* (1907), 409.

ACADÉMIE DES SCIENCES.		COMMISSION TEMPORAIRE.	LOI du 18 GERMINAL.	ARRÊTÉ du 13 BRUMAIRE AN IX	VALEURS.
Décade					1,000,000 mètres.
Degré		Grade ou degré			100,000
Poste			Myriamètre	Lieue	10,000
Mille	Millaire	Millaire	Kilomètre	Mille	1,000
Stade			Hectomètre		100
Perche			Décamètre	Perche	10
Mètre	Mètre	Mètre	Mètre	Mètre	1
Palme	Décimètre	Décimètre	Décimètre	Palme	0.1
Doigt	Centimètre	Centimètre	Centimètre	Doigt	0.01
Trait	Millimètre	Millimètre	Millimètre	Trait	0.001
Tonneau	Muid	Cade	Kilolitre	Muid	Mètre cube
Setier	Décimuid	Décicade	Hectolitre	Setier	
Boisseau	Centimuid	Centicade	Décalitre	Boisseau, Velte	
Pinte	Pinte	Cadil	Litre	Pinte	Décimètre cube.
			Décilitre	Verre.	
			Centilitre.		
Millier	Millier	Bar		Millier.	
Quintal		Décibar		Quintal.	
Décal		Centibar	Myriagramme.		
Livre	Grave	Grave	Kilogramme	Livre	Décimètre cube d'eau distillée.
Once	Décigrave	Décigrave	Hectogramme	Once.	
Drâme	Centigrave	Centigrave	Décagramme	Gros.	
Maille	Milligrave	Gravet	Gramme	Deniers.	
Grain		Décigravet	Décigramme	Grain.	
		Centigravet	Centigramme.		
		Milligravet	Milligramme.		
		Are	Hectare	Arpent	10,000 mèt. carrés
		Déciare			1,000 *idem.*
		Centiare	Are	Perche carrée	100 *idem.*
					10 *idem.*
			Centiare	Mètre carré	1 *idem.*
			Stère	Stère	Mètre cube
			Déci-stère	Solive.	
Unité monétaire.					Décagr. d'argent,
			Franc	Franc.	5 grammes d'arg.
			Décime	Sol.	
			Centime	Denier.	

FIGURE 1 The Republican weights and measures. From J. Delambre, *Base du système metrique décimal* (Paris, 1801).

the reform was to replace the oppressive religious subdivision of time with one that expressed the ideology of the new era. Significantly, the calendar of the French republic was based on the decimal system and relied on the principles of scientific naturalism. Each month was in fact divided into three parts of ten days each (*décades*). The months and the days were renamed after the products of nature (vegetables, animals) or activities characteristic of the period of the year when they took place (see Fig. 2).[31] In the caption of a famous allegorical representation of the new calendar, Philosophy dictates the names of the new months and days after consulting the book of nature (see Fig. 3).[32] It is not surprising, then, that the feast instituted by Robespierre to honor the Supreme Being (Vendémiaire 10) was also intended to honor nature.

The names of the ten days of the *décade* were *primidi, duodi, tridi, quartidi, quintidi, sextidi, septidi, octidi, nonidi, décadi.* The names of the months were taken directly from nature:

Autumn: *vendémiaire, brumaire, frimaire.*
Winter: *nivôse, pluviôse, ventôse.*
Spring: *germinal, floréal, prairial.*
Summer: *messidor, thermidor, fructidor.*[33]

Despite efforts to spread the use of the revolutionary calendar, it was abolished at the beginning of 1806.

The attempted reform of the calendar, which was unique in modern history, was followed by several minor reforms of civil and social terminology. In 1793 the Abbé Henri Grégoire published a memoir in which he proposed a reform of city street names.[34] In it he outlined a "système de nomenclature" taking terms from agriculture, trade, arts, natural sciences, and history. The names of the streets of French cities had to be as short, clear, and expressive as those used in the sciences.

---

[31]On the reform of the calendar during the French Revolution, see *Almanach National de France* (Paris 1794–1795), 1–10.

[32]The caption of Fig. 3 says: "Sur un sommet élevé, la philosophie assise sur un siège de marbre décoré des images de la Nature féconde et ayant pour diadème le bonnet de la Liberté, foule à ses pieds les gothiques monuments d'erreur et de superstition sur lesquels était fondée l'ancienne division des temps, et puise dans le grand livre de la Nature les principes et les dénominations du nouveau calendrier que près d'elle un génie attentif à sa voix trace sous sa dictée. De l'autre côté, sont ses divers attributs, tels que le livre de Morale, la Règle et un niveau, emblème de l'Egalité."

[33]See Deloche and Leniaud (eds.), *La culture des sans-culottes* (1989), 78–79.

[34]Henri Grégoire, "Système de dénominations topographiques pour les places, rues, quais, etc., de toutes les communes de la République" (1793?), *ibid.*, 119–133.

VENDEMIAIRE.

Premier mois de l'année Républicaine & de l'Automne, prend ſon étymologie des vendanges qui ont lieu pendant ce mois.

Sol. à la ♎ le 1, à 9 h. 2 m. du ſ.

ERE RÉPUBLICAINE.		Signes des départs&arrivées des Poſtes & Diligences.	Jours de la Lune.	Lunaiſons ſuivant l'ère républicaine.
1 Primedi.	Raiſin.	l.	29	
2 Duodi.	Safran.	ma.	30	
3 Tridi.	Châtaignes	me.	1	
4 Quartidi.	Colchique.	j.	2	Nouvelle lune le 3 à 5 h. 3 min. du matin.
5 Quintidi.	CHEVAL.	v.	3	
6 Sextidi.	Balſamine.	ſ.	4	
7 Septidi.	Carottes.	d.	5	
8 Octidi.	Amaranthe	l.	6	
9 Nonidi.	Panais.	ma.	7	
10 *Décadi.*	CUVE.	me.	8	
11 Primedi.	Pomme de terre.	j.	9	
12 Duodi.	Immortelle.	v.	10	Premier quartier le 11 à 6 h. 42 minutes du matin.
13 Tridi.	Potiron.	ſ.	11	
14 Quartidi.	Réſéda.	d.	12	
15 Quintidi.	ANE.	l.	13	
16 Sextidi.	Belle-de-nuit.	ma.	14	
17 Septidi.	Citrouille.	me.	15	
18 Octidi.	Sarrazin.	j.	16	
19 Nonidi.	Tourneſol.	v.	17	Pleine lune le 18 à o heur. 37 min. du mat.
20 *Décadi.*	PRESSOIR.	ſ.	18	
21 Primedi.	Chanvre.	d.	19	
22 Duodi.	Pêche.	l.	20	
23 Tridi.	Navet.	ma.	21	
24 Quartidi.	Amarillis.	me.	22	
25 Quintidi.	BŒUF.	j.	23	Dernier quartier le 24 à 7 h. 9 min. du ſoir.
26 Sextidi.	Aubergine.	v.	24	
27 Septidi.	Piment.	ſ.	25	
28 Octidi.	Tomate.	d.	26	
29 Nonidi.	Orge.	l.	27	
30 *Décadi.*	TONNEAU.	ma.	28	

FIGURE 2 The first month of the Republican calendar. From *Almanach National de France* (Paris, 1794–1795).

FIGURE 3 The Republican calendar after a contemporary engraving. From Dominique Julia, *Les trois couleurs du tableau noir de la Révolution* (Paris, 1981).

We know that much of the topographical nomenclature of Paris was radically changed, although the process lasted only until the end of the century.

The linguistic and naturalistic contagion of eighteenth-century France even reached the revolutionary parliament on several occasions. In September 1791, Talleyrand, inspired by the accuracy achieved in scientific language, declared to the Convention that the French language had to eliminate synonyms, and he appealed for the creation of "a universal grammar" of ideas capable of reflecting the natural order of human sensations.[35]

Whether they failed or succeeded, these attempts to create naturalistic nomenclatures were rooted in the political and philosophical intention of the French Enlightenment to create an accessible, naturalistic, and universal system of linguistic and conceptual coordinates within which to interpret the world.

THE GRAMMAR OF THE ARTS

Having considered the historical significance of the description of nature, we may now understand better how important language became. To describe a phenomenon thoroughly implied an appropriate designation of its parts. Accordingly, language was intended to be the fundamental, if not the only, means of definition of natural objects.

The question of a philosophical reform of scientific language had undoubtedly been discussed during the Renaissance and the seventeenth century. During the Renaissance, for instance, the main linguistic problems besetting science were the reform of the classical vocabulary and the translation of scientific and technical terms into the vernacular.[36] The seventeenth century saw the development of interest in the subject and many studies of the origin and structure of language were published. As Paolo Rossi has pointed out,[37] Wilkins, Leibniz, Kircher, and many others discussed the possibility of creating a universal language, capable also of expressing scientific knowledge. However, these attempts were either made by metaphysicists to facilitate the read-

---

[35]On the reform of the French language on a scientific basis, see Dominique Julie, *Les trois couleurs du tableau noir. La Révolution* (Paris, 1981), 222–226.

[36]On the evolution of scientific language in the Renaissance, see the classic study by Leo Olschki, *Geschichte der neusprachlichen wissenschaftlichen Literatur,* 2 vols. (Heidelberg-Leipzig, 1919–1922).

[37]Paolo Rossi devoted many of his studies to this topic. See particularly *Clavis universalis. Arti della memoria e logica combinatoria da Lullo a Leibniz* (Bologna, 1983), and the essay "Lingue artificiali, classificazioni, nomenclature," in *Aspetti della rivoluzione scientifica* (Naples, 1971), 295–367; see also Maria Luisa Altieri Biagi, "Lingua della scienza tra Seicento e Settecento," in *Lettere Italiane,* XXVIII (1976), 410–461, and M.M. Slaughter, *Universal Languages and Scientific Taxonomy in the Seventeenth Century* (Cambridge, 1982). On the same subject, but mainly restricted to the English philosophers of the seventeenth century, see the important study by Lia Formigari, *Linguistica ed empirismo nel Seicento inglese* (Bari, 1970). See also Hans Aarsleff, *From Locke to Saussure: Essays on the Study of Language and Intellectual History* (London, 1982), 42–93 and 239–292; James Knowlson, *Universal Language Schemes in England and France, 1600–1800* (Toronto and Buffalo, 1975).

ing and interpretation of the Bible, or they relied on medieval scholasticism. In criticizing them Descartes declared that logic of this kind served more to explain things already known than to teach new ones,[38] and he denied the importance of language by basing his philosophy and science on the clarity with which the human mind comprehended ideas. Within a different philosophical framework, Locke, in the third book of *An Essay Concerning Human Understanding* (1690), warned philosophers of the impossibility of a universal language and asserted that the ultimate basis of knowledge was sensation, rather than words.

The eighteenth-century *philosophes,* widely influenced by both Descartes and Locke, were therefore both aware and critical of the metaphysical abuses of language and of the *Mathesis universalis* of the previous century.

Reading the writings of the *philosophes* on language and those of a century before, it is indeed difficult to find a continuity of purpose.[39] The society of the Enlightenment and the new institutional status of science in the eighteenth century presented scientists with more empirical and economic problems. The great quantities of data accumulated in the first half of the century needed to be ordered. The first tentative interaction among science, technology, and economics was a sign of an important turning point in the attitude to scientific language. In this respect the discussion of the mechanical arts offers an interesting example.

The developments of some branches of the mechanical arts, the consequent acceleration of the productive process, and the qualitative improvements and quantitative increase in manufactures made the *philosophes* more aware of the importance of generally reorganizing economics and of reevaluating the role of the arts and technology. In publishing the first volumes of the *Encyclopédie,* Diderot made the first systematic attempt to do this. With the plates describing all the arts

---

[38]"Mais, en les examinant, je pris garde, pour la logique, ses syllogismes et la plupart de ses autres instructions servent plutôt à expliquer à autrui les choses qu'on sait, ou même, comme l'art de Lulle, à parler, sans jugement, de celles qu'on ignore, qu'à les apprendre," René Descartes, *Discours de la Méthode* (1637), Texte et commentaire par Etienne Gilson (Paris, 1976), 17.

[39]According to Paolo Rossi, the rationalism of the Enlightenment became, from this point of view, a decisive turning point. "E' proprio il razionalismo illuministico che segna, da questo punto di vista, una svolta decisiva. Una serie di problemi cha avevano appassionato per secoli i cultori di logica e retorica, i teorici del discorso e gli studiosi del linguaggio, vennero eliminati per sempre dalle scene della cultura europea, perdettero significato e senso, apparvero manifestazioni delle folli aspirazioni di secoli che si erano posti sotto il segno delle empie ricerche astrologiche, magiche e alchimistiche, o sembrarono i relitti, ancora presenti nell'età della nuova scienza, delle tenebre medievali," in Rossi (1983), 19.

and trades, he successfully showed their inherent potential for exploitation on a larger scale.

According to Diderot, linguistic inadequacies were an important factor impeding a better understanding of the arts and of their progress:

> I have found the language of the arts to be very imperfect for two reasons: the scarcity of proper nomenclature and the frequency of synonyms. Some tools have different names while others have only the generic name "engine" or "machine," without any additional name to distinguish them [. . .]. One would wish for more attention to analogy of form and use. Geometers do not have as many names as they have figures, but in the arts a hammer, a pair of tongs, a bucket, a shovel, etc., have almost as many names as there are arts. A good part of the language changes from manufacture to manufacture. Yet I am convinced that the most unusual operations and the most complex machines could be explained by a rather small number of familiar, well-known terms, if it were decided to use technical terms only when they communicate a distinctive idea.[40]

The vast number of synonyms for technical terms inevitably contributed to general confusion and hindered improvement of the internal organization of manufactures. Economic betterment certainly depended on technological progress, on the quality of natural resources, on the division and organization of labor, etc., but without a rational nomenclature, improvement in these respects could not be enough.

As a solution to this impasse, Diderot saw language as the only means of communicating abstract notions, and he understood that the more complex the arts became, the more they required precise definition. To distinguish the simple elements of a science, language could provide an analytical method capable of indicating the differences between them:

> It would be desirable if a good logician, well versed in the arts, undertook to describe the elements of a "grammar of the arts." For a first step he would have to determine the value of the correlatives "big," "large," "average," "thin," "thick," "slight," "small," "light," "heavy," etc. *For this purpose one must seek a constant measure in nature or evaluate the height, width, and average force of a man, and relate to it all indeterminate expressions of quantity, or at least set up tables to which artists would be asked to make their language conform.*[41] [my italics]

[40]Diderot, article "Art" of the *Encyclopédie*, in *Encyclopedia: Selections* (Indianapolis-New York-Kansas City, 1965), 13.
[41]*Ibid.*, 14.

Diderot, like his contemporaries, wished to find definitions for the arts by taking coordinates from nature but, as emerges from his writings, he found this particularly tricky to put into practice. It is striking, however, to see the important part played by the naturalistic view of language in the birth of eighteenth-century science and technology.

The central nature of this theme is underlined in another famous article in the *Encyclopédie*, in which Diderot tried to be more precise:

> Sciences are born only from the application of our ideas to ideas previously conceived, and from the combination of our thoughts, our observations and our experiences with the thoughts, observations and experiences of other men. Without the double convention which attaches the ideas to the voice and the voice to the signs, everything would remain inside the man and would pass away; without the grammars and the dictionaries, which are the universal interpreters between peoples, everything would remain concentrated in one nation and disappear with it.[42]

While emphasizing its importance, Diderot seemed to underline the conventionality of language rather than its naturalness. Ultimately the gap between artifact and nature was bridged by Diderot's belief that "each science has its name; each notion in a science has its name; everything that is known in nature has been designated, as has everything that has been invented in the different arts, and the phenomena, and the operations and the instruments."[43]

If everything around mankind required a definition in order to be known, the names man used for this purpose could not be allowed to be mere arbitrary conventions. They had somehow to be connected to the object they intended to define; and the more closely the definition conformed to the nature of the object to which it referred, the more explanatory and descriptive the name was.

We may conclude by saying that Diderot was committed to creating stricter correspondence between ideas and nature, the two necessary conditions being: 1) to perceive reality, and 2) to represent it as it was. The classical antithesis between reason and sensibility that was such a feature of the philosophy of the Enlightenment was partially resolved by Diderot, but not only by him, by projecting naturalistic properties onto the intellectual faculties. Being the main natural tool of representation, language was seen as the bridge between the thoughts and the senses, between the ideas and the world.

---

[42]Diderot, article "Encyclopédie," in *Oeuvres complètes*, vol. VII (Paris, 1976), 189.
[43]*Ibid.*

"Eadem Natura, Eadem Nomenclatura"

"To admit therefore the study of botany, and to reject that of nomenclature, is a most absurd contradiction."[44] The author of this statement was one of Linnaeus' greatest admirers in France. Jean Jacques Rousseau's interest in botany was in fact inspired by the quality and the thoroughness of Linnaeus' work. In particular, Rousseau considered the system of naming devised by the Swedish naturalist one of the most important achievements in the history of botany: "he [Linnaeus] established at length a clear nomenclature, founded upon the true principles of the art which he had set forth. He preserved all the ancient genera, which were truly natural; he connected, simplified, united, or divided the rest as their true characters required."[45]

It is well known that not all opinions on Linnaeus' nomenclature, especially in France, were so positive. The controversy concerning method, system, and nomenclature was in fact one of the liveliest in eighteenth-century natural history. It has been the object of several studies, although with little consensus among the historians concerned.[46]

The principles upon which Linnaeus based his botanical system were rather remarkable. On the one hand, he presented naturalists

---

[44]Jean Jacques Rousseau, *Letters on the Elements of Botany Addressed to a Lady* (London, 1794), 18. The original edition of *Lettres élémentaires sur la botanique à Madame de L**** was published in Geneva in 1781. The passage quoted here from the first English edition belongs to the introduction to the Swiss edition of the *Fragments pour un Dictionnaire des termes d'usage en Botanique*, in *Collection complète des Oeuvres de J.J. Rousseau*, vol. VII (Geneva, 1781), 459–527.

[45]Rousseau, *Letters* (1794), 11.

[46]On the spread of Linnaeus's ideas on nomenclature and on the controversy that followed, see William Whewell, *History of Inductive Sciences*, 3rd ed. (London, 1857), vol. III, 256–276; Henri Daudin, *De Linné à Jussieu–Méthodes de la classification et idée de série en botanique et en zoologie* (1740–1790) (Paris, 1926); William T. Stearn, "Species Plantarum and the Language of Botany," in *Proceedings of the Linnean Society of London*, 165 (1955), 158–164; *idem.*, "An Introduction to the Species Plantarum and Cognate Works of Carl Linnaeus," in Carl Linnaeus, *Species Plantarum* (facsimile edition) (London, 1957), vol. I; John Lewis Heller, "The Early History of Binomial Nomenclature," in *Huntia*, 1 (1964), 33–70; François Dagognet, *Le catalogue de la vie. Etude méthodologique sur la taxinomie* (Paris, 1970); Frans A. Stafleu, *Linnaeus and the Linneans* (Utrecht, 1971), 79–114 and 267–339; James L. Larson, *Reason and Experience: The Representation of Natural Order in the Work of Carl von Linné* (Berkeley-Los Angeles-London, 1971), 122–153; Philip R. Sloan, "The Buffon-Linnaeus Controversy," in *Isis*, 67 (1976), 356–375; John Lewis Heller, *Studies in Linnean Method and Nomenclature* (Frankfurt am Main-Bern-New York, 1983); Gunnar Eriksson, "Linnaeus the Botanist," in Tore Frängsmyr (ed.), *Linnaeus: The Man and His Work* (Berkeley-Los Angeles-London, 1983), 63–107; Giulio Barsanti., *La Scala, la Mappa, l'Albero. Immagini e classificazioni della natura fra Sei e Ottocento* (Florence, 1992), 129–214.

with a taxonomy able to present the most complete inventory of the three kingdoms of nature, on the other hand, some parts of this system were based on philosophical and metaphysical principles that were rejected and criticized by the *philosophes.* Since his early works, the main thrust of Linnaeus' botanical research had been towards the finding of a system of classification capable of reflecting the true nature or essence of plants. The search for such a "natural method" had already perplexed the botanists of the Renaissance. Caesalpinus and, later, Gesner and Tournefort created their botanical systems of classification after criticizing the obsolete artificial systems of the past but were unable to formulate a "natural" method themselves. Linnaeus was more convinced than his predecessors that he had found the true natural method of botany in the sexual organs of plants, which he regarded as "the fragments of the natural method."[47] The principles of this method were simple but effective: classification (*dispositio*) and nomenclature (*denominatio*).

The system of botany was represented by divisions into classes, orders, genera, species, and varieties, with the aid of which it was possible to classify all plants. The system was to Linnaeus the *filum araindeum* that helped the naturalist to recreate the order of nature as it had been conceived by God. Without the system, nature would have remained a mystery, a realm of chaos and obscurity.

The genera and species of Linnaeus' system of nature were the work of both God and nature (*opus naturae*).[48] The basis for the classification of all plants, said Linnaeus, had to be their fructification. From this starting point, Linnaeus proceeded to the conclusion that the essential parts of the plant were therefore the sexual organs, followed by the fruits; for classification, the investigation of the sex of plants was *maxime necessaria.* These principles were developed and refined in the later works. The awareness that the formulation of a natural method depended on the precision of the classification and the nomenclature drove Linnaeus to state the basis of his philosophy of botanical language with particular care.

When *Fundamenta Botanica* did not enjoy the success he had hoped for and the names he used were generally rejected, Linnaeus wrote *Critica Botanica*[49] to clarify the role of and the innovations in the new botanical nomenclature.

---

[47]"Nos naturalis methodi fragmenta exhibere conabimur" (par. 77), Carl Linnaeus *Fundamenta Botanica* (Amsterdam, 1734), 7.

[48]"Naturae opus semper est species et genus" (par. 162), *ibid.*, 19.

[49]Linnaeus, *Critica Botanica* (1737), translated into English by Sir Arthur Hort (London, 1938).

The main change that occurred after *Fundamenta* was the introduction of the *differentia specifica*, which was supposed to be the *essential* character of the plant. "The *differentia* was a logically composed definitive element containing the attributes that distinguished the species thus named from all other species in the genus. This meant that the *nomen specificum* became something more than merely a name; it became a reflection of the essential nature of the species and threw a direct light on its relationship with other species."[50]

According to Linnaeus the *nomen specificum* had to consist of no more than 12 words, thus being considerably shorter than the long and verbose botanical classifications of the past.

Between 1751 and 1753, Linnaeus published his *Philosophia botanica* and *Species plantarum*, in which he made various additions to his system of naming, the most important being the introduction, or rather the systematic use, of the binomial nomenclature for designating the species. "However the introduction of short convenient specific epithets from everyday use (*nomina trivialia*) was never a primary object of the *Species plantarum*. To Linnaeus the true specific name (*nomen specificum legitimum*) was not an arbitrarily formed biverbal name but a descriptive name which distinguished the plant from all others of its own genus, by expressing differential characters imprinted by the creator."[51] The introduction of the *nomen triviale* and the use of binomial nomenclature were therefore an artificial response to an "economic" demand for synthesis.

Despite all the difficulties encountered by his natural method, Linnaeus never doubted his ability to classify nature with a system of names appropriate to the intrinsic nature of the objects. As it did to the *philosophes*, a mistake in the choice of designation meant to Linnaeus the impossibility of a proper understanding of the natural objects. James L. Larson emphasizes that many of Linnaeus' ideas involved "most of the assumptions of eighteenth-century naturalists concerning the system, the elements, and the representation of natural order."[52] It is therefore more appropriate to discuss some of Linnaeus' ideas, such as his system of nomenclature and his conception of nature, from the point of view of the values that he shared with the Enlightenment in general. This makes it easier to understand why, despite all the controversy aroused by his philosophical conclusions, his ideas were so successful.

From this interpretative perspective, Giulio Barsanti has recently

---

[50]Eriksson (1983), 101.
[51]Stearn (1957), 162.
[52]Larson (1971), 143.

published an article in which he reconsiders the dispute between Linnaeus and Buffon,[53] showing that the philosophical and scientific background to this famous *querelle* between the two naturalists was often one of shared rather than opposed traditions.

Undoubtedly the aim of Buffon was to criticize the method and the philosophy of nature proposed by Linnaeus, but many of his criticisms were based on a lack of understanding of Linnaeus' proposals.

According to Buffon, the methods of classification used by Linnaeus overlooked the infinite complexity and gradations of nature. His method claimed to categorize objects according to distinguishing characteristics, but in doing so, Linnaeus pretended to "evaluate the whole by considering just one part whereas nature proceeds by unknown degrees. Accordingly nature cannot be divided because it proceeds from one species to another and often from one genus to another by imperceptible steps."[54] When this quotation has been taken literally, Buffon has been seen as a continuist and Linnaeus, with his classifications and divisions of nature, as a discontinuist. However, Buffon based his conception of nature on the image of the *scala naturae.* Paradoxically the *scala naturae* displayed natural objects according to the hierarchy of attributes, proceeding from the simplest to the most complex and emphasizing the specific differences between natural bodies rather than their generic similarities.[55] "By contrast, the topographical classification of Linnaeus brings out the affinities between each body of a particular species and with all bodies contiguous to it."[56] Thus Linnaeus' picture of nature showed a universe in which every individual was linked to every other and in which it was impossible to trace any rigid demarcation. This striking alternation of positions is shown more clearly in the attempt made by the two naturalists to classify man. Buffon, the supposed *continuist,* provided man with a soul and placed him at an infinite distance from other animals; Linnaeus, the supposed *discontinuist,* classified man among the quadrupeds.[57]

Another historical myth is that of the contrast between Buffon the *nominalist* and Linnaeus the *realist.* We have already analyzed Linnaeus' position on language and nomenclature. Undoubtedly Buffon criticized the divisions and strongly disputed the existence of the genera, classes, and orders in nature. "In nature," he declared, "only indi-

[53]Barsanti (1984).

[54]Buffon, *Oeuvres* (1954), 10.

[55]Barsanti (1984), 89.

[56]*Ibid.*, 89–90.

[57]*Ibid.*, 90–91.

viduals exist."[58] In that case, the meaning and the role of species in *Histoire naturelle* were not so different from those proposed by Linnaeus in *Systema naturae.* In 1765, in the preface to the third edition of the *Histoire naturelle,* Buffon emphasized in fact that "the species were the only beings of nature."[59] With this declaration Buffon implicitly acknowledged the possibility of a natural classification. The French naturalist was in fact well aware that natural history could not avoid looking for a logical method of classification and nomenclature. "From all we have just said," he admitted, "it follows that there are two equally dangerous pitfalls in the study of natural history, the first being that of having no method, the second that of wishing to relate everything to a particular system."[60] The solution proposed by Buffon to this dilemma was a compromise between conceiving nature as a whole and introducing classificatory divisions that lost sight of its primitive unity. The one and only natural method, therefore, consisted in linking the things that were similar to each other and separating those that were different. As the species were the only products of nature, only a descriptive and a comparative method could be adequate:

> Among natural things, only what is exactly described is well defined; but in order to describe with exactitude, one has to see, see again, examine, compare the things one intends to describe.[61]

A meticulous attention to accuracy of description was to Buffon the only thing that could save the naturalist from lapsing into artificial and subjective systems of classification. But despite these programmatic declarations, Buffon obviously could not classify nature simply by describing it. The French naturalist was forced to use systems and divisions, and above all, he had to use terms appropriate to the nature of the objects concerned.

The distance between Linnaeus and Buffon was therefore not as immense as is commonly believed. Both had the same purpose: to classify natural things by presenting an homogeneous and realistic picture of nature. Both tried to find a natural method, although for Linnaeus the method involved things created as such by the God of

---

[58]"Car en général plus on augmentera le nombre des divisions des productions naturelles, plus on approchera du vrai, *puisqu'il n'existe réellement dans la nature que des individus,* et que les genres, les ordres et les classes n'existent que dans notre imagination" [my italics], in Buffon, *Oeuvres* (1954), 19.

[59]*Ibid.,* 35.

[60]*Ibid.,* 13–14.

[61]*Ibid.,* 14.

the Bible, whereas the God of Buffon was almost synonymous with nature itself. Other important differences obviously remained. Buffon conceived genera of a nominalist character, whereas Linnaeus believed in their existence in nature. Buffon supported a philosophy of nature dominated by the idea of a dynamic chain of natural beings, whereas Linnaeus had a more static vision of nature. As Frans Stafleu has pointed out, one of the most important innovations of Buffon's natural history was the introduction of time in nature.[62] With the publication of the *Histoire naturelle* "we might also say that botany moved from 'definition' to 'description'."[63] Buffon did not intend by this shift of emphasis to diminish the value and scientific importance of words, but more interestingly, he took the discussion forward from where Linnaeus left off. Moreover, his attempt to give pride of place to pure description in the defining of natural objects overwhelmed him with an incoherent multiplicity of objects and created an even stronger urge for a system of classification.

After 1760 the success of the binomial nomenclature in Europe seemed to be establishing the supremacy of Linnaeus once and for all. However, there were some attempts, especially in France, to put the naturalistic appeals of Buffon into practice. The most consistent efforts were made by Adanson, Lamarck, and Jussieu, who developed Buffon's ideas in their own different ways.

Michel Adanson, an experienced botanist, tried to apply Buffon's instructions regarding a *pure* and *natural* description of plants literally. In his *Famille des plantes*, published in 1763, Adanson rejected all the principles of classification proposed by Linnaeus, adopting an "ultra-conservative nomenclature."[64] In order to preserve the natural character of his method, Adanson's descriptions of plants were purely inductive and the naming of them followed neither a classificatory system nor a hierarchy of his observations. In Adanson's view, the flowers, the stems, the pistils, etc. could not be taken as more than preliminary indications for the classification of plants. The results of his over-descriptive approach were pedantic, wordy, and, nevertheless, incomplete descriptions of the plants. Adanson hoped to be able progressively to collect enough empirical data for a systematic comparison that would have led to a complete and natural inventory of all plants and their mutual relations. The enterprise was obviously impracticable, and the

---

[62]Stafleu, *Linnaeus* (1971), 309.
[63]*Ibid.*, 310.
[64]*Ibid.*, 311.

modest impact of Adanson's work testifies more realistically to the fate of Buffon's ideas on method and nomenclature.

In 1778 Jean Baptiste Lamarck published his three-volume *Flore Française*,[65] a work that was highly praised by his master Buffon. In the long *Discours préliminaire*, Lamarck was critical of Linnaeus' systematics and his artificial fragmentation of nature. In Lamarck, too, the intention to create the true natural method was present. In classifying plants, he rejected, like Adanson, all subordination to one or more *essential* characters. "It is necessary," he stated as his first principle, "to make use of all the characters a plant can offer."[66] Accordingly, the exact determination of a plant could be achieved only by what Lamarck called the natural order (*l'ordre naturel*),[67] which was a system incorporating all the relationships between plants and in principle similar to the one sought by Adanson. Considering botany as a science of continuous gradation from the most complex to the simplest, Lamarck insisted that it would be possible to arrange the vegetable kingdom of France according to such a natural order. A comparison of all the characters of similar plants "would offer for our consideration an immense collection of objects, among which each species is distinguished from the others by a perceptible and constant difference; the gradation of these differences is the foundation of the order we are proposing."[68] The determination of genera, for instance, had to be regarded rather as an investigation of the relations between plants than as a means of knowing them and indicating them without error.[69]

As Pietro Corsi has emphasized, "the ideal of a natural classification expressed in the *Discours préliminaire* could not be fulfilled in practice, although the author insisted that the true natural order consisted

---

[65]Jean Baptiste Lamarck, *Flore Française ou description succinte de toutes les plantes qui croissent naturellement en France, disposée selon une nouvelle méthode d'Analyse. . . .*, 3 vols. (Paris, 1779). On the botanical works of Lamarck, see Daudin (1926), 187–203, and the recent study by Pietro Corsi, *The Age of Lamarck: Evolutionary Theories in France 1790–1830* (Berkeley-Los Angeles-London, 1989), 40–55.

[66]Lamarck, *Flore . . .*, vol. I, lx.

[67]"Le but d'un ordre naturel au contraire est d'enchaîner toutes nos idées, de nous faire saisir tous les points communs par les quels les êtres se tiennent les uns aux autres, de n'offrir aucun objet à nos régard, sans nous montrer en même temps tout ce qui existe en-deçà à au-delà, et de nous exercer par ce moyen à ces grandes vues qui parcourent toute la sphère d'un sujet, et qui sont pour ainsi dire le coup-d'oeil du génie," *ibid.*, lxxxvii–lxxxviii.

[68]*Ibid.*, xc–xci.

[69]"La formation des genres par les botanistes modernes doit être plutôt regardée comme une recherche sur les rapports des plantes, que comme un moyen de les connoître et de les indiquer sans erreur," *ibid.*, xxi.

of an imperceptible gradation of form."[70] In fact the projected *Théâtre universel de botanique*, which was to have included an exact analysis of all known plants together with the descriptions and the natural order that such an analysis indicated, was never completed. The infinite variety and multiplicity of plants could not be classified without a rigid and methodical nomenclature. A few years later, when Lamarck was called upon to compile the first botanical volumes of the *Encyclopédie méthodique*, he finally acknowledged the necessity of classificatory methods and nomenclature, declaring: "names as we know are only the signs of our ideas; however, these [. . .] arbitrary signs acquire a *real value* by constant use"[71] [my italics]. Reluctantly the French naturalist was forced to admit that the use of systematic language in science was imperative.

The last attempt to formulate a natural method was made by a descendant of the botanical dynasty of Jussieu, Antoine-Laurent.[72] The purposes and the general principles of his work did not differ from those of Lamarck:

> The plants seem [. . .] to form with each other a continuous chain, of which the two extremes are the smallest herb and the largest tree. By means of a very smooth gradation one goes from one to the other, in the process assigning a place near each other to those of which the affinity is marked by a greater number of mutual relations.[73]

This continuous chain was established by nature, and its order was consequently natural. However:

> The impossibility of bringing together, or even knowing, all the plants that must constitute the general chain will always be an obstacle that cannot be overcome. There will be empty spaces, difficult to fill, but even though nature has dispersed the material necessary for the reconstruction of this order, it enables us at least to get an inkling of the principles on which it is founded. Among the number of characters provided by plants there are a few essential ones which are general and invariable and which apparently must form the basis of the required

---

[70]Corsi, *The Age of Lamarck* (1984), 44.

[71]Article, "Nomenclature," in *Encyclopédie Méthodique: Botanique* (Paris, an IV), vol. 4, 498–499.

[72]"En 1789 "a paru" le *Genera plantarum* de M. de Jussieu, ouvrage fundamental en cette partie, et qui fait, dans les sciences d'observation, une époque peut-être aussi importante que la *chimie* de Lavoisier dans les sciences d'experience," Georges Cuvier, *Rapport sur les Sciences naturelles* (Paris, 1989), 305.

[73]Passage quoted by Stafleu (1971), 326.

order. They are not arbitrary but based on observation, and are to be found only by going from the particular to the general.[74]

The character chosen to this end by Jussieu was the seed, which he considered the heart of the plant. Unlike the pistil, the petals, the stamens, or other reproductive organs, the seed had the practical advantage of constantly retaining its form and consequently its morphological identity. The sexual organs were also regarded as essential characters. By analyzing these organs, Jussieu established which were constant and which varied in each family of plants, classifying the more constant ones hierarchically in order of importance. Finally the families of plants were classified by "a calculation of the importance of the organs and its application to different vegetables."[75] Using this method, Jussieu was able to classify more than 100 primitive families of plants. Within his system, which was indeed very effective, was contained Linnaeus' sexual system, together with its nomenclature. The fight against Linnaeus' linguistic realism and his artificial divisions of nature had not, in the end, any real winners. After the death of Buffon in 1788, Linnaeus was no longer the main target of French botanists. Thirty years later, his ferocious critic Lamarck was able to regard him as "a man of superior genius and one of the greatest known naturalists, who, having collected the facts, taught us precision in the determination of the characters of all the orders."[76]

Condorcet was even more explicit and asserted that Linnaeus' sexual system inspired a revolution in botany,[77] whereas Buffon was "great only within the limits of his talent,"[78] i.e., as a literary figure rather than a scientist.

Another curious episode that shows even better the underlying affinities between the Swedish naturalist and the French academic establishment was the founding of the Société linnéenne de Paris in December 1787 by Pierre Marie Auguste Broussonet, a pupil of a Lin-

---

[74] *Ibid.*

[75] Cuvier (1989), 306.

[76] Lamarck, *Philosophie zoologique* (1809), 2nd ed. (Paris, 1873), vol. I, 131.

[77] "Ce système [of Linnaeus] fit une révolution dans la botanique; la plupart des écoles de l'Europe s'empressèrent de la suivre, et de publier les catalogues de leurs plantes, rangées d'après la méthode de Linnaeus," Condorcet, "Eloge de Linné," in *HARS* [1778] (1781), 72.

[78] "Bernoulli, Linnaeus, Haller, etc. [. . .]," Condorcet wrote to Mme. Suard, "are all far superior to the Comte [Buffon] because he is great only within the limits of his talent, whereas the others have conquered Europe with their discoveries." Quoted in R. Rashed (ed.), *Sciences a l'époque de la révolution française* (Paris, 1988), 462.

naean botanist from Montpellier, Antoine Gouan.[79] Buffon was still alive, but he did not possess enough authority to hinder the creation of a society explicitly celebrating the scientific works of his old enemy. During the 42 meetings held in 1788, the members of the society thoroughly discussed every possible aspect and application of Linnaeus' system and nomenclature. It is interesting to note that the Société linnéenne differed in aim from the Linnaean Society founded in London by James Edward Smith in 1788. The Linnaean Society was in fact "a body of naturalists associated for the purpose of cultivating the Science, *not to list themselves as the followers of any person whatever, any further than truth directs them.*"[80] By contrast, for full membership of the Société linnéenne de Paris, allegiance to the doctrines of Linnaeus was a precondition. Another relevant difference between the two societies was in composition. With the exception of Joseph Banks (honorary member) and Thomas Martyn (fellow), few of the 51 members of the Linnaean Society were more than amateur naturalists. The Société linnéenne de Paris, on the other hand, could list among its members the leading French scientists of the day: Lavoisier, Fourcroy, Guyton de Morveau, Bayen, Seguin, Vauquelin, among the chemists; Lamarck, Gouan, Duhamel, Olivier, Lacépède, Desfontaines, Millin, Bosc, Faujas, Pelletier, Parmentier, Thouin, Daubenton, among the naturalists; Haüy, Sage, Hassenfratz, Monnet, de Lamétherie, among the mineralogists, etc. Equally significant is the presence of many of Buffon's pupils, the only significant omission being Adanson. Unlike the Linnaean Society, the Société linnéenne had a short life, and in 1790 it changed its name to Société d'histoire naturelle de Paris. Despite this change, the name of Linnaeus was still regarded with the greatest respect. On 23 August 1790, this admiration became a cult when a bust of Linnaeus was erected in the Jardin des Plantes in Paris (see Fig. 4). In his reconstruction of French natural history, published in 1792,[81] Millin stressed the debt of French naturalists to the work of Linnaeus. Etienne L. Geoffroy, d'Argenville, and Romé de l'Isle were all influenced by the methods of classification used by the Swedish naturalist. As far as the criticism he received from botanists was concerned, "national pride, esprit de corps and the reluctance of experi-

[79]On the origin and progress of the society, see my "The *Société linnéenne de Paris* (1787–1827)" in *Svenska Linnésällskapets årsskrift* (1990–1991), 151–175.

[80]Quoted in A.T. Cage, *A History of the Linnean Society of London* (London, 1938), 11–12.

[81]Millin, "Discours sur l'origine et les progrès de l'histoire naturelle en France," in *Actes de la Société d'histoire naturelle*, vol. 1 (1792), iii–xvi.

Figure 4 The bust of Linnaeus in the Jardin des Plantes in Paris. From *Actes de la Société d'histoire naturelle*, vol. 1 (Paris, 1792).

enced botanists to abandon the ideas to which they were attached, were the causes of this indifference."[82]

During the first half of the century, only Bernard de Jussieu respected and admired Linnaeus. As a zoologist, Buffon was even more critical than the botanists, but, said Millin, "in recent years [. . .] and especially since his death, the new generation, leaving the former one with its old mistakes and prejudices, has espoused the true principles, those of the Linnaean school."[83] Millin clearly exaggerated the impact

[82] *Ibid.*, xi.
[83] *Ibid.*, xiii.

of Linnaeus, but his words are still significant when considered alongside the traditional picture of a Linnaeus unanimously rejected on French soil. The Aristotelian faith in a correspondence between the definition and essence of objects was mainly criticized for its debt to theological commitment rather than its linguistic structure. In fact the question of the existence of species and genera raised philosophical issues, such as creationism and the structure of nature. Nevertheless, the insistence on realism and on the fundamental importance of scientific language and nomenclature were absorbed and developed in their own way by the French naturalists.

The Birth of Medical Philosophy

Eighteenth-century medicine could not avoid being affected by the *contagion nomenclative*, with both its theoretical and its practical implications. Of the Baconian sciences, the one remaining in the most antiquated state was probably medicine. Except in a few establishments,[84] medical matters were taught without a direct knowledge of practical medicine or surgery. The university medical faculties claimed superiority to the arts of pharmacy and surgery, although these arts enjoyed the possibility of conducting empirical investigations, and the main examples of experimental progress were in the preparation of pharmaceutical remedies and in the work of a few capable surgeons. But even here, the institutional situation still reflected the heritage of the Middle Ages. It was not until 1743, for instance, that a royal decree created a legal distinction in France between barbers and surgeons and between pharmacists and druggists.[85] This archaic state of medical legislation enables us to understand the skepticism of medical faculties with regard to active collaboration with the medical arts. Furthermore, the kind of medicine that was taught within the universities was the preserve of a guild that resembled those of the Middle Ages in its structure and system of privileges. In the eighteenth century, just as in Rabelais' time, the faculties of medicine, law, and theology represented the conservative, hidebound side of the learned world. In this context it is instructive to examine the story of the big-

[84]The Leyden school of medicine led by the charismatic figure of Boerhaave made the development of practical medicine its main concern.

[85]On the institutional status of medicine in eighteenth-century France, see Pierre Huard, "L'enseignement médico-chirugical," in Taton (ed.) (1986), 171–257.

gest hospital in Paris, the Hôtel-Dieu. At the end of 1787, Jacques René Tenon, professor of medicine at the *Collège de chirurgie*, presented a series of memoirs[86] on the state of the hospital to the Academy of Sciences in Paris. The memoirs contained a huge amount of data, statistics, comparative mortality tables, etc., revealing the disastrous situation in French public hospitals, and followed the setting up in 1786 by the Academy of a commission to propose improvements to the Hôtel-Dieu. The commission comprised Daubenton, Bailly, Lavoisier, Laplace, Darcet, Coulomb, and Tenon himself.[87] Most of the *Rapport* of the commission consisted of a description of the state of the hospital. "The greatest disadvantages of the Hôtel-Dieu, and one of the sources of its insalubrity, is the mixing in the same building, often in the same room, of patients who are contagious with those who are not."[88] "Madmen spilled over into the surgical wards. Sufferers from smallpox and syphilis mingled with expectant mothers. Everywhere everything had to be done right there in fetid company: calls of nature answered, bandages changed, infections drained, veins bled, meals provided, operations performed etc."[89] The air of the surgery room was infected almost permanently.[90] The same bed could be shared by three, or even more, pregnant women![91] More generally the contamination of the air caused the spread of all sorts of infection. Air needed to be changed continuously, otherwise adulteration by the patients' breathing, together with the natural presence of nitrogen, would create a suffocating environment.[92] According to the commission, these macroscopic problems explained the mortality rate of the Hôtel-Dieu (one patient in four), which was one of the highest in Europe. No improvement had been achieved since 1721. Of a hospitalized popula-

---

[86]The memoirs were published under the title of *Mémoires sur les hôpitaux de Paris* (Paris, 1788).

[87]"Rapport des commissaires chargés par l'Académie de l'examen du projet d'un nouvel Hôtel-Dieu," in LO, vol. 3, 604–707.

[88]*Ibid.*, 633.

[89]Charles C. Gillispie, *Science and Polity in France at the End of the Old Regime* (Princeton, 1980), 253.

[90]LO, vol. 3, 639.

[91]*Ibid.*, 641.

[92]"L'homme respire sans cesse l'air dont il est entouré, il ne peut s'en passer un instant; c'est cet air qui entretient la vie. Mais la masse entière de l'air n'est pas consacrée à cet emploi; chaque portion d'air est composée, pour les trois quarts environ, d'une fluide nommé *mofette atmosphérique*, dans lequel les animaux ne pourraient vivre, et pour l'autre quart d'un air éminemment respirable, destiné à entretenir la vie, et qui par cette raison a été nommé *air vital*," *ibid.*, 647. These observations were obviously drawn from Lavoisier's studies on the respiration of animals.

tion of 1,108,741 between 1721 and 1773, 244,720 patients had died, a regular mortality rate of around 25 percent.[93]

These incredible figures testified to the general inefficiency of the French medical system. A complete reorganization—political, institutional, and scientific—was therefore urgently needed. According to Tenon, the first requirement was to classify the diseases and to place patients showing different pathologies in different rooms.[94] The classification of diseases demanded a change in hospital building design. The French architect Bernard Poyet designed a hospital that would comply with the contents of the *Rapport* (see Fig. 5).[95] The plan was innovative in ordering and classifying the rooms according to the type of disease, sex, and age of the patients. These apparently simple ideas were a turning point in the treatment of infectious disease and in the organization of medicine as a whole. In no other science had the order and the separation of its objects such *vital* importance. Thus, the classification of diseases was a crucial step both in theoretical medicine and in the social reorganization of the health care system.[96]

The first significant nosological classification had been produced by François Boissier de Sauvages, who in 1731 published a work entitled *Nouvelles classes de maladies,*[97] in which all known diseases were

---

[93]*Ibid.*, 653–654.

[94]J. Tenon, *Mémoires* (1788), 457.

[95]On the ideas of order and classification in the building of eighteenth-century hospitals, see the interesting essay by Thomas A. Markus, "Buildings and Ordering of Minds and Bodies," in Peter Jones (ed.), *Philosophy and Science in the Scottish Enlightenment* (Edinburgh, 1988), 169–224.

[96]The importance of the connection between scientific classification of diseases and public health is well expressed in the following passage of the third report of the hospital commission: "Chaque pavillon sera un hospice, et l'hôpital sera un assemblage de douze hospices; et le système de bâtiments que nous proponons, a tous les avantages de cette espèce d' hôpitaux sans en avoir les inconvéniens. Le plus grand de ces inconvéniens est de ne pouvoir qu'exclure certaine maladies, sans pouvoir les distinguer et les séparer. Ici, elles sont toutes reçues et toutes classées; chacune aura son departement, fermé s'il le faut [. . .].

Le soin de classer les maladies est en effet important, et on peut y satisfaire au moyen de nos subdivisions qui sont plus nombreuses qu'il ne faut. La connoissance du nombre des maladies que peut fournir chaque espèce de maladies, seroit utile pour savoir d'avance combien de subdivisions on doit leur attribuer." "Troisième rapport des commissaires chargés par l'Académie de l'examen des projets relatifs à l'établissement de quatre Hôpitaux," in *HARS* [1786] (1788), 37.

[97]F.B. de Sauvages, *Nouvelles classes de maladies, qui dans un ordre semblable à celui des botanistes, comprennent les genres et les espèces de tous les maladies, avec leurs signes et leurs indications* (Avignon, 1731). On the influence of Sauvages on the progress of systematic classification in medicine, see the study by Fredrik Berg, "Linné et Sauvages, Les rapports entre leurs systèmes nosologiques," in *Lychnos* (1956), 31–54.

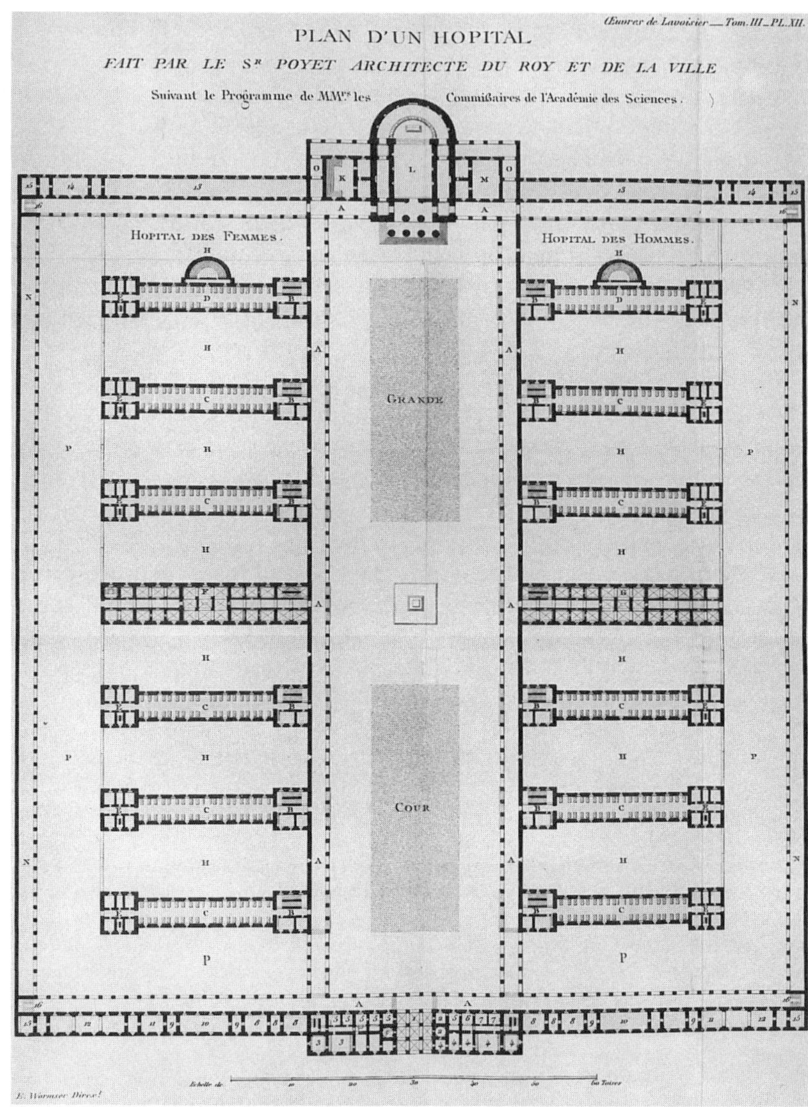

Figure 5 Plan of the hospital projected in 1786 by the French architect Bernard Poyet according to a new classification of diseases. From LO, vol. 3.

subdivided into genera and species in the manner adopted by the botanists. However, in the stagnant atmosphere of medicine in the first half of the eighteenth century, this work failed to have any real influence, being ignored for more than 30 years.

The reform of medicine had to overcome the institutional conservatism of the universities if it was to be successful. One important step in this direction was taken by the anatomist Félix Vicq d'Azyr when he founded the Société Royale de Médicine in 1776.[98] The aim of the Société was to promote systematic research into practical medicine and anatomy and, above all, into epidemics, epizootic diseases, and other illnesses of particular social importance.[99] This brought Vicq d'Azyr into conflict with the medical faculties, whose interests remained circumscribed by the narrow horizon of abstract medical disputes. The most important innovation of the Société was to introduce instruction in surgery as an integral part of the teaching of medicine,[100] which had the immediate consequence of causing the physicians of the Société to place a new emphasis on experience and observation. But the greatest merit of Vicq d'Azyr, as of Lavoisier, Condorcet, and Turgot, was in studying the general structure of nature and society from the point of view of his particular discipline and specialization. It was this general commitment that gave direction to his medical reforms. At the same time, he attempted to use methods borrowed from philosophy and other Baconian sciences. The problem of the internal organization of medicine was tackled by studying various themes, such as the classification of diseases, their nomenclature, the relation between the physical and the moral, and the need to agree on a normative method of dealing with scientific problems. This last matter is the one that concerns our historical investigation most closely, but from what is outlined above, the emergence of a close interaction between the political and institutional sides and the theoretical and philosophical sides of medicine should be apparent.

The preconditions of an effective political reform of medicine included an ordered nosology and a practical system for naming dis-

[98]The best study on Vicq d'Azyr and more generally on the attempt to reorganize medicine in eighteenth-century France is by Sergio Moravia, "Filosofia e medicina in Francia nel XVIII secolo," in *idem., Filosofia e scienze umane nell'età dei Lumi* (Florence, 1982), 109–192; on the problem of classification in medicine, see the excellent study by William Randall Albury, *The Logic of Condillac and the Structure of French Chemical and Biological Theory, 1780–1800* (Baltimore, 1972), 92–109.

[99]Moravia (1982), 111–112.

[100]*Ibid.*, 115.

eases. Vicq d'Azyr, among others, took on the daunting task of producing them, which he saw as follows:

> To restore order in the human understanding applied to the study of the various sciences it is necessary to remake their languages. What, in fact is it to study a science? It is to acquire ideas of all the parts which compose it. It is to associate those ideas in such a way that their impressions are reproduced by themselves and succeed one another without effort or labor. It is to order them so that some of them—formerly individual but now become general—are subdivided into classes, genera and species; while the rest, isolated, await new filiations. It is to take care, while going from the known to the unknown, both for the exactitude of the facts in observation and for the chain of intermediate judgments in reasoning. Finally, it is to learn to put the entire activity of the mind to work in fixing the nature and relations of one's thought by means of words and signs.[101]

As Condillac had stressed in his philosophical writings, language was not only a means of communication, but also a vehicle for achieving new knowledge and enriching our capacity for understanding nature. By the analysis of language, man associated ideas with the corresponding facts; a well-ordered language and nomenclature simplified not only the apprehension of a science, but also scientific endeavor itself. Even if medicine was still very complex in the number and quality of its objects, Vicq d'Azyr believed in a systematic inventory of the human body and its diseases. The idea of language as an analytical method of displaying all the scattered parts of any body of knowledge as a whole was also borrowed by Vicq d'Azyr from the analytic philosophy of Condillac.

The sciences of geometry, physics, natural history, and, finally, chemistry succeeded in applying the analytical method to their nomenclature. It was now time to do the same thing for medicine. With this project in mind, Vicq d'Azyr set down preliminary conditions for the construction of an anatomical nomenclature:

> It would be, perhaps, a useful enterprise to substitute for the old nomenclature of anatomy an entirely new one in which the names in the different classes would have a regular correspondence in respect to their genera, their composition and their suffixes—a nomenclature submitted to constant rules, the methodical distribution of

---

[101]F. Vicq d'Azyr, "Deuxième discours sur l'anatomie," in *Oeuvres* (Paris, 1805), vol. 4, 211; English translation by Albury (1972), 101.

which would be such that the mind would easily grasp its plan and the memory effortlessly preserve it.[102]

The ease with which the human mind and memory would retain such a nomenclature was not due only to the efficiency of its linguistic devices but also, and primarily, to its identity with the "natural succession of ideas."[103] Vicq d'Azyr, like his contemporaries, attached great weight to the naturalistic aspects of analysis and a nomenclature calculated to reduce nature to reason and vice versa. It was no coincidence, therefore, that the members of the Société called themselves *medicins-philosophes.*[104]

Vicq d'Azyr set forth four general principles regulating the nomenclature of anatomy:

1. Every organ which one proposes to describe ought to be treated as a geometric solid, of which one shall first examine the faces, edges, and angles of the exterior, and afterward the same divisions of the interior.
2. In the denominations which one gives to the faces, edges and angles of these organs one shall employ only names which one can apply to all the animals having these organs; and these names shall be composed from the most remarkable parts of the organs, or from those of their surrounding regions, or from their uses when these are well established and so easy to grasp that there cannot be any ambiguity in that regard.
3. There are no expressions which can replace the words *anterior, posterior, superior* and *inferior* as characters of a general division for the whole extent of the body of man and of the animals [. . .]. Thus it will be necessary to substitute for these expressions which are peculiar to each of the great regions of the animal body [. . .].

   I shall call *syncipital* that [face of the geometric figure] which is directed toward the summit of the frontal bone, or synciput; *basal* that which answers to the base of the skull; *facial* that which is turned toward the face; and *occipital* that which is turned toward the occiput. One sees that this nomenclature can be extended to all the animals which have a bony head, since in all of them the synciput is opposite the base of the skull and the face is opposite the occiput.
4. Not only ought the corresponding regions of the same organ to be designated in the same manner, but these organs themselves ought to bear the

[102]Vicq d' Azyr (1805), 212; English translation by Albury (1972), 106.

[103]"Dans l'ordre des nos recherches il faut choisir les mots propres à la formation des noms génériques et specifiques avant de définir; et il faut définir avant d'analyser. L'analyse ou la division est, au fond, la même opération de l'esprit: c'est dans la succession naturelle des idées, c'est dans la manière dont on les acquirent et dont les enchaîne qui il faut chercher les élémens de cette méthode" Vicq d'Azyr (1805), 215.

[104]Moravia (1982), 183.

same name in all the animals; without which the comparisons which our work requires could never be carried out.[105]

Towards the end of the century, with Lavoisier's nomenclature triumphant and Condillac's philosophy canonical, the physicians intensified their efforts to reform medical nomenclature and bring it closer to nature. In 1796 Constant Duméril published an essay on a reform of anatomical nomenclature by which he hoped to establish "the means to compose expressive terms taken from the *nature* of the objects they are meant to describe"[106] [my italics]. In the same year, Philippe Pinel published his *Nosographie philosophique*,[107] one of the most important medical treatises of the second half of the century. Pinel, like Vicq d'Azyr, was aware that to solve the theoretical problems that burdened medicine first required a systematic classification of the diseases, but he was unimpressed by the previous attempts of Boissier, Adam Nietzki, William Cullen, Linnaeus, Christian Gottlieb Selle, and Jean Baptiste Michael Sagar, who had tried to classify diseases by ordering them in classes, genera, and species as if they had the same properties as plants or minerals.[108] These attempts were arbitrary, and the methods were artificial. The only method based on observation and on the structure of nature was the analytical one. Hippocrates was the first to put this method into practice, Condillac the last to provide the conceptual framework for using it in the classification of diseases.[109] These being the sources, it was obvious that "a methodical and controlled distribution presupposes a *permanent order* in its object, which is subject to certain general laws"[110] [my italics]. Accordingly, "the diseases, studied in relation to their affinity, form a natural chain of ideas. Like all the other objects of natural history, they are classified by their external characters and they are finally given *exact and invariable denominations*"[111] [my italics].

---

[105]Vicq d'Azyr (1805), vol. IV, 222–224; English translation by Albury (1972), 104–105.

[106]Constant Duméril, "Projet d'une nomenclature anatomique basée sur la terminaison," in *Magazin Encyclopédique*, 2 (1796), 455.

[107]P. Pinel, *Nosographie philosophique ou la méthode de l'analyse appliquée à la médicine*, 2 vols. (Paris, 1796). The title of the work illustrates the success and the influence of Condillac's analytical method.

[108]*Ibid.*, vol. 1, IV–V. On this, see Moravia (1982), 175–177.

[109]Pinel (1796), xi–xii.

[110]*Ibid.*, vi.

[111]Pinel, *La médicine clinique rendue plus précise et plus exacte par l'application de l'analyse*, 2nd ed. (Paris, 1804), ix.

The traditional nosologies, which confused the different stages of one and the same disease with different diseases, had to be replaced by a natural method, capable of recognizing similarities and of overlooking inessential details. However, the analytic method and observation alone could not solve all the difficulties involved in creating a rational nosology. The terminology had to be reformed as well. Pinel argued that the only way in which medicine, like natural history, could deserve to be called a science was by the reform of its language, together with its classificatory criteria.[112] "In the exposition and categorization of the history of diseases, my constant aim . . . was to put a strict exactitude into the designations."[113]

By the end of the eighteenth century, the link between language, philosophy, and nature was being asserted by an increasingly large number of physicians. Charles-Louis Dumas paid particular attention to the problems in most of his works on medicine. According to Dumas, in medicine as in all other branches of knowledge:

> Things at first make themselves known by sensation, then by appropriate designations, and the name of each thing must be shaped to the idea which we have of it and which sensation has given us.

Thence:

> Let us conclude that a language becomes more applicable to sciences the more it agrees both with nature and with our order of ideas. The goal of any reform of the language of a science must be to connect words with things, by attaching to them the impression of the ideas we have conceived.[114]

The study of anatomy, too, was determined by nature itself:

> *This truly analytical method is based on the natural one [. . .] and I have adhered to it with advantage in my methodical system of nomenclature and classification of the muscles of human body* when I related all the known

---

[112]*Ibid.*, 325–326.

[113]"Le but constant que je me suis proposé ici dans l'exposition et la distribution de l'histoire des maladies, celui que doivent avoir sans cesse en vue ceux qui cultivent la médicine philosophique, ou qui ne la font point consister dans la prescription aveugle de vaines formules, c'est de mettre exactitude sévère dans les dénominations: car le premier objet à remplir dans une science quelconque est de s'entendre; ce qui devient impossible lorsque la vraie signification des termes n'est point fixée," *ibid.*, 377.

[114]See C.L. Dumas, *Système méthodique de nomenclature et de classification des muscles du corps humain* (Montpellier, 1797), 2 and 5.

> muscles to forty-seven regions, [. . .] presented in a natural sequence.[115] [my italics]

The attitude of Pierre Cabanis to this problem, which was similar to that of Buffon in natural history, did not appear at first sight to be in tune with the general enthusiasm for medical philosophy among his colleagues.[116] He seemed to be reluctant to attribute any epistemological value to scientific language and to a systematic approach to nosology. He explained his skepticism by pointing out that the evolution of each disease had its own individual and specific stages that defied rigid and unambiguous classification. According to Cabanis, a rigid classification entailed a danger of superficial clinical observation and diagnosis. But despite his opposition in principle to methodical nomenclature, Cabanis himself proposed a philosophy of medical language. In effect,

> the classifier and the empirical philosopher, when equally talented, do not follow such different paths as we might suppose. Nature leads them both, as if by the hand; she shows them the objects in their true colors and she impresses these same objects on their memory by means of certain striking traits, she classifies them by real analogies or dissimilarities. . . . This method of Nature is as simple as it is eternal and fertile.[117]

The main characteristic of this natural method was the observation of the symptoms of all the stages of a single disease. By noting them and, in a second process, comparing them with the symptoms of other diseases, it was possible to obtain a satisfactory picture of the natural gradation of all diseases. Thus, symptoms were the essential characters of medical knowledge, the features on which a natural nosology could be based. "The symptomatic method is the work of nature herself."[118] Within this naturalistic framework, the function of words was not to be

---

[115]Dumas, *Principes de physiologie ou introduction à la science expérimentale, philosophique et médicale de l'homme vivant,* 4 vols. (Paris, 1800), vol. 1, 31.

[116]On this aspect of Cabanis' thought, see Dagognet (1970); Sergio Moravia, *Il pensiero degli Idéologues: Scienza e filosofia in Francia (1780–1815)* (Florence, 1974), 13–23; *idem.* (1982), 183–213. The valuable work by Martin S. Staum, *Cabanis: Enlightenment and Medical Philosophy* (Princeton, 1980), is restricted mainly to an analysis of Cabanis' medical doctrine.

[117]On this aspect, see Pierre Cabanis, *Du degré de certitude de la médicine* (Paris, 1798), 76–77. This work had been ready for publication in 1789, but with the events of the French Revolution, Cabanis had other priorities.

[118]*Ibid.*, 81.

conventional. Being the effects of our perceptions, "words, so to speak, catch the sensations, they summarize and fix them."[119]

These views were not very different from those of Vicq d'Azyr, Pinel, and Dumas.

To summarize this chapter, two central themes emerge as the main characteristics of the scientific thought of the French Enlightenment:

1. The study of a particular discipline was always undertaken within the more general framework of the study of nature and each science provided a particular angle from which to view nature. Naturalistic arguments were used to legitimate scientific activities objectively and institutionally.

2. Directly connected to this, the second general theme of the century was the development of the nomenclatures, methods, systems, and logic used to classify nature. The differences between the examples did not obscure the common naturalistic purpose. Condillac and Diderot, Linnaeus and Buffon, Vicq d'Azyr and Cabanis, though considering different fields, sought to devise a natural method of classifying and naming the objects of their respective sciences.

It is a truism to argue that none of these methods was really natural or that the picture of nature produced by the Enlightenment was too vague and abstract. But what is a truism today is often the product of earnest discussion and heated debate in the past.

---

[119]Cabanis, *Coup d'oeil sur les révolutions et sur la réforme de la médicine* (Paris, 1804), 162.

CHAPTER THREE

The Birth of Chemical Language (1500–1782)

"Mugghiava con la voce dell'afflitto,
Se che, con tutto che fosse di rame,
Pure el parea dal dolor trafitto;
Così, per non aver via nè forame
Dal principio nel foco, in suo linguaggio
Si convertian le parole grame."

Dante Alighieri, *La Divina Commedia, Inferno,* Canto XXVII, 10–15

"La logique est inutile pour inventer et donner de la science."

Jean-Baptiste Van Helmont, *Oeuvres* (Lyons, 1670), 43.

"Utilitas expressit nomina rerum."

Lucretius, *De Rerum Natura,* Liber Quintus, 1029.

The Question of the Sources

As we saw in Chapter Two, a need arose during the eighteenth century for the Baconian sciences to devise a method of classifying and naming the objects of their study. If the general response to this need was to try to devise a naturalistic classification of phenomena, the solutions to the specific problem of a technical nomenclature obviously varied widely between the different sciences with their different assumptions and conditions.

Notwithstanding their linguistic and epistemological inferiority to mathematics and astronomy, each Baconian science had its own specific lexical tradition to reform and the chemical one was particularly complex. An abundance of chemical terms of all kinds and etymologies was the burdensome medieval inheritance of the humanists of the early Renaissance when they began to take a closer interest in philology.

Before the seventeenth century, there is no evidence of any serious attempt to reform the metallurgical, mineralogical, and chemical nomenclatures[1] or to establish new ones with systematic rules and methods.[2] The question of a "scientific" language seems to arise during the Renaissance, when the Greek and Latin sources were translated into the vernacular[3] and also particularly when the classical

---

[1]I relate here the terms *chemical*, *metallurgical*, and *mineralogical* to classical antiquity only for convenience. It is obvious that a distinction between chemistry and mineralogy is feasible only in the seventeenth and eighteenth centuries. The chemical and mineralogical knowledge of the ancients was an integral part of their philosophy or of the *materia medica*. As has been recently shown in the important works by Dietlinde Goltz, *Studien zur Geschichte der Mineralnamen in Pharmazie, Chemie und Medizin von den Anfängen bis Paracelsus* (Wiesbaden, 1972), and Robert Halleux, *Le problème des métaux dans la science antique* (Paris, 1974), a correct approach to the mineralogical and chemical knowledge of the ancients has to rely on a *philological* analysis of the sources rather than on a *historical* one.

[2]This does not mean that the ancients named minerals totally without rhyme or reason. But the names they used denoted many different characterisitcs, such as color, therapeutic properties, geographical provenance, etc., and there was no method or consistency in their practice.

[3]By scientific language I thereafter mean the language established by the Aristotelian philosophy of nature and accepted as the vernacular of science in most Renaissance universities. On the linguistic aspect of Renaissance science, see the classical study by Leo Olschki, *Geschichte der neusprachlichen wissenschaftlichen Literatur*, 2 vols. (Heidelberg-Lepzig, 1919–1922).

language could not provide an adequate term to denote a new substance, disease, engine, etc.

The task of assessing the importance of language in science during the sixteenth and seventeenth centuries has posed questions which have found little consensus among historians of Renaissance science and culture. Historians confronted with this problem have differed profoundly in their evaluation of the relationship between magic and experiment, science and pseudoscience, and in their perception of a continuity or discontinuity between alchemy and chemistry. Lynn Thorndike was among the first historians of science to take up a definite position on this issue, and as early as 1923, he declared that "magic and experimental science have been connected in their development; [. . .] magicians were perhaps the first experimenters."[4] In 1923 this famous statement was an isolated voice raised against the more common belief in the experimental and humanist origin of modern science expressed by George Sarton. Today these positions are completely reversed. In reconstructing the history of modern science, it now becomes more and more difficult to maintain a sharp distinction between such different intellectual traditions as magic and experimental sciences. This trend has been useful in challenging the unilateral and positivist tendency to consider the history of science within the rigid interpretative framework of a progression and accumulation of positive experiments. However, some historians have followed it to the point of conjuring up a world of fancy in which magical formulas and hermetic faith alone inspired the first steps towards the abandonment of the Aristotelian world view and the establishment of modern science. The unfortunate effect of this confused and confusing interpretation can in particular be seen in studies of early chemistry. Historians who have examined the state of chemical knowledge during the Renaissance have looked for the source of early chemical language in the alchemical, hermetic, and magical traditions.[5] The success of Paracelsus's works and doctrines has probably been one of the main causes of this interpretative trend. Starting from Paracelsus's popularity during the Renaissance,

---

[4]Lynn Thorndike, *A History of Magic and Experimental Science* (1929), vol. 1 (New York, 1949), 2.

[5]See, for instance, F.A. Yates, *Giordano Bruno and the Hermetic Tradition* (London, 1964); A.G. Debus and R.P. Multhauf, *Alchemy and Chemistry in the Seventeenth Century* (Los Angeles, 1966); F.A. Yates, "The Hermetic Tradition in Renaissance Science," in C.E. Singleton (ed.), *Art, Science and History in the Renaissance* (Baltimore, 1968), 255–274. The limitations and the theoretical dangers of these interpretations of Renaissance science have been convincingly criticized by Paolo Rossi in his "Tradizione ermetica e rivoluzione scientifica," in *idem, Immagini della scienza* (Rome, 1977), 149–181.

and following a very peculiar chronology, historians have claimed to find the origins of the early chemical lexicon in the writings of Hermes Trismegistus and in medieval alchemy. This interpretation is so established that the development of chemical nomenclature as the result of a slow and difficult process of emancipation from alchemy is regarded almost as proven.

Undoubtedly the lexicon of seventeenth and early eighteenth-century chemistry did indeed owe many of its colored expressions to the Paracelsians and to the alchemical and magical tradition that sustained his works. However, a realistic portrayal of the development of the language of chemistry has surely to take into account the metallurgical and mineralogical traditions.

When we compare the language of Renaissance alchemy with that of seventeenth-century chemical treatises, we can see that only a part of the nomenclature of the latter derives from the Hermetic or Paracelsian tradition. An enormous number of terms were taken from metallurgy and mineralogy. The names of many minerals, gems, stones, rocks, earths, and compounds and of some inflammable substances were the legacy of a naturalistic tradition that was often explicitly opposed to alchemy.

Despite the importance of practical knowledge in the mining industry of the early Renaissance, its role in the development of chemistry has only sporadically been studied. On the contrary, by stressing the fact that the Renaissance alchemists pursued their research in chemical laboratories and borrowed many mineralogical notions, most historians of chemistry have emphasized the historical influence of magic and alchemical ideas on the birth of seventeenth-century experimental chemistry very heavily. In reality, early chemistry is a synthesis of two conflicting traditions. The social contexts from which alchemy and metallurgy emerged and the aims they pursued were, in fact, completely different. We observe, on the one hand, the first attempt to classify minerals and certain chemical substances in a logically ordered lexicon, capable of efficiently meeting the practical needs of miners and accessible to the largest possible audience, and on the other hand, the use of metaphors, obscure symbols, and hermetic expressions devoid of any form of rational and scientific communicative content except to the adept of the supreme art. The highly developed division of labor in Renaissance mines, the rich system of laws that regulated their organization, and the close collaboration between theoreticans and the *Bergmeister*, the smelter, the assayer, the melter, the washer, the digger, the lawyer, the local authority, etc., made metallurgy a shared body of knowledge essential to the prosperity of entire communities. By contrast, "the alchemist [. . .] worked alone. He formed no school. This rigorous solitude, together with his preoccupation with the endless obscurities of the work, was sufficient to activate the unconscious and, through the power of imagination, to bring into being things that apparently were not before."[6] The difference between metallurgy and alchemy was not merely confined to their languages and social background; Hélène Metzger and Robert Halleux have convincingly pointed out that whether the alchemist used the same notions and instruments as the metallurgist or not, his epistemological purpose did not rely on the decisive role of *experience* and empirical verification as did that of the miner and the metallurgist.[7] In fact, the alchemists did not submit their research into the transmutation of metals to empirical confirmation because the essence of their beliefs transcended the empirical authority of facts. According to Paracelsus:

> Magic has power to experience and fathom things which are inaccessible to human reason. For magic is a great secret wisdom, just as reason is a great public folly.[8]

Thus, the experience of the alchemist is completely different in character from the accessible experience of vulgar reason.

The assertion of this philosophy of nature was, in a more general context, typical of the continuing opposition during the sixteenth century between humanism and the *Counter Renaissance*, between the revival of the classical tradition and the reaction against it. With this cultural dichotomy in mind, I have tried to depict the complexity and variety of sixteenth-century chemical knowledge.

There is a risk of interpreting the progress of early "chemistry" either as the result of the magical aspirations of the alchemical tradi-

---

[6]C.G. Jung, "Paracelsus as a Spiritual Phenomenon," in *idem*, *The Collected Works*, vol. 13 (New York, 1967), 179. Jung has been one of the few scholars to understand that a correct and historical understanding of alchemy has to investigate its spiritual, mystical, cosmological, and philosophical sources rather than its pretended experimental continuity with chemistry.

[7]"En alchimie, le laboratoire n'a pas de rôle crucial. La fonction de la pratique est, d'abord, d'illustrer la vérité de la théorie. La réussite d'un procédé démontre à l'opérateur qu'il a bien compris les anciens. La qualité de la pratique est une conséquence directe du niveau de compréhension de la théorie.

Car si l'expérience échoue, l'échec n'infirme pas la théorie," from Robert Halleux, "Pratique industrielle et chimie philosophique de l'Antiquité au XVIIe siècle," in *L'actualité chimique* (Janvier-Février, 1987), 19. See also the outstanding contribution by H. Metzger, "L'évolution du règne métallique d'après les alchimistes du XVIIe siècle," in *Isis*, 4 (1922), 466–482.

[8]Paracelsus, *Selected Writings* (1951), 3rd ed. (Princeton, 1988), 137.

tion or as the sober and practical expression of the metallurgical and mineralogical one. A study of early chemical names shows the importance of both of these traditions. My main aim is to avoid understating or exaggerating the influence of either tradition and to follow the prudent warning of Charles B. Schmitt:

> Just as the positivists have projected their own criteria of scientificity onto the past and thus failed to understand the context from which seventeenth-century scientific thought emerged, so too those who have followed the line of approaching "science" through the Renaissance tradition of magic often fell into the other extreme by failing to apply precise intellectual categories to historical information.[9]

"Die Wissenschaft des Seigerns begriffen"

During the sixteenth century, the mining industry went through a period of rapid growth, particularly in Germany, France, Sweden, and Italy. The sudden need for iron, for armaments, and for primitive manufactures, inspired new efforts to improve and systematize metallurgical knowledge.[10] Accordingly, the quality and speed of progress in the mining industry, which had relied upon the individual ability of craftsmen until the end of the fifteenth century, now began to depend increasingly on a "scientific" elaboration of the miners' technological knowledge.

In this situation, where there was an urgent need for metals and a lack of appropriate scientific knowledge of them, the first attempts to systematize metallurgy had to fulfill the practical and economic purposes of the emerging mining industry as well as satisfy the scientific curiosity of the learned reader. A metallurgical treatise had to reveal clearly the mysteries hidden in the subterranean veins to both the miner and the naturalist.

Georgius Agricola understood perfectly this double aspect, practical and scientific, of metallurgy, and he realized that the valuable empirical knowledge of the miners needed to be underpinned by a corpus of scientific theory. Agricola accomplished this task by incorporating all the mineralogical notions of the ancients, from Aristotle and Theo-

---

[9]Charles B. Schmitt, "Some Considerations on the Study of the History of Seventeenth-Century Science: Lessons from Hélène Metzger," *Corpus*, n. 8–9 (1988), 26.

[10]On the connection between the development of metallurgy and sixteenth-century warfare, see Lewis Mumford, *Technics and Civilization* (1934) (London, 1955), 65–77.

phrastus to Pliny and Dioscorides, in a new compilation of observations and experiments collected in the field in more recent times. To understand fully the cultural background to this project, we have to bear in mind that Agricola was a friend of Erasmus and a humanist himself; the attention he paid to the classical tradition was therefore rather an indication of his belief in its scientific value than a display of dry erudition. Greek and Latin sources showed humanists the right way to investigate nature and accordingly the humanists tried to reconcile the classical sources with more recent observations and knowledge. This liaison with the classical tradition was so strong that when accommodation with the sources was impossible, they still made the mental effort to consider the new phenomena or the new theories from a classical viewpoint. In these circumstances it is understandable that the composition of a scientific treatise presupposed a philological exploration of the classical sources and a careful decoding of their lexicon. Indeed, the question of language became equally relevant to the development of any science. Only after tracing a word back to its classical derivations could a humanist feel satisfied as to what it truly meant. What Paracelsus considered vain and arid semantic quibbles were in reality slow but important steps towards a detachment from the tradition to which the terms belonged. The impossibility of describing what they could now see with the limited range of the Greek and Latin lexicon inspired the humanists to create a new scientific nomenclature.

Agricola is one of the most notable representatives of this scientific humanism; before undertaking his reform of the metallurgical knowledge of antiquity, he had published Latin and Greek grammars and a commentary on the medical works of Galen.[11]

In his first metallurgical work, published in 1530, he drafted a mineralogical lexicon based on the classical one,[12] giving the name and a description of each metal and mineral. To understand the difficulties he had to face, it is enough to recall that the word *metal* still had a very general meaning in the sixteenth century. The Greeks used the

---

[11]On Agricola's non-mineralogical works, see the still valuable bibliographical remarks of H.C. Hoover, in G. Agricola, *De re metallica* (1912), 2nd ed. (New York, 1950), 650. For a wider reconstruction of metallurgy in the seventeenth century, see the surveys by Hans Prescher, "Georg Agricola," supplementary volume of the reprint of Agricola's *Von Bergwerck XII Bücher* (1557) (Leipzig, 1985), and by Pamela O. Long, "The Openness of Knowledge: An Ideal and Its Context in 16th-Century Writings on Mining and Metallurgy," in *Technology and Culture*, 32 (1991), 318–355.

[12]Georgius Agricola, *Bermannus, sive de re metallica* (Basel, 1530). I have here used the excellent edition edited by Robert Halleux (Paris, 1990).

word *métallon* to designate a mine or a vein rather than any specific mineral, and the Latin *metallum* did not modify its meaning.[13] Agricola managed to be more precise, and in 1546 he stated:

> A metal is a natural mineral body which is either liquid or solid and will melt in fire. The molten metal, on cooling, again becomes hard and returns to its original form. In this it differs from a stone that melts in a fire. The molten stone becomes hard when cool but does not return to its original form and appearance.[14]

Interestingly, Agricola's definition of a metal was derived from the technical operation of smelting. Although still very general, it marked an important step forwards. This advance, as Robert Halleux has pointed out, was due to the technical knowledge of metals obtained during the Renaissance. The introduction of smelting furnaces and of more advanced pumping and hauling machinery added to the miners' knowledge of minerals and enabled them to observe their specific characteristics, such as fusibility and expansion; it is significant that the Latin synonym used by Agricola for metallurgist (*excoctor*) was derived from smelting.[15] Furthermore, the comparison of the classical sources with the new mining industry inevitably posed new problems, such as the discovery of metals unknown to the ancients. The astrological association of seven metals with the seven known planets was seriously called in question by the identification of bismuth (called *bisemutum* by Agricola) as a new metal.[16] Discoveries, such as this, led Agricola to recognize that sometimes the knowledge of the ancients was incomplete:

> Now I realize that Pliny, though diligent and learned, had ignored many things, especially in the mineral kingdom. He only translated into Latin what he read in the works of the Greeks.[17]

These gaps in the knowledge of the ancients were discovered by Agricola in the course of his philological analysis of their texts and by constant comparison of the Greek and Latin names with the experience of the miners. The lexicographical lacunae raised new questions with regard to the defining of newly discovered substances or compounds. How to derive the new names? What kind of language to use? Which criterion to adopt: the philological or the etymological?

---

[13]Robert Halleux (1974).

[14]Agricola, *De natura fossilium* (1546), English translation (New York, 1955), 18–19.

[15]Agricola (1950), 34.

[16]Agricola (1990), 67.

[17]*Ibid.*, 71.

The publication of *Bermannus* encouraged other naturalists into metallurgical research. In 1540 Vannoccio Biringuccio's *De la Pirotechnia,*[18] a technological treatise covering all the metallurgical knowledge of the sixteenth century, was published posthumously. According to Lynn Thorndike, however, the work has to be considered no more than "a sixteenth-century version in Italian of what one might find in Latin works of the three previous centuries."[19] This unfair judgment is consistent with Thorndike's view of the experimental science of the Renaissance as sterile, or even reducible to its magical or esoteric background.[20] In reality Biringuccio made important and original contributions to metallurgy, such as new methods of making steel and glass and of producing saltpeter,[21] all products that became important with the technological development of the sixteenth-century mining industry and with the renewed needs of warfare. Furthermore, the detailed description of the chemical laboratory, of machinery for extracting metals, of smelting furnaces, etc., made the treatise a highly successful editorial enterprise.[22] Another important and original aspect of Biringuccio's work was the criticism of alchemy and of belief in the transmutation of metals:

> The more I look into this art of theirs, so highly praised and so greatly desired by men, the more it seems a vain wish and fanciful dream that it is impossible to realize unless someone should find some angelic spirit as patron or should operate through its own divinity. Granted the obscurity of its beginnings and the infinite processes and concordances that it needs in order to reach its destined maturity, I do not understand how anyone can reasonably believe that such artists can ever do what they say and promise.[23]

Biringuccio was equally derisive concerning the existence of the *potable* gold, the *philosopher's stone,* and the *quintessence,* which, despite all the efforts of their inventors, did not bring any of them back to

---

[18]Vannoccio Biringuccio, *De la pirotechnia* (Venice, 1540), translated and reedited in English by Cyril Stanley Smith and Martha Teach Gnudi under the title *The Pirotechnia* (1942), reprinted (Cambridge, Mass., 1959).

[19]Thorndike (1949), vol. 5, 544.

[20]Even less satisfactory is Thorndike's treatment of Agricola, whose historical significance and influence have been overlooked.

[21]See the introduction by C.S. Smith to Biringuccio, *The Pirotechnia* (1959), xvi–xvii.

[22]The book was reprinted in Italy in 1550, twice in 1559, and again in 1678. It was translated into French in 1556, and reprinted in 1572 and 1627. A part of it was translated into Spanish in 1569 and into English in 1555 and 1577. See *ibid.*, 458–461.

[23]*Ibid.*, 35–36.

life.[24] Biringuccio's main reason for rejecting alchemy lay in the cultural legacy on which he relied. His conception of matter was explicitly shaped by the Aristotelian doctrine of four elements. The impossibility of transmuting metals was argued with an appeal to the authority of Aristotle rather than from a particularly innovative experimental awareness. Biringuccio believed that by altering the structure of natural phenomena the alchemists were trying to substitute themselves for God. Along with the Church, he considered their audacious enterprises blasphemous and unnatural:

> They [the alchemists] could indeed ridicule Nature, as when they say that with their medicine they wish to correct her defects and faults, reducing imperfect metals to that perfection which she, in her weakness, has not been able to reach . . . .
>
> And finally, taking all the alchemistic principles and comparing them with the processes of Nature, and pondering on the procedure of the one and of the other, it seems to me that there is no proportion between their powers, granting that Nature operates in things from within and causes all of her basic substances to pass wholly one into the other; while art, very weak in comparison, follows Nature in an effort to imitate her, but operates in external and superficial ways, and it is very difficult, even impossible, for her to penetrate things.[25]

Biringuccio's naturalism was typically Aristotelian, and if we take into account the fact that he wrote *De la Pirotechnia* under the pressure of specific practical and technological needs, such as the production of saltpeter and fireworks, the manufacture of nitric acid, etc., we can better understand his distaste for the abstract speculations and the remote promises of the alchemists.

It is, in addition, interesting to note that Biringuccio's criticism of alchemy had important implications for the technical nomenclature he himself used. The Italian metallurgist preferred, for instance, to change the name *mercury* used by the alchemists to the more descriptive *quicksilver.* The alchemists "called it mercury in consideration of its properties, perhaps because of its resemblance to the planet of him whose actions place him between gods and men. . . . Thus these investigators think that it has a similar place among perfect metals, since it is the primary mineral matter." In consequence, alchemists "wish to

---

[24]"But in spite of all this, the fathers and inventors of the art who exalted it with such high praise are all dead and have not enjoyed even one period of youth, to say nothing of two or three; and as far as I know they are not yet raised from the dead as they promised," *ibid.*, 39.

[25]*Ibid.*, 40 and 37–38.

prove at all costs that quicksilver is truly a defect of Nature so that they can hope to remedy it with their art. For this reason they are continually agitated in mind and body in their effort to aid it and to supply the necessary thing that Nature lacked in order to bring it to perfection."[26] As these passages clearly show, Biringuccio rejected both the methods and the nomenclature of the alchemists. He questioned the legitimacy of the association between planets and metals and rejected any heavenly influence on nature but that of God. Instead of the symbolical tables of the alchemistic treatises, 83 woodcuts depicting the metallurgical laboratory and its implements adorned Biringuccio's work.

However, being written in a difficult and archaic Italian, *De la Pirotechnia* was soon surpassed in clarity and elegance by the works of Agricola, whose constant care to define the terms he used philologically made metallurgy far more intelligible to learned men not acquainted with the practical art.

In 1546 Agricola published *De natura fossilium*, an important treatise in which he outlined the first classification of the mineral kingdom.[27] Agricola divided minerals according to the following scheme:

- Mineral bodies
  - Homogeneous bodies
    - Simple minerals
      - Earths
      - Solidified juices
      - Stones
      - Metals
    - Compound minerals
  - Mixtures[28]

With this classification the mineral kingdom began for the first time to assume its modern features. This general division was followed by the first attempt to outline a *method* of classifying minerals. "Mineral substances," Agricola wrote, "vary greatly in color, transparency, luster, brilliance, odor, taste and other properties which are shown; [. . .] in order to show the differences in minerals I shall begin by classifying them according to color, then I shall describe the nature of each form."[29] The classification of minerals was thus determined from their external characteristics. This method was not completely new: it had been used by Pliny and eventually by Albertus Magnus. Agricola

[26] *Ibid.*, 79 and 80.

[27] Georgius Agricola, *De natura fossilium* (1955). In the same year (1546), Agricola wrote a letter to Wolfgand Meurer in which he presented a Latin-German lexicon with 480 metallurgical terms. See Agricola, *Ausgewählte Werke*, vol. 3 (Berlin, 1956), 10–43.

[28] This scheme is taken from Agricola (1950), 3.

[29] Agricola (1955), 5.

could, however, claim the distinction of being the first to provide a systematic taxonomy based on external characteristics and accurate description.

Related to this, another problem was to find an appropriate nomenclature:

> In the discussion of minerals learned men, traders and miners have been of a great assistance to me. As a rule, in connection with each mineral, I shall mention the places where it has been found or where it is found today. It is well known that certain regions produce famous earths, others solidified juices, stones, gems, marbles and metals. Some of these substances lack names as previous writers have not mentioned them and it will be necessary for me to give them new names. As a rule I will give them Greek names as they cannot be named aptly in Latin.[30]

Agricola's mention of the assistance of "learned men, traders and miners" is important, because it underlines both the humanistic and the practical sources of the metallurgical lexicon. With the technical advice and expertise of miners, Agricola was able to identify new minerals; from the reports of traders, he came to know of the existence of mineral specimens from remote geographical areas; with the knowledge of the learned, he was able to name the new species of minerals, following the philological paths of classical authorities from Theophrastus to Pliny.

With this combination of theory and practice, Agricola transliterated the German lexicon of the miners into Greek and Latin and tried constantly to introduce the new names into the scientific tradition established by the ancients. Agricola's humanistic endeavor was intended to bring all the different sources (Greek, Latin, medieval, and contemporary) together in as coherent a lexicon as possible. The result of his efforts was *De natura fossilium,* in which he included 573 Latin and 115 Greek terms, avoiding, as far as possible, any vernacular

[30] *Ibid.*, 3. The original quotation *per extenso* reads as follows: "Sed cum nostrae venae non gignant omnis generis res fossiles, eas quae nobis desunt, non modo a Germania regioinibus quae eis abundant, verum ab omnibus fermè, a quibusdam Asiae et Africae, apportandas curavi. in quibus negociis conficiendis mihi et docti domines, et mercatores, et metallici operam navarunt. quo circa ad quamque rem fossilem regiones in quibus quondam nata sit, aut nunc nascitur, soleo adiicere. aliae enim ferunt terras insignes, alias succos concretos, quaedam lapides, gemmas quaedam, aliae marmora, aliae metalla. *ac quoniam earum rerum partim carent nominibus, quod de eis nihil scripserunt veteros, necessarium fuit, ut ipsis nova ponerem nomina: et quidem saepius Graeca, quod Latina tam apta poni non possent*" (my italics), Agricola, *De natura fossilium* (Basel, 1558), 164–165.

contaminations.[31] This was a considerable enlargement of the technical vocabulary when compared with the total of approximately 270 mining terms presented in the *Bermannus.* Indeed, within 16 years Agricola had been able to add enormously to his knowledge of the mineral kingdom. His surveys of German and Bohemian mines after 1530 acquainted him with many new minerals, gems, stones, and earths.

With the publication of his major work *De re metallica* in 1556, Agricola reached scientific maturity, displaying a confident mastery of the metallurgical and chemical knowledge available at the time. To consider Agricola's work solely from the standpoint of the history of metallurgy or of technology is to overlook the many *chemical* notions, operations, substances, and instruments described in *De re metallica.* Furthermore, the boundaries among metallurgy, mineralogy, and chemistry during the sixteenth century are hardly distinguishable.

Just as he had classified minerals, Agricola now provided, for instance, a classification of salts (Books VII and XII); he improved the technical apparatus of the chemical laboratory and greatly increased the accuracy of chemical operations, such as assaying. The tremendous success enjoyed by *De re metallica*[32] was also due to the many clear and detailed woodcuts describing the processes, instruments, engines, machinery, and implements in use in the sixteenth century, and it remained a primary source, both in mineralogy and chemistry, until the first half of the eighteenth century.

From our perspective the publication of *De re metallica* represents a particularly important landmark in the history of chemistry because its attempt to reform chemical and metallurgical nomenclature was inspired by a dissatisfaction with alchemy. This is an eloquent sign of the existence of a metallurgical tradition very distinct from the alchemical one. From the very title of his work, Agricola underlined his explicit intention of providing his readers with a *clear* and *descriptive* language.[33]

---

[31]See the introduction to Agricola (1955), vii.

[32]Georgius Agricola, *De re metallica* (Basel, 1556). The Latin edition was reprinted in 1561, 1607, 1621, and 1657. The work was translated into German in 1557 and reprinted in 1580, 1607, and 1621, and it was translated into Italian in 1563. I have here used the English translation made by Hoover in 1912 and reprinted in 1950.

[33]"Georgii Agricolae de re metallica libri XII. Quibus Officia, Instrumenta, Machinae, ac omnia denique ad Metallicam spectantia, *non modo lucullentissime describuntur, sed & per effigies, suis locis insertas; adiunctis Latinis, Germanicisque appellationibus it ob oculos ponuntur, ut clarius tradi non possint*" (my italics) (Basel, 1556).

But let us follow these arguments as Agricola developed them. Having listed the sources he relied on,[34] Agricola continued:

> Seeing that there have been so few who have written on the subject of the metals, it appears to me all the more wonderful that so many alchemists have arisen who would compound metals artificially, and who would change one into another. . . .
>
> All these alchemists[35] employ obscure language and Johannes Aurelius of Rimini alone has used the language of poetry. There are many other books on this subject, but all are difficult to follow, because the writers upon these things use strange names, which do not properly belong to the metals, and because some of them employ now one name and now another, invented by themselves, though the thing itself change not.[36]

The connection established between the progress of metallurgy and its language was a vital contribution. Agricola was perfectly aware of the existence of an alchemical tradition, and in relying on the metallurgical tradition, from Pliny onwards, he rejected the obscure language of alchemy and its pretentious claims regarding the transmutation of metals. Furthermore, Agricola, like Biringuccio, neither accepted the association between metals and planets nor shared the common belief in the influence of heavenly forces on the subterranean realm. He "rejected absolutely the Biblical view which [. . .] was the opinion of the vulgar; further, he repudiated the alchemistic and astrological views with great vigour. There can be no doubt, however, that he was greatly influenced by the peripatetic philology."[37]

Agricola saw himself as carrying on the classical tradition, and he tried to identify the emergence of a rational and systematic exploitation of mineral deposits with a philological reconstruction of metallurgical knowledge. This humanistic aim, both practical and learned, did not have room within its scientific horizon for the spurious sources of alchemy, which in the eyes of a classical scholar were rooted in the

---

[34]Agricola acknowledged Theophrastus and Pliny among the ancients, and Pandolfus Anglus, the anonymous works *Ein nützlich Bergbüchlin* (ca. 1500), and *Probier Büchlein* (1510), and Biringuccio's *Pirotechnia* among the moderns as sources of his works. "Beyond these books," Agricola declared, "I do not find any writings on the metallic arts," *De re metallica* (1950), xxvi–xxvii. In reality Agricola knew and used in his treatises Aristotle, Dioscorides, Galen, and Albertus Magnus, who all dealt at some time with metallurgy.

[35]Among them Agricola included Hermes, Zosimos, Geber, Lull, etc. See Agricola (1950), xxvii.

[36]*Ibid.*, xxvii–xxviii.

[37]*Ibid.*, 46 (remark by Hoover).

Middle Ages rather than in antiquity. To the German humanist, the alchemists were not *Auctoritates,* and most of their traditions, far from being as old as they claimed, went back only to the Middle Ages, an epoch to which the neglect of classical languages had brought confusion rather than scientific clarity. Moreover, their obscure language and their "invented" terms made their works virtually incomprehensible and certainly quite useless in the daily work of the *Bergmeister.*

It is also interesting that Agricola, in keeping with classical medical thought, was reluctant to use mineral and chemical substances for therapeutic purposes and regarded the new medical terms introduced by Paracelsus as *inaudita vocabula.*[38]

In reality the question of names proved a difficult one, even for such a diligent humanist and philologist as Agricola, although he was well aware of the problems:

> The things dealt with in this art of metals sometimes lack names either because they are new, or because, even if they are old, the record of the names by which they were formerly known has been lost. For this reason I have been forced by a necessity, for which I must be pardoned, to describe some of them by a number of words combined, and to distinguish others by new names.[39]

As has been pointed out by Robert Halleux, each of these new names could be a redeployment of a classical term, a restriction of the general meaning established during antiquity, a periphrasis, a transliteration of a German word, or a word from the vernacular with the addition of a Latin suffix, e.g., *bitumen, quartzum, cobaltum, spatum,* and *bisemutum.*[40] These general rules were still too vague to engender an effective and unequivocal nomenclature. The limited state of mineralogical and chemical knowledge in the sixteenth century did not yet allow of a more precise solution. However, despite this objective difficulty, the many empirical facts and observations collected by Agricola were presented to the reader as a coherent body of notions, and in a terminology that could easily be identified.[41] Indeed, the nomenclature presented in *De re metallica* was extremely analytical and descriptive, and acting on his belief that metallurgy needed "one who teaches others to give exact names to everything," Agricola coined many new

---

[38]Quoted by Halleux in Agricola (1990), xxix.

[39]Agricola (1950), xxxi.

[40]See the comments of Halleux, in Agricola (1990), xxvii–xxix.

[41]For a detailed analysis of Agricola's mineralogical lexicon, see the extensive comments by Hoover on the fifth book of *De re metallica* (1950), 101–115.

terms. He used mineralogical terms that primarily evoked the external characters of the substances to which they referred. In the case of silver minerals, for instance, he coined the following names:

Latin (1556)	*German (1546 and 1557)*	*Modern Term*
Argentum purum in venis reperitur	*Gedigen silber*	Native silver
Argentum rude plumbei coloris	*Glas ertz*	Argentite
Argentum rude rubrum	*Rot gold ertz*	Pyrargyrite
Argentum rude rubrum translucidum	*Durchsichtig rod gulden ertz*	Proustite
Argentum rude cineraceum	*Gedigen graw ertz*	Part Cerargurite/
Argentum rude nigrum	*Gedigen schwartz ertz*	Part Stephanite[42]
Argentum rude purpureum	*Gedigen braun ertz*	

By 1556, out of 11 known silver minerals, seven had been discovered and named by Agricola. The same descriptive nomenclative criteria were used for all the other metals and minerals. Another of Agricola's valuable contributions to the specialization of metallurgical nomenclature was in the creation of terms related to mining. Agricola identified and named five different kinds of vein[43] and 20 different soil strata.[44] He also designated a long list of mining instruments, revealing the richness and variety of technical devices used in sixteenth-century mines.[45] The classification of surveying instruments, such as the compass (*instrumentum cui index*), the tripod (*tripus*), and the hemicycle (*hemicyclium*), described in the fifth book, created the

[42] *Ibid.*, 108–109.

[43]

Latin (1556)	*German (1546 and 1557)*	*English*
Vena profunda	Gang	Fissure vein
Vena dilatata	Schwebender gang	Bedded deposit
Vena cumulata	Gesuchte or Stock	Impregnation
Fibra	Klufft	Stringer
Commissurae saxorum	Absetzen des gesteins	Seams or joints

From Agricola (1950), 42.

[44] *Ibid.*, 126–127.

[45] *Ibid.*, 137 and 151.

foundations of a new science: mining surveying (*geometria subterranea*). Last but not least, Agricola devoted most of the seventh book to the definition and conversion of Latin and German weights and measures.

It is worth mentioning in passing that a French-German dictionary published at the end of the sixteenth century by the humanist and mathematician Levinus Hulsius[46] listed several mining terms, showing that they were considered a part of the general vocabulary of the language.[47] By contrast, there are hardly any alchemical terms to be found in it, confirming the decidedly esoteric character of alchemical language. In effect, the "openness" of metallurgical nomenclature was one of its main characteristics. This openness "did not necessarily refer to wide public dissemination of knowledge. Rather it could signify the art of writing down orally disseminated craft knowledge, making it accessible to an unskilled learned and noble audience. It could mean (as it did for Agricola) the development of a clear technical vocabulary."[48] The vital importance of mining during the period 1500–1800 to the development of the military and economic strategies of most European countries well explains the necessity of Agricola's effort to provide metallurgy with a clear nomenclature.

Agricola did not remain an isolated figure in the early history of metallurgy. The success of his works encouraged many *Bergmeister* from different mining districts to organize their rich technical knowledge systematically. The copious literature on mining and assaying produced between the publication of *De re metallica* and the end of the eighteenth century includes more than 100 treatises, many of which enjoyed several editions.[49] These texts, which should not be confused with the works of mineralogy, dealt primarily with technical methods of exploiting mining ores and were usually published in Germany or Sweden. Except for the writings of Agricola, the scientific background and historical impact of this considerable output have been overlooked by historians of chem-

---

[46]Levinus Hulsius, *Dictionnaire François-Allemand & Allemand-François*, 3rd ed. (Frankfurt, 1607).

[47]Specific and technical mining terms, including some coined during the sixteenth century, such as *Berggang, Berghawer, Bergknapp, Bergmeister, Bergsalz, Untergraber, Erzegrube*, etc., were included in Hulsius' dictionary.

[48]Long (1991), 353.

[49]In their "Agricola Bibliographie 1520–1963," published as vol. 10 of Agricola's *Ausgewählte Werke* (Berlin, 1971), Rudolf Michaëlis and Hans Prescher listed some 75 works dealing with mining, assaying, and the science of surveying mines (*Markscheidekunst*), published between 1556 and 1800 and explicitly following Agricola's approach to metallurgy. Despite the immense value of this bibliography, many of the metallurgical works printed in Sweden during the same period are missing.

istry, who have too often contented themselves with tracing the origin of modern chemistry with reference only to its alchemical, medical, and pharmaceutical sources.[50] In reality, the impact of the metallurgical tradition on the development of eighteenth-century chemistry was tremendous. It may suffice here to mention that Stahl wrote two books on assaying and metallurgy, that Wallerius and Bergman were active in the Swedish Board of Mines (*Bergskollegium*), and that Lavoisier accompanied Guettard on his mineralogical survey of France before turning to chemical research, to indicate that the influence of the metallurgical tradition was not confined to the mines. But the importance of the metallurgical literature goes beyond its historical diffusion. Following the path pointed out by Agricola, his successors made every effort to equip their science with a clear and practical nomenclature. Christopher Entzelt and Lazarus Ercker illustrated their works with descriptive woodcuts depicting the technique of mining and with bilingual (Latin-German) dictionaries in which the technical nomenclature was explained in detail.[51] This terminology was derived from Agricola, and no concession was made to alchemy, which was openly criticized. In the early seventeenth century, Löhneyss and Schindler developed Agricola's ideas and enriched metallurgical science with the first description of the extraction of zinc and with a new method of smelting metals (see Fig. 6).[52] Significantly, at a time when Paracelsism was exerting a great influence on the physicians and scholars of Europe, these same metallurgists followed their master Agricola in rejecting

[50]With the exception of a few specific studies of German, Swedish, and French mining districts, the secondary literature on metallurgy is poor. Partington (1969), for instance, has grouped metallurgical and mining literature with pharmaceutical and mineralogical works, thus losing sight of the epistemological unity and consistency of the metallurgical tradition. Archibald and Nan Clow, *The Chemical Revolution: A Contribution to Social Technology* (London, 1952), dealt with every conceivable chemical industry but mining.

On specific mining districts, see Sten Lindroth, *Gruvbrytning och Kopparhantering vid Stora Kopparberget intill 1800–talets början* (Uppsala, 1955), 2 vols.; for a shorter analysis, see Hugo Olsson, *Kemiens historia i Sverige intill år 1800* (Uppsala, 1971), 40–101; Svante Lindqvist, *Technology on Trial: The Introduction of Steam Power into Sweden. 1715–1736* (Uppsala, 1984), 5–107 and 213–231; Anders Lundgren, "The New Chemistry in Sweden: the Debate that Wasn't," in *Osiris* (second series), 4 (1988), 147–155; Bertrand Gille, *Les origines de la grande industrie métallurgique en France* (Paris, 1947); Otfried Wagenbreth and Eberhard Wächter (eds.), *Die Freiberger Bergbau: Technische Denkmale und Geschichte* (Leipzig, 1986).

[51]Christopher Entzelt, *De re metallica, hoc est, de origine Varietate & Natura Corporum metallicorum Lapidum, Gemmarum . . . Libri III* (Frankfurt, 1557); Lazarus Ercker, *Beschreibung Aller fürnemisten Mineralischen Ertzt unnd Bergkwercksarten . . .* (Prague, 1574).

[52]C.C. Schindler, *Metallische Probier Kunst—Bericht vom Ursprung und Erkentniss der Metallische Erze* (Dresden, 1607); Georg Engelhard von Löhneyss, *Grundtlicher und aussfürlicher Bericht vom Bergwerck* (1617), fifth ed. (Stockholm-Hamburg, 1690).

FIGURE 6 Engraved frontispiece of the mineralogical work by Georg Engelhard Löhneyss, *Bericht vom Bergwerck* (Zellerfeld, 1617), illustrating the entrance of a mine.

the belief in the transmutation of metals or the influence of planets on the subterranean world, in denying the efficacy of the *virgula divinatoria*, and in deriding alchemical nomenclature. The exoteric character of the metallurgical tradition is also illustrated by its institutional organization. Unlike alchemy, which was often connected with secret or closed societies, mining industry was naturally supported by local government and was an integral part of the life of the community in which it was situated. As an activity involving the labor force of entire villages, the mining of the sixteenth and seventeenth centuries was strictly regulated by a code of laws. Furthermore, ever since Agricola's time, mining had been characterized by an elaborate specialization of labor. The mining prefect, the *Bergmeister*, the jurate, the notary, the manager of the mine, the manager of the tunnel, the digger, the shoveller, the level worker, the washer, the buddler, the sifter, the smelter, the master of the mint, etc.[53] were all essential parts of a remarkably complex industrial structure in which it was obvious that knowledge both practical and theoretical had to be easily available to everyone involved.

During the eighteenth century, the decline of alchemy accelerated, and knowledge of metallurgy was enhanced in the works of Rössels, Schlüter, Cramer, Beyern, and Rinman,[54] and by detailed descriptions of the mining districts of Germany, Bohemia, Sweden, France, Spain, and England.[55] Making the metallurgical knowledge even more accessible, three mining dictionaries were published during the eighteenth century and reprinted several times (see Fig. 7).[56] Much of the nomenclature pro-

[53]Agricola (1950) lists 25 different kinds of duties in the mines. During the seventeenth and eighteenth centuries, several *Bergkurthel* and *Bergkrecht*'s treatises were published, regulating the organization of labor in the mines and its relation to the local tax system and civil authorities.

[54]Balthasar Rössel, *Speculum Metallurigiae Politissimum, oder: Hell-polierter Berg-Bau-Speigel* (Dresden, 1700); Christoph Andreas Schlüter, *Gründlicher Unterricht von Hütt-Werken*, 2 vols. (Braunschweig, 1738); Johannes Andreas Cramer, *Anfangsgründe der Metallurgie*, 3 vols. (Blaukenburg-Quedlinburg, 1774–1777); August von Beyern, *Gründlicher Unterricht vom Bergbau nach Anleitung der Markscheidekunst*, 2nd ed. (Altenburg, 1785); Sven Rinman, *Anledningar til Kunskap om den gröfre Jern och Stålförädlingen och des förbättrande* (Stockholm, 1772).

[55]Johann Charpentier, *Mineralogische Geographie der Chursächsischen Lande* (Leipzig, 1778); Johann Gottlieb, *Gesammelte Nachrichten von Schlesischen Bergwerken* (Breslau-Leipzig, 1775); Johann Friedrich Züecker, *Die Naturgeschichte und Bergwerksverfassung des Ober-Harzes* (Berlin, 1762); Gabriel Jars, *Voyages Métallurgiques . . . en Allemagne, Suède, Norvege, Angleterre, & Ecosse*, 3 vols. (Paris, 1774–1781); Johan Jacob Ferner, *Physikalisch-Metallurgische Abhandlungen über die Gebirge und Bergwerke und Ungarn* (Berlin-Stettin, 1780). See also Schlüter (1738).

posed by Agricola in 1556 survived in these lexicons and the new terms which were derived followed the linguistic principles of the German humanist.

It is difficult to be certain about the kind of influence metallurgy had on the birth of chemical nomenclature, but we know that at least in a few cases it played a very important role. In Sweden, most chemists were trained in the metallurgical tradition, and their expertise in the art of assaying certainly owes much to the long experience accumulated during centuries of mineworking. This practical background had an important incidental effect on Swedish chemical language, which was in fact deeply dependent on the metallurgical lexicon. In Germany, too, much metallurgical terminology was embodied in the chemical lexicon, although it had to coexist there with the alchemical language derived from the Paracelsian and medical tradition. Even in France the metallurgical tradition considerably affected the way chemists expressed themselves. In the works of Hellot and Macquer, for instance, most of the metallurgical terms can be traced back to the work of Schlüter.[57]

Thus, a terminology that was inspired by humanistic and philological principles, that was in open and explicit opposition to the language of alchemy, and that grew up and developed in the mining districts of Europe was absorbed into the language of eighteenth-century chemistry.

THE CHARACTERS OF MINERALOGY

Despite the initial impetus given by the work of Agricola, mineralogy became during the Renaissance an integral part of natural history, rather than of metallurgy. The investigations of minerals that were published during the sixteenth and seventeenth centuries were the work of naturalists who established an intimate connection between the three natural kingdoms (mineral, vegetable, animal).[58] Following the method used in botany and zoology, the classification of minerals

---

[56][Johann Caspar Zeisig,] *Neues und wohlein-gerichtetes Mineral- und Bergwerks-Lexicon* (1730), 2nd ed. (Chemnitz, 1743); *Bermännisches Wörterbuch* (Chemnitz, 1778); Sven Rinman, *Bergwerks Lexicon,* 2 vols. (Stockholm, 1788–1789).

[57]See, for example, the articles "Métaux" and "Mine" in vol. 2 of Macquer's *Dictionnaire de chymie* (Paris, 1766).

[58]Among these works, the following are particularly representative: Andreas Cesalpino, *De Metallicis Libri tres* (Rome, 1596); Ferrnate Imperato, *Dell'Historia Naturale* (Naples, 1599); Anselmus Boëtius de Boodt, *Gemmarum et Lapidarum Historia* (Hanover, 1609); Michele Mercati, *Metallotheca,* composed during the second half of the sixteenth century but published only much later (Rome, 1717–1719); Ulisse Aldrovandi, *Musaeum Metallicum* (Bologna, 1648).

FIGURE 7 Engraved frontispiece of Johann Caspar Zeisig's *Mineral- und Bergwerks-Lexicon* (Chemnitz, 1743), illustrating a miner dictating to the god Mercury the names of the mining instruments. The engraving emphasized the descriptive and didactic aims of the metallurgical literature.

was based on their external and observable features. This approach was a definite departure from that of Agricola in that it deliberately ignored the physical and chemical qualities of minerals. Agricola acquired most of his mineralogical knowledge in the *officinae* and based his definitions and identification of minerals on the results of a wide range of technical and chemical operations. The nature of the abundant illustrations in these studies helps us to understand the difference. In Agricola's and other metallurgical works, the metals and minerals were always shown in mining and assaying contexts, emphasizing the technological background to their definition. By contrast, the metals and minerals presented by representatives of the naturalist tradition were depicted in curiosity cabinets (see Figs. 8 and 9), far removed from their origins. In these treatises much emphasis was placed on the peculiar and curious shapes of minerals, on the wonders of particular specimens. Often (especially in Mercati and Boëtius de Boodt) the minerals illustrated were either mythological or did not belong to the mineral kingdom. Thus one might find shells, unicorn's horns, crab bones, toad stones, and other marvels of nature treated as minerals and described in detail. Although its character changed radically, mineralogy remained within the domain of natural history until it became an autonomous science at the beginning of the nineteenth century. There were two main reasons for the persistence of such a tradition. The first concerned the fact that mineralogy was very much a matter of description and classification.[59] Linnaeus, in particular, included the systematization and classification of minerals within his wider taxonomic interests. In his *Systema naturae*, published in 1735, the Swedish naturalist successfully established epistemologically the domain of natural history within the three kingdoms of nature. Although he did not pay much attention to the mineralogical side, his division of natural history and his method of classifying minerals enjoyed an enormous influence during the eighteenth century (see Fig. 10). The second reason concerned the belief, rooted in Paracelsism but denied by Agricola, that minerals grew in the same way plants and animals did. An internal fluid, it was assumed, circulated inside the minerals and caused their changes of form.

Like all other Baconian sciences, eighteenth-century mineralogy was characterized by a huge increase in the number of new observa-

[59]On this theme, see Evan M. Melhado, *Jacob Berzelius: The Emergence of His Chemical System* (Uppsala, 1981), 102–104.

FIGURE 8 Engraving illustrating the interior of the Vatican collection of minerals, stones, earths, and gems. From Michele Mercati, *Metallotheca* (Rome, 1717–1719).

tions, data, and objects. The fastest growth was in Sweden and Germany, followed by England, France, and Italy.

Inevitably, this considerable increase in interest all over Europe led to differences in approach and in the method of analyzing minerals. The problem of classification was probably the most controversial question. The increasing number of mineral species and the diversity of their forms, colors, densities, and purities, gave rise to a need for systematic classification similar to that attempted in botany. The traditional custom of classifying minerals simply by describing their visible characteristics brought increasing confusion. The same mineral might be assigned to two different species, a compound was often considered as a primitive substance, and so on. The solutions found to these problems were at first merely practical and economical; it is therefore no coincidence that the first mineralogists to establish a new system of classification were the Swedes. Using Linnaeus' taxonomy as his model, Johan Gottschalk Wallerius divided the mineral kingdom into classes, orders, genera, and species and, in 1747, produced the first systematic inventory of the products of the earth. In his definition of

FIGURE 9 One of the 19 *armaria metallorum* of the Vatican mineralogical cabinet, containing salt and niter. From Michele Mercati, *Metallotheca* (Rome, 1717–1719).

REGNI MINERALIS
CLASSIS SEXTÆ, ORDO PRIMUS
METALLA NOBILIA.

Des Mineral-Reichs
Sechster Classe, Erste Ordnung.
Die Edlen Metalle.

29

Nomen genericum. Geschlechts-Name.	Characteres generici. Kennzeichen des Geschlechts.	Differentiæ specificæ. Unterschied der Arten.	Synonima specifica. Gleichgeltende Namen der Arten.	
	Metallum			
I. AURUM. Chym. Sol. Gold. ☉	Firmum, tenax, luteum, in aqua forti non solubile. Dicht, zähe, gelb, löset sich im Scheide-Wasser nicht auf.	Nudum nativum. Rein gewachsen. Minera varia vestitum. (32.) In mancherley Ertz versteckt.	*Aurum nativum.* Gediegen Gold. *Minera aurifera.* Goldhaltig Ertz.	 Lapis solaris. Güldische Bergart.
II. ARGENTUM. Chym. Luna. Silber. ☽	Firmum, tenax, album, in aqua regis non solubile. Dicht, zähe, weiß, löset sich im Königs-Wasser (♈) nicht auf.	Nudum nativum, forma varia. Rein gewachsen, in mancherley Gestalt.	*Argentum nativum.* Gediegen Silber.	 Bauren-Ertz.
		Minerali alio mixtum. Mit andern Bergarten vermischt.	*Mineræ Argenti.* Silber-Ertze.	
		Plumbei coloris, splendens, malleabile. Bleyfarbig, glänzend, so sich hämmern läßt.	*Min. argenti vitrea.* Glas-Ertz.	 Silber-Glas-Ertz.
		Rubrum diaphanum & opacum. Roth durchsichtig und undurchsichtig.	*Min. florenorum rubra.* Roth Gulden-Ertz.	Min. Lunæ rubra. Roth gültig Ertz.
		Albo-griseum splendens, cupro mixtum. Licht grau glänzend, so auch Kupfer hält.	*Min. florenorum alba.* Weiß Gulden-Ertz.	Min. Lunæ alba. Weiß gültig Ertz.
		Amorphum, minera varia vestitum. Ohne bestimte Gestalt, in mancherley Ertz versteckt.	*Min. argentifera.* Silberhaltig Ertz.	Lapis lunaris. Silbrige Bergart.

D 3 REGNI

Figure 10 The classification of minerals according to Linnaeus's taxonomic system. From Johann Lucas Waltersdorff, *Systema Minerale in quo regni mineralis producta omnia systematice per classes, ordines, genera et species proponuntur* (Berlin, 1748).

minerals, he finally distinguished them, for the first time, from living bodies.[60] He studied the formation and morphology of crystals and provided the first *chemical classification* of minerals, stressing the importance of chemical analysis in mineralogy. Despite providing all these useful pointers to the future path of mineralogy, Wallerius did not come down in favor of any specific classificatory system and confined himself to presenting a description of each mineral in alphabetical order. He also adopted the traditional nomenclature with only a few changes.

The rich inventory and the methodological indications proposed by Wallerius showed naturalists a vast spectrum of possible ways of developing new systems of mineralogy. In demonstrating all the alternatives, Wallerius focussed attention on the significance of the role to be played by classification.[61]

The first method to be developed was the chemical one. Johan Friedrich Henckel, Chriestlieb Ehrgott Gellert, and Johann Heinrich Pott claimed that only by studying the internal chemical properties of minerals was it possible to classify them properly. However, the German mineralogists never succeeded in providing a sufficient collection of chemical analyses of minerals incorporating their principles. With the imperfect instruments then in use for chemical analysis, they only succeeded in applying the chemical approach to a few classes of substance, leaving most of the mineral kingdom unconsidered.

Nevertheless, this line of attack was continued by Axel Fredrik Cronstedt, who in 1758 published a more determined defence of the superiority of the chemical approach in his celebrated *Försök till mineralogie.*[62]

> But as every country had a different name for these bodies "minerals," they often gained more names than there were real species. . . .

---

[60]"Il y a des naturalistes qui prétendent que les minéraux ont une vie semblable à celle dont jouissent les végétaux: mais personne n'ayant encore pû jusqu'à présent remarquer, même à l'aide des meilleurs microscopes, que ces substances eussent un suc contenu dans les fibres ou veines. . . . On ne voit point sur quel fondement on attribueroint une vie aux minéraux, à moins qu'on ne volut appeller vivant tout ce qui a la faculté de croitre et de s'augmenter," Wallerius, *Minéralogie, ou description générale du regne minéral* (The first edition was published in Swedish in 1747.) (Paris, 1753), vol. I, 1–2.

[61]On Wallerius's mineralogy, see Metzger, *La genèse de la science des cristaux* (1918), 2nd edition (Paris, 1969), 57–60; Ferdinando Abbri, *Le terre, l'acqua, le arie: La rivoluzione chimica del Settecento* (Bologna, 1984), 55–62; *idem.*, "Le teorie chimiche," in Paolo Rossi (ed.), *Storia della scienza moderna e contemporanea*, vol. I (Turin, 1989), 556–561.

[62]A.F. Cronstedt, *An Essay Towards a System of Mineralogy* (London, 1770), translated by Gustaf von Engeström.

> To remove and alter these inconveniences, they have in later and more enlightened times endeavoured to fix proper names to the subjects of the mineral kingdom, according to their external marks, as in regard to Figure, Colour and Hardness; *but these characters afterwards having been found not sufficient, it was necessary to discover others more solid by the result of chemical experiments, which added to the former ones would make a complete system.*[63] [my italics]

Cronstedt, dissatisfied with the results achieved by a simple application of botanical and external classifications in mineralogy, claimed that only chemical analysis could reveal the *essential characters of minerals.* Using the blowpipe,[64] Cronstedt was indeed able to produce chemical analyses of a much larger number of mineralogical species.

Thus, the minerals were assigned to classes, genera, and species according their chemical composition and properties. Although Cronstedt did not touch on the controversial theme of the ultimate composition of matter, he denied the primitive status of certain classes, such as the earths, by using systematic chemical analysis as his investigative method.

This implicit principle came to the surface with the works of another important Swedish scientist, Torbern Bergman.[65] Bergman was by profession a chemist, but he was also well acquainted with mineralogy and natural history. In the early years of his scientific career, he collaborated with Linnaeus and produced a herbarium, of which he later wrote in his autobiography:

> My herbarium, collected in Skara, was of much use in the countryside: in it I had the names of many herbs, whose shapes I could compare with Arch. Linnaeus' descriptions, and thus gradually learn the meaning of terminology in science.[66]

This early collaboration inspired Bergman to approach mineralogical topics within the more general framework of natural history and

---

[63] *Ibid.*, viii.

[64] On the importance of the blowpipe in the early history of chemical and mineralogical analysis, see the essay by William B. Jensen, "The Development of Blowpipe Analysis," in John T. Stock (ed.), *The History and Preservation of Chemical Instrumentation* (Dordrecht-Boston-Lancaster-Tokyo, 1986), 123–136; see also Gustav von Engeström, *Description of a Mineralogical Pocket Laboratory* (London, 1770).

[65] On Torbern Bergman, see my essay, "T.O. Bergman and the Definition of Chemistry," in *Lychnos* (1988), 37–67, parts of which I have used here.

[66] "Mitt i Skara samlade herbarium kom mig nu på landsbygden til pass: jag hade der namn på en hel hop örter, huilkas skapnad jag kunde jämföra med Arch. Linnés beskrifningar och således efter hand lära mig termernes betydelse uti vetenskapen," Torbern Bergman, "Själfbiografi," in *Äldre Svenska biografier,* vol. 3 (Uppsala, 1916), 87.

to take an active interest in the problems of scientific naming and nomenclature.

In 1782 Bergman published his *Sciagraphia regni mineralis secundum principia proxima digesti,*[67] which had an immediate impact among European chemists and naturalists alike. The book is in the Linnaean tradition. According to Bergman, the main task of mineralogy was to establish the character of inorganic substances. To this end, it was necessary to find an effective method: "as in the vegetable kingdom different methods have been formed upon the roots, the leaves, the flowers, the fruits etc., so also in mineralogy many methods may be devised."[68] Yet the classificatory methods commonly used took into account only the external qualities of minerals, such as color, hardness, and texture, although "they are generally, too, of different kinds, rare and dense, figured and shapeless, admitting of every possible variety. This general view of the subject shows us how little external characters can be depended on."[69]

In fact these secondary qualities did not show any difference, for instance, between calcareous and other kinds of earth, or between many other mineralogical substances that were very different but to outward appearances identical. For this reason, "classes, genera and species are . . . to be formed upon internal nature and composition."[70] Unlike Cronstedt, Pott, and Marggraf, Bergman was convinced that chemical analysis could be used in all mineralogical experimentation. However, even the most rigorous analysis failed to resolve all the problems when it came to establishing the genus and species of a compound. The main difficulty was in determining the variability of the proportions of the chemical constituents. Bergman therefore had to provide a more flexible solution:

> In methodizing fossils, compounds should rank under the most abundant ingredient. Thus let a and b represent the component parts; if the former be the heavier the compound must be placed under the genus of that.[71]

But this rule admitted several exceptions, since it was possible for an ingredient present in a smaller quantity to react more powerfully,

---

[67]Bergman, *Outlines of Mineralogy* (Birmingham, 1783). This work was also translated into French, German, and Portuguese.

[68]*Ibid.*, 6.

[69]*Ibid.*, 7.

[70]*Ibid.*, 9

[71]*Ibid.*, 10.

thereby constituting the characterizing element. In this case only the external characterization should determine the genus of the compound.

Therefore, chemical analysis led to many exceptions to the general rules and showed the impossibility of relying only on internal qualities. Despite these problems, the *Sciagraphia* contained one of the first outlines of a general reform of the classification and nomenclature of minerals. By subdividing the minerals into four classes—salts, earths, inflammable substances, and metals—and submitting them to the same rules of identification, Bergman made the significant advance of attempting to link the investigation of minerals with that of chemical objects.

Two years later, in 1784, Bergman published another important mineralogical work, entitled *Meditationes de systemate fossilium naturali*,[72] which had in fact a greater impact on chemical than on mineralogical nomenclature.[73]

Bergman argued that each class of mineral had to be designated by a unique name that identified it as salt, earth, metal, or *phlogisticatum*.[74] Also the genus, wherever possible, had to be denoted with a single word. The salts were subdivided into acids (*acidum*) and alkalis (*alkala*). Earths were subdivided into *calcareum*, *magnesiacum*, *argillaceum*, *siliceum*, and *barytes*. Therefore, the earths were designated by the suffix *-um*, with the exception of barytes, a name accepted by Bergman that had been derived by Guyton de Morveau from the name *terra ponderosa*. The names of the 14 known metals also ended with the suffix *-um* and consisted of one word.[75] Bergman also tried to use the binomial nomenclature systematically. He was firmly convinced of the efficacy of the *nomen triviale* used by his mentor Linnaeus:

> Although these names may have assumed from the inventor, some virtue, ancient appellation, property or accidental circumstance respecting the species, yet they should generally be limited to one word, and very seldom indeed to two. They may be considered as surnames, distinguishing the individuals contained in the same genus.[76]

---

[72]Published in *Nova Acta Regiae Societatis Scientiarum Upsaliensis*, 4 (1784), 63–128. On the relevance of this memoir, see W.A. Smeaton, "The Contributions of P.J. Macquer, T.O. Bergman and L.B. Guyton de Morveau to the Chemical Nomenclature," in *Annals of Science*, 10 (1954), 87–106, and Crosland (1978), 144–167.

[73]I discuss the content of this work in more detail later in this chapter.

[74]Inflammable substance.

[75]The genera of metals were: *aurum*, *argentum*, *hydrargyrum*, *plumbum*, *cuprum*, *ferrum*, *stamnum* , *vismutum*, *niccolum*, *arsenicum*, *cobaltum*, *zincum*, *antimonium*, and *magnesium*.

[76]Bergman (1784), 125.

The focus on chemical analysis and properties made the mineralogical nomenclature outlined by Bergman more popular among chemists than among naturalists, even though the whole system was essentially constructed on the principles of Linnaeus' botanical nomenclature and classifications. Most naturalists of the second half of the eighteenth century still preferred to classify minerals according to their external characters. The persistence of the traditional taxonomy was attributable to the fact that the chemical approach failed to isolate and identify a large number of minerals. As late as 1814, when Berzelius tried to outline a new system of chemical mineralogy, "the analytical techniques were not yet available to accomplish the full subordination of mineralogy to inorganic chemistry."[77]

The most successful method based on external characters had already been outlined by Linnaeus, who stressed the importance of the crystalline structure of salts in their identification.[78] The French naturalist Jean-Baptiste Romé de l'Isle developed this piece of intuition into a mineralogical system.[79] Through a series of observations of different crystals of minerals, de l'Isle was able to formulate the first law of crystallography, which stated that the inclination of the angles of the faces of each crystal species remained constant, although the dimensions of the crystal might vary considerably. De l'Isle intended to classify all minerals in accordance with this law, claiming that in using the differences in the forms of the crystals he was classifying minerals by a natural method. This classification was presented in a table that displayed the minerals according to the shape of their crystals.[80] De l'Isle's approach was undoubtedly successful in providing naturalists with a general system for the classification of minerals, but it still suffered from some significant disadvantages.

In 1784, René Haüy published his *Essai d'une théorie sur la structure des cristaux*, in which he disputed the possibility of making crystallography the basis of the distribution of minerals,[81] on the grounds that the forms of crystals of different minerals were often identical. Haüy went on to argue that crystallography was an autonomous science and that

---

[77]Melhado (1981), 295.

[78]Metzger (1969), 68.

[79]J.B. Romé de l'Isle, *Essai de cristallographie ou description des figures géometriques propres à differens corps du Regne Minéral connus vulgairement sous le nom de Cristaux* (Paris, 1772); *idem, Description méthodique d'une collection de minéraux du cabinet de M.D.R.D.L.* (Paris, 1773).

[80]See Fig. 11.

[81]"A l'égard du premier de ces objets, il est certain d'abord que jamais on ne pourra faire de la Cristallographie la base d'une distribution méthodique des minéraux," Haüy (1784), 5.

its object was the geometrical study of the relations and structural differences between crystals.[82]

At the end of the eighteenth century, the most common method of classifying minerals remained that based on their external characters. In 1772 Wallerius published the most detailed descriptive mineralogical inventory of the century.[83] A year later, the German mineralogist and geologist Abraham Gottlob Werner[84] published an essay[85] that set out to define the theoretical principles of the external approach. Werner was not convinced of the feasibility of the chemical approach on a large scale and also rejected crystallography, claiming that the "physics of Minerals" was a matter of mathematics and geometry rather than of mineralogical classification.[86] Botanical classification was considered too abstract. The only natural way to describe minerals was to classify them by their external characters, without any exceptions or hierarchy. Applying this approach, Linnaeus and Wallerius had made many useful observations, but they were not altogether consistent, referring also to other methods of classification, such as the chemical or the crystallographical.[87]

According to Werner, the species of a mineral should be identified by the observation or the assay of its color, form, dimension, transparency, hardness, solidity, flexibility, adhesion to the tongue, smell, taste, coldness, sound, unctuosity, weight, etc. Accordingly, "the name given to character is considered appropriate if it fully expresses the nature of the character and distinguishes it from the others of its species or genus."[88] The purpose of this kind of external approach was to provide sound practical advice for the mineralogist in his daily work, whereas the other methods, chemical and crystallographical, required too abstract a theoretical knowledge and were useful only to a much lesser degree.

Werner's position was shared by Buffon, who in 1783 published an *Histoire naturelle des minéraux*, in which he criticized the methods of classification in use among the naturalists and the chemists for

---

[82]Metzger (1969), 81–84.

[83]Wallerius, *Systema mineralogicum, quo corpora mineralia in ordines, genera et species . . . describuntur*, 2 vols. (Stockholm, 1772).

[84]On Werner, see the recent biography by Martin Guntau, *Abraham Gottlob Werner* (Leipzig, 1984).

[85]A.G. Werner, *Von den äusserlichen Kennzeichnen der Fossilien* (Leipzig, 1774), translated into English by Albert V. Carozzi under the title *On the External Characters of Minerals*, (Urbana, 1962).

[86]Werner (1962), 54.

[87]*Ibid.*, 7–16.

[88]*Ibid.*, 19.

TABLEAU CRISTALLOGRAPHIQUE.

Nos.	FIGURE & PROPORTIONS du Priſme.	FIGURE & PROPORTIONS des Pyramides.	NOMBRE & FIGURE des côtés du Priſme.	NOMBRE & FIGURE des côtés des Pyramides.	NOMBRE total des côtés.	NOMS des Sels.	NOMS des Pierres.	NOMS des Minéraux.	Nos. de la Planche & de la Figure qui repréſentent ces Criſtaux.	FIGURES données par M. le Chevalier Von Linné.
1	Priſme oblong, hexaëdre, dont les faces ſont égales.	2 Pyramides égales, hexaëdres, plus courtes que le priſme.	6 Rectangles longitudinaux.	12 Triangles iſoſceles, 6 = 6.	18	Le Tartre vitriolé.	Criſtal de roche. Améthiſte. Fauſſe hyacinte.	Mine de Plomb verte.	Pl. I. fig. 1. — —	Syſt. nat. ed. XII. fig. 1. Syſt. nat. ed. IX. fig. 4. Amœn. ac. t. 16. f. 6.
2	Priſme court, hexaëdre, dont les faces ſont égales.	2 Pyramides égales, hexaëdres, de la longueur du priſme.	6 Quarrés. — — —	— — *Idem.* — —	18	*Idem.*	*Idem.*	— — — —	Pl. I. fig. 2.	
3	Priſme très-court, hexaëdre, dont les faces ſont égales.	2 Pyramides égales, hexaëdres, plus longues que le priſme.	6 Rectangles tranſverſes, quelquefois linéaires.	— — *Idem.* — —	18	— — —	*Idem.* Grenat.	— — — —	Pl. I. fig. 6. — —	Syſt. Nat. XII. fig. 2. Muſ. Teſſ. t. 2. fig. 1.
4	Priſme oblong, hexaëdre : deux faces oppoſées plus larges que les autres.	2 Pyramides hexaëdres : 2 faces oppoſées plus larges & tronquées dans chaque pyramide.	6 Rectangles longitudinaux d'inégale largeur.	8 Triangles & 4 Trapèſes. 6 = 6.	18	— — —	Criſtal de roche jaune.	M. de plomb blanche.	Pl. I. fig. 3. — —	Syſt. Nat. XII. fig. 4.
5	Priſme oblong, hexaëdre : deux faces oppoſées plus étroites que les autres.	2 Pyramides hexaëdres : dans chacune d'elles deux faces oppoſées plus étroites.	— — *Idem.* — —	12 Triangles, dont 2 très-étroits dans chaque Pyramide. 6 = 6.	18	Tartre vitriolé.	Criſtal de roche.	— — — —	Pl. I. fig. 4.	
6	Priſme oblong, hexaëdre, dont les faces ſont égales.	2 Pyramides triëdres, courtes, égales.	3 Rectangles longitudinaux, alternes avec 3 hexagones allongés.	6 Pentagones, 3 = 3. —	12	— — —	Criſtal de roche.	— — — —	Pl. I. fig. 5.	
7	— — O — —	2 Pyramides égales, hexaëdres, jointes baſe à baſe.	— — O — —	12 Triangles iſoſceles. 6 = 6.	12	Tartre vitriolé.	Criſtal de roche. Fauſſe Hyacinte.	— — — —	Pl. I. fig. 7. — —	Syſt. Nat. XII. fig. 3. — Ed. IX. fig. 6.
8	— — O — —	Les mêmes, tronquées au ſommet.	— — O — —	12 Trapèſes & 2 petits hexagones. 7 = 7.	14	*Idem.*	— — — —	— — — —	Pl. I. fig. 17.	
9	— — O — —	2 Pyramides hexaëdres ; tous les angles formés par l'union des 2 baſes ſont tronqués.	— — O — —	12 Pentagones irréguliers, & 6 petits rhombes.	18	*Idem.*	— — — —	— — — —	Pl. I. fig. 8.	
10	— — O — —	Les mêmes, dont les ſommets ſont auſſi tronqués.	— — O — —	12 Hexagones irréguliers, 2 petits hexagones & 6 rhombes.	20	*Idem.*	— — — —	— — — —	Pl. I. fig. 9.	
11	— — O — —	2 Pyramides hexaëdres, jointes baſe à baſe ; l'une beaucoup plus longue que l'autre.	— — O — —	12 Triangles inégaux. 6 = 6.	12	Vitriol arſénical.	Spath calcaire.	— — — —	Pl. I. fig. 12. — —	Syſt. Nat. IX. fig. 5. Am. Ac. t. 16. fig 9.
12	— — O — —	2 Piramides hexaëdres, inégales, jointes baſe à baſe, la ſupérieure tronquée.	— — O — —	10 Triangles inégaux & 3 tétragones. 7 = 6.	13	— — —	Spath calcaire.	— — — —	Pl. VI. fig. 5.	
13	— — O — —	2 Pyramides hexaëdres, inégales, tronquées, jointes baſe à baſe.	— — O — —	3 Hexagones & 4 Triangles, d'une part, 1 hexagone & 6 trapèſes de l'autre. 7 = 7.	14	Vitriol martial.	Spath calcaire.	— — — —	Pl. II. fig. 16.	
14	— — O — —	2 Pyramides égales, hexaëdres, engagées par leur baſe en ſens contraire.	— — O — —	12 Scalènes oppoſés 2 à 2 ; les deux les plus proches ſont égaux. 6 = 6.	12	— — —	Spath calcaire.	— — — —	Pl. I. fig. 13. — —	Syſt. Nat. XII. fig. 31. — IX. fig. 7.
15	— — O — —	Les mêmes ; tous les angles formés par la jonction des deux baſes ſont tronqués.	— — O — —	12 Pentagones irréguliers, & 6 petits rhombes.	18	— — —	Spath calcaire.	— — — —	Pl. II. fig. 21.	
16	Priſme oblong, hexaëdre, dont les côtés ſont égaux.	2 Pyramides égales, hexaëdres, dont les plans ſont alternes avec ceux du priſme.	6 Hexagones allongés. — —	12 Rhombes aigus. 6 = 6.	18	Tartre vitriolé.	Spath calcaire.	— — — —	Pl. II. fig. 18. — —	Syſt. Nat. XII. fig. 40.
17	Priſme oblong, hexaëdre, dont les côtés ſont inégaux.	— — *Idem.* — —	6 Hexagones ; les 3 alternes larges.	12 Trapèſes inclinés 2 à 2. 6 = 6.	18	Tartre vitriolé.	Spath calcaire.	— — — —	Pl. II. fig. 19.	
18	Priſme court, hexaëdre, dont les côtés ſont égaux.	— — *Idem.* — —	6 Hexagones courts. — —	12 Rhombes. 6 = 6. —	18	— — —	Grenat *ſuivant M. Linné.*	— — — —	Pl. II. fig. 17. — —	Syſt. Nat. XII. fig. 31.
19	Priſme oblong, hexaëdre, dont les côtés ſont égaux.	*Idem*, tronquées au ſommet.	6 Hexagones allongés. — —	12 Pentagones & 6 Rhombes. 9 = 9.	24	— — —	Spath calcaire.	— — — —	Pl. II. fig. 20.	
20	Priſme très-long, hexaëdre, dont les côtés ſont égaux, & ſouvent ſtriés.	2 Pyramides hexaëdres, très-courtes relativement au priſme.	6 Rectangles longitudinaux, fort étroits.	12 petits Triangles. 6 = 6.	18	Le Nitre.	— — — —	M. de Plomb blanche.	Pl. V. fig. 17.	
21	Priſme long, hexaëdre, tronqué de biais.	Sans Pyramide ; un des bouts adhérant ou irrégulier.	6 Rectangles ou trapèſes allongés & 1 hexagone incliné.	— — O — —	7	Le Nitre.	Criſtal de roche.	— — — —	Pl. I. fig. 18 & 19. Pl. V. fig. 18 & 19.	
22	Priſme oblong, hexaëdre, dont les côtés ſont égaux.	Sans Pyramide ou tronqué aux deux bouts.	6 Rectangles longitudinaux & 2 hexagones horiſontaux.	— — O — —	8	Sel de Seignette. Tartre vitriolé.	Spath calcaire ; Emeraude du Pérou. Mica ; Pierre de Croix.	M. de Plomb verte. Molybdène.	Pl. II. fig. 1. & Pl. III. fig. 23. — —	Syſt. Nat. XII. fig. 5. — IX. fig. 9. Amœn. Ac. fig. 16.
23	*Idem*, dont les deux bouts ſont									

Figure 11 Romé de l'Isle's classification of minerals. From J. B. Romé de l'Isle, *Essai de cristallographie* (Paris, 1772).

their inability to reflect the complexity of nature.[89] This typical argumentation did not prevent him from using the chemical nomenclature of Guyton de Morveau to classify minerals. As will be shown later, Guyton's nomenclature was based largely on Linnaeus' and Bergman's binomial principles.[90]

Despite the wide range of approaches, eighteenth-century mineralogists had in common a deep interest in devising a new and more accurate classification and nomenclature for use in dealing with minerals.

"OBSCURUM PER OBSCURIUS"

It is not within the scope of this study to investigate the internal structure and significance of the nomenclature of alchemy.[91] Such an undertaking would raise questions that are extraneous to the historical problem with which I have chosen to deal. However, it would be a mistake to underestimate the influence of alchemical thought on the early stages of chemistry. Although this influence has sometimes been exaggerated, it would be wrong to deny that during the century 1550–1650 Paracelsus' chemical writings were a constant source of inspiration to many natural philosophers.

For this reason I shall consider those areas of alchemical thought and nomenclature that had a historical influence on the lexicon and on the experimental laboratory of early chemistry. It might be argued that in doing so I am arbitrarily presupposing a distinction between alchemy and early chemistry that did not actually exist in the sixteenth century. But in fact, as I showed in the discussion of Agricola's thought, there were emerging during the Renaissance, if not *disciplines*,[92] differing cultural traditions that dealt with chemical and mineralogical objects in

[89]"Souvent les naturalistes, et plus souvent encore les chimistes, lorsqu'ils ont observé quelques rapports communs entre deux ou plusieurs substances, n'hésitent pas de les rapporter à la même dénomination; c'est là l'erreur majeure de tous les méthodistes, ils veulent traiter la Nature par des genres et espèces," Buffon, *Histoire naturelle des minéraux* (Paris, 1783), vol. I, 74.

[90]*Ibid.*, vol. II, 160–163.

[91]On alchemical nomenclature, see Marcelin Berthelot, *Les origines de l'alchimie* (Paris, 1885); Carl Gustav Jung, *Psychologie und Alchemie* (1944), English translation (London, 1953); and the classic work by Crosland (1978), 3–62.

[92]The notion of a discipline being a contemporary one, it is obvious that it does not fit the complexity of sixteenth-century science *historically*. I therefore prefer, when possible, to rely on the broader concept of *tradition*, which implies a certain body of knowledge without a strict categorization of it.

completely different ways, and, pervasive as the influence of Paracelsus was, the appearance of his works did not eliminate the Aristotelian interpretation of matter taught in the European universities overnight. On the contrary, people like Paracelsus, Patrizi, Telesio, and Bruno:

> failed to overthrow the Aristotelian tradition in natural philosophy, not because they were persecuted, or because their opponents preferred vested interests and habits of thought to the truth, but because their impressive doctrines were not based on a firm and acceptable method. Aristotelian natural philosophy, rich in subject matter and solid in concepts, could not possibly be displaced from the university curriculum as long as there was no comparable body of teachable doctrine that could take its place.[93]

Furthermore, as was shown by Agricola and Biringuccio, consistent scientific progress could also be made within the "limits" of Aristotelian natural philosophy.

The historical function and importance of Paracelsus' criticism of Aristotle and the classical tradition do not lie in their scientific results but in the effectiveness with which Paracelsus and his disciples were able to erode the absolute self-confidence of Aristotelian science. As has been pointed out by Kristeller and Alexandre Koyré,[94] the cultural roots of Paracelsianism implied opposition not only to the methods and contents of classical philosophy but also to the very structure of the *scientific method.* The frequent appeals to experience and practical experiment made by Paracelsus did not stem from an empirical conception of philosophy similar to either the Aristotelian or the humanistic model. His notion of experience was in fact conceived within a cosmological and mystical framework.

In the specific case of chemistry, the opposition between Aristotelian science and a mystical philosophy of nature has been well illustrated by Owen Hannaway in his interesting survey of the emergence of chemistry during the sixteenth and seventeenth centuries.[95]

After the death of Paracelsus in 1541, many of his pupils, Oswald Croll and Gerard Dorn in particular, spread and developed his philo-

---

[93]Paul Oskar Kristeller, *Renaissance Thought and Its Sources* (New York, 1979), 47–48.

[94]Kristeller (1979); Alexandre Koyré, "Paracelse," in *idem, Mystiques, spirituels, alchimistes* (Paris, 1955). On Paracelsus, see also Jung (1967), and the classic work by Walter Pagel, *Paracelsus: An Introduction to Philosophical Medicine in the Era of the Renaissance* (Basel, 1982).

[95]Owen Hannaway, *The Chemists and the Word: The Didactic Origins of Chemistry* (Baltimore and London, 1975).

sophical ideas in a chemical context, although their efforts met with strong resistance from the universities and the Church.

Oswald Croll,[96] who published his *Basilica chymica* in 1609, was the more consistent in applying Paracelsus' cosmological ideas to alchemy. This application, however, was far from being a compromise between the *theory* and *practice* of chemistry: "the transformation of mere matter did not concern him; his goal was the control and manipulation of the spiritual, cosmic forces of nature which lay encapsulated within the useless dross of matter."[97] The experimental phase of chemistry represented for Croll a mere means of intercepting the hidden communications between microcosmos and macrocosmos. The elements composing matter were not at all the constituents of concrete bodies and substances but "the cosmogonic regions, or wombs, in which the fruits of the Astra are generated."[98] Accordingly, alchemy as a whole played the role of a metaphysical and therapeutic interpretation of the signs of the universe. As might befit this superior and philosophical mission, the "chemical" nomenclature visualized by Croll was particularly abstract and obscure:

> The caelestial Medicine onely, or the WORD of God (which is the Firmament of all physick, without which no drug will doe good) is that which healeth all things, and by the efficacy of the WORD (in which lyeth hid, and from which proceedeth all force beyond any naturall actions) all Medicins become powerfull; *as the bark is not the kernell, so herbs are not the medicines, but a signe onely of the Word signified.*[99] [my italics]

Thus the Word had a metaphorical value rather than a descriptive one. This theological connection between the Word and God explicitly undermined the practical manipulation of remedies as well as their classification and naming. Moreover, Croll's metaphorical conception of language led him to oppose the descriptive character of botanical nomenclature, and he continued:

> Oh that the Botanists of our time, who being ignorant of the internal Form of plants, know only their matter, substance and body, would devote as much care to the descernement of the Signatures of Plants as they do to their manifold and frequently frivolous disputes about

[96]On Croll, see Hannaway (1975), 22–72.

[97]*Ibid.*, 4.

[98]*Ibid.*, 35.

[99]Quoted *ibid.*, 48.

> the accurate naming of them, it would render a much richer and more beneficial service to medicine.[100]

This extremely important passage reveals a common attitude of the alchemists (not only Paracelsian) to empirical observation and descriptive nomenclature. In considering the study of the *matter*, *body*, and *substances* of plants, in other words their observable features, useless and insufficient for their *real* identification, Croll criticized the empirical and Aristotelian structure of sixteenth-century botany. When he called the naming of plants *frivolous*, Croll was not only anti-Aristotelian but also "anti-scientific" in the sense that he was opposing all the characteristics of scientific language, beginning with its descriptiveness; "the opposition to the taxonomic enterprise [was] absolute."[101]

To the Paracelsians, nature was a chain of correspondences and analogies permeating the universe, and its flux involved the naturalist himself in its cosmological and organic unity. Within this vast and vital universe, the *true* physician had to reveal the hidden relations between microcosmos and macrocosmos, and interpret the signatures concealed by God in each single body. It is obvious that the Aristotelian approach to nature, which gave a definition and a name to all things according to their empirical and therapeutic properties, was not acceptable to the Paracelsians. Their use of the word had a higher meaning that superseded the mere function of appropriately describing the form of an object. "The Word in nature which the Paracelsian sought to articulate was, in the last analysis, ineffable."[102]

The vitalistic correlation between all natural kingdoms could only be vaguely symbolized, and certainly not captured in a systematic and descriptive nomenclature. Most alchemical symbolism reveals the immense effort of the alchemist to express the whole system of correspondences and analogies, and the ancient association of planets and metals fitted in during the sixteenth and seventeenth centuries with this new metaphysical meaning (see Appendix). The metaphorical and symbolic structure of alchemical language is demonstrated by Martin Ruland's attempt to compose a lexicon of alchemical terms based on Paracelsian principles. There we find that copper had 25 distinct ac-

---

[100] *Osvaldi Crollii Tractatus de signaturiis Internis Rerum, seu de Vera et viva Anatomia Majoris et Minoris Mundo* (Frankfurt, 1609), quoted and translated *ibid.*, 63.

[101] *Ibid.* My use of terms, such as scientific and anti-scientific, refers exclusively to the body of scientific knowledge accepted as such within Renaissance universities.

[102] *Ibid.*, 62.

cepted meanings,[103] and that Gold was considered a microcosmos and a heavenly substance at the same time.[104] In this flood of obscure notions, the classical terminology used by the Greeks and Romans was criticized or ignored. It is not surprising, then, that the Paracelsians' criticism of Aristotle's conception of matter was associated *tout court* with the criticism of scholastic nomenclature and the definition of natural objects.

From this point of view, the position of Jean-Baptiste Van Helmont is significant. According to him, "logic [was] useless for inventing and making science,"[105] and he criticized Aristotelian syllogisms and definitions on the grounds that the only authority for this kind of question was the Bible.[106] Holding these ideas, Van Helmont conceived an alchemical language permeated with biblical, metaphorical, and cosmological references. The term *gas*, for instance, far from denoting the elastic fluids with which we associate it, signified the vital principle placed by God in all natural bodies. The dissociation of *gas* from a specific body was considered by Van Helmont as a sign of the transmutation of that body into another.

> Hence, *gas* may be defined as the material or watery vector of object-specificity, the spiritual carrier of the specific life-plan of an object. Air and water are general media shared by all objects; *gas* is a privilege reserved for the individual. It expresses the close interlocking of matter with seminal spirit [. . .].
>
> The development of *gas* indicates the "beginning of transmutation" of the whole object. Grapes with their skin intact dry out; deprived of their skins they soon "conceive the ferment of ebullition," the "cause of fermenting," and so do fruit and vegetables of all sorts once they are crushed. Again this is not the liberation of something *from* the object; rather, the *gas* that develops is the object itself, whole and transmuted.[107]

[103]"Aes non semper cuprum denotat sed quandoque aurum, vel argentum a natura sine alteri metallis, aut lapidis per mixtione . . ., Aes caldarium, aes abstractum, aes residuum, aes liquefactum, etc.," from Martin Ruland, *Lexicon Alchemiae sive dictionarium alchemisticum, cum obscuriorum Verborum, et rerum Hermeticarum, tum Theophrast-Paracelsicarum phrasium, planam explicationem continens* (Frankfurt, 1612), 9.

[104]"Gold ist microcosmus/ hat drey principia und vier Element und ist ein himlische substanz/ himmel und stralen der Sonnen/ darum bestehet es im Fewer/ und ist dei höchste medicin, hat inn sich alle sternen dess himmels/ und alle kräuter der Erden," *ibid.*, 93.

[105]Jean Baptiste Van Helmont, *Oeuvres* (Lyon, 1670), 43.

[106]On Van Helmont, see Metzger, *Les doctrines chimiques en France du début du XVIIe siècle à la fin du XVIIIe siècle*, 2nd ed. (Paris, 1969), 166 and ff.

[107]Walter Pagel, *Joan Baptista Van Helmont: Reformer of Science and Medicine* (Cambridge, 1982), 63.

The notion of gas, far from denoting a mere elastic fluid, embodied a cosmological system of matter that found its basis in its power of spiritual explanation.

Although Van Helmont shared the metaphysical and theological background and, like alchemists generally, constantly opposed Aristotelianism, he placed much more emphasis on experimentation than Croll and the other Paracelsians did. It is doubtful, however, whether the meaning he attached to the word *experiment* had much in common with the one in use a century later. In her outstanding and definitive study of seventeenth-century chemistry, Hélène Metzger stressed the danger concealed in the positivist attitude of separating the experimental endeavors of naturalists, such as Van Helmont, from their religious and philosophical beliefs, and convincingly demonstrated that the real historical importance of alchemy lies not in its experimental achievements but rather in the complexity of the philosophy behind it.

Despite the widespread diffusion of Descartes' *mechanism* among natural philosophers, chemists in France still displayed a completely different approach to nature, very similar to that of the Paracelsians and Van Helmont.

Nicaise de Febvre, who was deeply influenced by both these philosophical doctrines, published his *Traité de la chymie* in 1660.[108] The cosmological belief in the transmutation of matter and in the existence of a Universal Spirit led de Febvre to challenge the function of a precise chemical nomenclature:

> After having worked with the compounds, chemistry finds in its last resolution, five substances which it admits as elements and principles. On these [chemistry] builds its doctrine. . . . They are water, or phlegm, spirit or mercury, sulphur or oil, salt and earth. Some chemists give other names because it is allowed to all to denote the elements in the way they find most appropriate. This question of names is of little interest as long as there is agreement on the nature of things.[109]

But the nature of chemical "things" was far from being revealed, and the ambiguous terminology used by de Febvre was a sign of the scientific fragility of his treatise.[110] The practical notions of the *Traité* were confined to a confused collection of distillations, calcinations, precipitations, etc. of animal and vegetable substances. By contrast,

---

[108]See Metzger (1969), 62 and 82ff.

[109]Nicaise de Febvre, *Traité de la chymie* (1660), 2nd ed. (Paris, 1669), vol. 1, 19.

[110]Metzger (1969), 80–81.

little space was devoted to metallurgy, the one field in which, thanks to Agricola, there was a clear and quite coherent body of knowledge.

To conclude, we may say that from a practical point of view the chemical treatises of the Paracelsians had the merit of undermining the dogmatism and the rigidity of the Aristotelian conception of matter and its inflexible nomenclature, which was still mostly embedded in the Greek and Latin classics. The *anti-classical* element in sixteenth-century alchemy, if not contributing anything particularly original to chemical knowledge, helped to question the legitimacy of Aristotelian natural philosophy, and the reactions against Paracelsism, such as those of Libavius and Glaser, bear witness to the historical importance of alchemy's *negative* role.

Andreas Libavius attacked alchemical and Paracelsian thought from an Aristotelian standpoint.[111] "The true chemist does not disguise his elementary preparations but offers them for what they are—oil, waters and the like. The object of chemistry is not to destroy the material matrix of natural substances," he said.[112] This argument, similar to the one advanced by Agricola 50 years before, was directed at the alchemists' wish to render nature more perfect by purifying metals. Nature was already perfect and harmonious and could only be observed and codified into a scientific language (taxonomy). Hence the importance that Libavius attached to the lucid expression of chemical knowledge. As Hannaway has observed:

> This man-made language of which Libavius speaks operates in a wholly different way from the language of Nature which Croll evokes. It defines, divides, distinguishes and establishes criteria from judgment—a judgment which separates things. It seeks to discriminate knowledge, whereas Croll's language sought to reveal and express the affinities and resemblances of things.[113]

Following the precepts of Ramist philosophy and dialectics, Libavius divided nature into several realms and sub-realms. The chemical sciences accompanying the *Alchemia* (1597) substituted an ordered Latin setting for the pictorial and allegorical world of the alchemists.[114] Thus Libavius took a completely different view of the role played by language in chemistry:

---

[111]On Libavius, see Hannaway (1975), 75–151.

[112]Quoted in *ibid.*, 87.

[113]*Ibid.*, 108.

[114]*Ibid.*, 73–74.

> The true chemist does not like neologisms. The Paracelsian likes nothing more. To hide perplexity of words like the Delphic Demon is the best that these artful dodgers are capable of. But this should not be taken to absurd extremes and [those words] which have been taken over from the Arabs and the Egyptians have been made more pliable now by frequent use. Let the Latin-speaking practitioner speak more latinly when he can. He dare not change everything, however, for change of this kind obscures the subject-matter.[115]

However, this projected reform of nomenclature had modest results indeed. Libavius' work was still heavily contaminated with alchemical and non-classical terms, and his strong criticism of Paracelsus did not lead to a foundation for chemistry nor to the systematization of its nomenclature.

In another well-known treatise, the *Traité de la chymie* by Cristopher Glaser, the attack on Paracelsism and the obscurities of alchemical nomenclature again ran into both practical and theoretical difficulties. This, more than any other seventeenth-century work, shows us the unstable bases of the chemical knowledge of the period.

In the *Preface* to his work, Glaser immediately pointed out that:

> Writers on chemistry differed in knowledge and attitude. Hence came their different ways of writing [about it]. Those who engaged in the superior chemistry, penetrating its greatest mysteries, were satisfied to grasp some notion of it, while those who appeared to be more communicative had actually written with such obscurity that they might well have seen ghosts instead of bodies, thorns instead of fruits.[116]

The uncertainty of alchemy and the obscurity of its language made Glaser extremely skeptical of any speculative interpretation of chemical matter. In his opinion, chemistry was a practical art useful to the physician, the pharmacist, the dyer, and the painter.[117] He demonstrated this pragmatic and technological approach in the particular care he took in the precise classification and description of chemical operations. Glaser listed no fewer than 48 operations,[118] a number that in relation to the actual technological and experimental status of sixteenth-century chemistry was definitely excessive.

---

[115]Libavius, *Rerum chymicarum Epistolica* . . . (Frankfurt, 1595), translated and quoted *ibid.*, 119.

[116]Glaser, *Traité de la chymie* (1663), 2nd ed. (Paris, 1676), Preface.

[117]*Ibid.*, 3–5

[118]*Ibid.*, 11–12

In fact, the contents of the *Traité de la chymie* were not particularly original and the emphasis on vegetable and animal analysis was a repetition of a Paracelsian leitmotif. The nomenclature used by Glaser was also derived from the Paracelsian lexicon.[119] Early chemistry was largely dependent on medicine, and it was quite natural that even those like Glaser who rejected the approach and the language of the Paracelsians were forced to compromise with them. If they wanted to work on the therapeutic qualities of chemical remedies, the literature to which they had to refer was necessarily Paracelsian. The only possible alternative to Paracelsism, as we have seen, was to pursue chemical investigation within the metallurgical framework offered by Agricola.

The history of chemistry between 1550 and 1650 was one of lively intellectual ferment and fierce polemic and debate. The arguments about alchemical principles and chemical elements took place in a framework of cosmological and metaphysical controversy, and their practical consequences were comparatively modest.

During this period "chemical" nomenclature developed into a Tower of Babel, and all sorts of terms, synonyms, codes, and symbols made their first appearance in European languages. However, from this chaotic context, something new was emerging. It might perhaps be called an *intuitive perception* of a scientific domain that was neither tied to the alchemical tradition nor dependent on the Aristotelian and scholastic one.

The emergence of this "prescientific" tradition, which did not become chemistry until the end of the eighteenth century, was extremely slow and still bore the imprint of various heterogeneous ideas.

The Emergence of Chemical Questions

During the second half of the seventeenth century, an increasing number of naturalists engaged themselves in chemical investigations. The growth of this branch of natural philosophy did not yet imply the emergence of chemical science as an independent discipline, but it laid down the experimental basis for its further development. In this connection the work of Robert Boyle undoubtedly represented a significant step towards the definition of chemistry as a science. However, most histori-

---

[119]Glaser again proposed the association between planets and metals and obscure Paracelsian terms, such as *Magister Jupiter, arcane coralline, pierre infernale, saturne celeste,* etc.

ans have studied the meaning and structure of Boyle's corpuscular philosophy with relation to its successors rather than its sources. As a result, the cultural and scientific context of Boyle's work has been ignored, and his contributions have inaccurately been emphasized as "modern."

As has recently been shown in an interesting essay by Antonio Clericuzio,[120] the traditional identification of Boyle's corpuscular philosophy with a mechanical conception of matter needs a reassessment.[121] Boyle's criticism of spagyric chemistry did not in fact imply the adoption of a mechanistic view. On the contrary, Boyle acknowledged on several occasions his debt to Van Helmont, and he accorded some credence to certain alchemical beliefs, such as in the transmutation and the "growth" of metals,[122] and also in the existence of *seminal principles*[123] and of occult qualities between natural bodies.[124] Notwithstanding this debt to sixteenth-century and seventeenth-century alchemy, these sources were woven into an innovative and original presentation of chemical knowledge.

The importance of Boyle's historical role is to be found in his emerging awareness of a chemical domain different both from the one prospected by the "vulgar chemist" (the Paracelsian) and from that of the Aristotelians.[125] Boyle's corpuscular philosophy helped him to raise new questions for chemical investigation and to regard "chemistry" as a discipline independent of mechanics. He explained chemical phenomena in terms of corpuscles endowed with chemical, rather than mechanical, properties.[126] Seen from this perspective, Boyle's obsessive emphasis on experimentation and his skepticism towards any kind of chemical *philosophy* not supported by laboratory observations are understandable. This empirical approach was, ac-

---

[120]Antonio Clericuzio, "A Redefinition of Boyle's Chemistry and Corpuscular Philosophy," in *Annals of Science*, 47 (1990), 561–589. On Boyle's nomenclature, see also Jan V. Golinski, "Chemistry in the Scientific Revolution: Problems of Language and Communication," in D. Lindberg and R.S. Westman (eds.), *Reappraisals of the Scientific Revolution* (Cambridge, 1990), 367–396.

[121]This view has been supported by Thomas Kuhn, "Robert Boyle and the Structural Chemistry in the Seventeenth Century," in *Isis* (1952), 12–36, and by Marie Boas, *Robert Boyle and Seventeenth-Century Chemistry* (Cambridge, 1958), 75–107.

[122]Robert Boyle, *Observations About the Growth of Metals*, in Boyle, *The Works*, T. Birch (ed.), 2nd ed. (London, 1772), vol. IV, 79–84.

[123]On this, see Clericuzio (1990), 583–587.

[124]Boyle, *Suspicions About Some Hidden Qualities in the Air*, in *idem, The Works* (1772), vol. IV, 85–96.

[125]For the position of Boyle on the subject of the *vulgar* and Aristotelian chemists, see *ibid.*, vol. I, 661.

[126]Clericuzio (1990), 563.

cording to Boyle, certainly not by chance, connected to the question of chemical language:

> I should immediately proceed to the proof of my assertion, but for the confidence wherewith chymists are wont to call each of the substances we speack of by the name of sulphur or mercury, or the other of the hypostatical principles, and the intolerable ambiguity they allow themselves in their writings and expressions [. . .].
>
> For I find that even eminent writers (such as Raymond Lully, Paracelsus and others) do so abuse the terms they employ, that as they will now and then give divers things, one name; so they will oftentimes give one thing, many names; and some of them (perhaps) such, as do much more properly signifie some distinct body of another kind; nay even in technical words or termes of art, they refrain not from confounding liberty; but will, as I have observed, call the same substance, sometimes the sulphur, and sometimes the mercury of a body.[127]

Compared with the comments of Libavius, de Febvre, and Glaser, Boyle's criticisms of the language of the Paracelsians were more precise and were repeated several times in the *Sceptical Chymist.*[128] His objections were prompted not only by a desire for clarity, but also by a totally different conception of chemical science. According to Boyle, a name should express, or rather *describe*, the concrete substance to which it referred. This need was particularly acute for the identification of controversial substances, such as mercury, that might denote several different substances with completely different properties. In this case Boyle was even more critical of the "vulgar chemists," since though "they seem to tell us, what they mean by the principle [mercury], they are wont to do it in terms so loose and so ambiguous, that the representations they make of it, are more like to panygirics, and sometimes to riddles, than to *clear definitions, or such as good descriptions*" [my italics].[129] Undoubtedly the sarcasm directed at alchemical terminology was not new, but in Boyle we find a more precise description of its ambiguity and the first attempt to dismantle the hypothetical and metaphysical structure underlying it. Moreover, Boyle was the first to link the *clarity* of chemical terminology to the empirical category of *description*, a connection that pointed towards a definite reform of the nomenclature. These apparently insignificant changes in content reflected in fact a radical change in attitude to

---

[127]Boyle, *The Sceptical Chymist*, in *idem, The Works* (1772), vol. I, 520.

[128]See also the passages in the *Introductory Preface*, *ibid.*, vol. I, 460 and 522–523.

[129]*Experiments and Notes About the Producibleness of Chymical Principles Being Parts of an Appendix, Designed to the Sceptical Chymist*, *ibid.*, vol. I, 629.

the methods of chemistry. Boyle was talking about a more experimental approach to chemical observation; experiments were the basis of all scientific investigations and the *experimental philosophy* their theoretical guide.[130] Although this was not as innovative and modern as many historians have believed, the notion of *experimental philosophy* involved a new interpretation of chemical matter and the permanent introduction of *experimentation* and *description* in the chemical laboratory. Boyle's empiricism, although still supported with many metaphysical and theological arguments, had important consequences in the delimitation of the domain of chemistry and in the reform of chemical nomenclature.

The inconsistency and incongruity of Paracelsian language and method were also criticized by Boyle elsewhere.[131] However, he was perfectly aware that even if chemistry could no longer be the obscure science depicted in the treatises of the "vulgar chemists," it was still far from attaining the perfection and the expressive rigor of physics. The experimental status of seventeenth-century chemistry did not yet allow it to satisfy Descartes' idea of a mathematicized science. From this perspective it is interesting to recall the celebrated controversy between Boyle and Spinoza on the nature of niter.[132] The polemic was occasioned by the publication in 1661 of *A Physico-Chymical Essay, Containing an Experiment touching the different Parts and Redintegration of Salt-Petre,*[133] in which Boyle decomposed niter and recomposed it into spirit of niter and fixed niter without basing his experiment "on the primary and mechanical affection of particles,"[134] but by considering only the *chemical* properties of the substances. Spinoza, who believed that the only possible approach to science was that of Cartesian mechanism, reacted to Boyle's experiment as follows:

> One will never be able to prove this by chymical or other experiment, but only by reason and calculation. For by reason and calculation we divide bodies infinitely, and consequently the forces which are re-

---

[130]"I told you already (saies Carneades) that there is great difference betwixt the being able to make experiments, and the being able to give a philosophical account of them," *ibid.*, vol. I, 522. The philosophical account referred to is neither a Paracelsian nor a mechanical one. Boyle was most likely thinking of his *experimental philosophy.*

[131]See, for instance, the *Appendix to the First Section of the Second Part of Some Considerations Touching the Usefulness of Experimental Philosophy, ibid.*, vol. II, 232 ff., where Boyle mentions the "unintelligibility" of pharmaceutical nomenclature.

[132]On the Boyle-Spinoza controversy, see A.R. Hall and M. Boas Hall, "Philosophy and Natural Philosophy: Boyle and Spinoza," in R. Taton and F. Braudel (eds.), *Mélanges Alexandre Koyré* (Paris, 1964), vol. II, 241–256; Clericuzio (1990), 573–577.

[133]Boyle, *The Works*, vol. I, 35–76.

[134]Clericuzio (1990), 575.

> quired to move them; but we shall never be able to prove this by experiments.[135]

This argument supported the Cartesian mechanistic view of nature grounded on the analysis and mathematical interpretation of matter, rather than on the "fallacy and arbitrariness" of the senses. Accordingly, the composition of niter could be determined only by a mechanical interpretation of its nature (reason + calculation). Boyle firmly rejected this approach, which he rightly considered inadequate for penetrating the nature of chemical substances. Chemistry could be reduced neither to mathematical language nor to mechanical principles as Spinoza claimed, observed Boyle in a manuscript:

> [The Cartesian philosophers] are so charm'd with ye clearness and pleasure of Theorys and explications, yt are deriv'd immediately from metaphisical and mathematical notions and theorems; yt they oftentimes give forced and unnatural accounts of things rather than not to be thought to have deriv'd them immediately from these highest principles. And, wch is much worse, they despise, and perhaps too condemn or censure all yt knowledge of the works of nature yt Physicians, Chymists, and others pretend to, because they cannot be clearly deduc'd from the doctrine of Attoms, or ye Catholick Laws of motion. The practice of these virtuosi is like, in my opinion, to prove so great an impediment to ye advancement of real learning. . . .[136]

This comment could not have been clearer in stating Boyle's criticism of mechanical philosophy and, more generally, of the application of mathematics to chemistry. Boyle understood that the complexity of the subject, the difficulty of the experimental preparation and analysis of chemical substances, the imperfectly known nature of elements and principles could not be expressed in purely mechanical terms. Boyle's attitude to chemical nomenclature implicitly acknowledges the ensuing dilemma. If, on the one hand, Boyle repudiated the nomenclature of alchemy and considered it an impediment to the progress of science, on the other hand, he could not foresee a new one, since the experimental and epistemological conditions for building a methodical and systematic nomenclature did not exist. Boyle's solution to the problem was to use long descriptive sentences to designate chemical substances and operations. This left him free to use any term without falling into the concise obscurities of alchemy, but at the price of being forced to resort to very long-winded descriptions to eliminate ambiguity.

---

[135] *The Correspondence of Henry Oldenburg*, vol. I (Madison and Milwaukee, 1965), 463.
[136] Manuscript by Boyle, quoted in Clericuzio (1990), 574.

As long as this was the nature of the discussion of the language of chemistry, it is obvious that the need for a systematic and methodical nomenclature did not arise.

Even when we consider the case of the French apothecary Nicolas Lemery, seen, like Boyle, as another example of a *mechanistic* chemist, we encounter a similar difficulty. In Lemery's works, too, the commitment to mechanism was programmatic rather than indicative of a concrete experimental endeavor. Like Boyle, Lemery was aware of the experimental difficulties to be overcome before a chemical science based upon the Cartesian notions of *motion, figure,* and *number* could even be conceivable. These discrepancies are evident if one compares the general and theoretical part of Lemery's *Cours de Chymie* (1675) with the experimental part of the work. At the beginning of his *Cours,* Lemery tried to explain acidity by referring to the shapes of the particles:

> Since there is no better way of explaining the nature of a thing as hidden as that of a salt than by attributing to its component parts shapes that correspond to all the effects it produces, I shall say that the acidity of a liquid consists in pointed parts of salt, which parts are in agitation; and I do not believe anyone will dispute that acid has points, since all experiences show it [. . .].
>
> As for alkalis, they can be recognized by pouring acid on them, for immediately or shortly thereafter there occurs a violent effervescence which lasts until the acid finds no more bodies to rarefy. This effect can reasonably lead to the conjecture that the alkali is a material composed of unyielding and brittle parts, the pores of which are shaped in such as way that when the acid points have entered into them, they break and separate anything that opposes their movement. . . .[137]

This apparently advanced mechanical explanation of the composition of salts in terms of the shapes of the particles, eventually gave way to a classical preparation of remedies and to a conception of chemical reaction largely based on the notion of the *mixt.* Furthermore, Lemery reproposed a scheme of the composition of matter based on the traditional five principles: Water, Spirit, Oil, Salt, and Earth.[138] The mechanistic commitments were an attempt to overthrow both the Paracelsian and the Aristotelian conception of matter and to incorporate, in a very

---

[137]Nicolas Lemery, *Cours de Chymie, contenant la maniere de faire les operations qui sont en usage dans la Médicine par une Méthode facile* (1675), 11th edition (Paris, 1757), 17–18. English translation in Émile Meyerson, *Explanation in the Sciences* (Dordrecht-Boston-London, 1991), 219.

[138]*Ibid.*, 4.

generic way, chemical phenomena in a *scientific* theory. But in reality they were of little use in Lemery's actual chemical experiments.

This more pragmatic attitude to scientific theory becomes clearer when we read the pharmaceutical preparations in the *Cours*, which hardly ever refer to a general theory of matter. The importance of Lemery's work is therefore to be found in his criticism of alchemy and Paracelsism and in his effort to define chemistry in a new light. In this connection Lemery said ironically of the old nomenclature:

> Most of the authors who have spoken of chemistry have written of it with such obscurity that they seem to have done their best not to be understood; and in this we might say that they have been too successful because this science has been hidden, and known only to a few people, for centuries.[139]

The criticism of alchemical language was not purely destructive; throughout the *Cours* Lemery often tried to explain the meanings of the terms he used with the aid of detailed descriptions of the external characters or the therapeutic properties of the chemical substances to which he was referring. At the beginning of the *Cours*, Lemery added a dictionary of the terms he used, together with the explanations.[140] This glossary, which included only 59 terms, is extremely interesting, and although many of the words had an alchemical etymology, their meaning was restricted by Lemery by a detailed description of their empirical qualities,[141] which simplified the identification of the substance or operation to which they referred. Thus, Lemery's nomenclature was the result of laboratory experience rather than of a theoretical and methodical attempt at reform. A part of the lexicon used by the French apothecary still had its origins in alchemical nomenclature, although the meaning he attached to each word was a manifestation of a more precise and completely different conception of chemistry.

Here mention may be made of Lemery's criticism of the astrological association planets-metals, which was based on the *empirical* observation that the knowledge of celestial bodies was too scanty to support any hypotheses of their influence on terrestrial bodies:

> But in truth there's nothing to confirm their opinion, and we find it every day plain enough, that the Faculties and Virtues are utterly

[139] *Ibid.*, xi.
[140] "Explication de plusieurs terms dont on se sert dans la chymie," *ibid.*, 32–36.
[141] Among the alchemical terms used by Lemery, we can list *Athanor, Menstrum, Magistere, Transmutation, Jupiter, Saffran d'or*, etc.

> false, which they attribute to the *Planets* and *Metals*; the *Metals* indeed are good use in *Physick*, and excellent *remedies* may be drawn from them; but their effects may be better explained by the causes nearer at hand than *Stars*.[142]

This skepticism probably accounts for the substitution of the term *vif argent* for mercury as the name of the metal.[143]

When he had to confront alchemy as a cultural tradition, Lemery was extremely severe, making no concession to its scientific pretensions: alchemy was "an art without art, whose principle is to lie, whose means is to manipulate, and whose aim is to beg."[144] Despite this dismissive remark, Lemery had to make many compromises with alchemical thought when he proposed his chemical nomenclature and described mineralogical remedies. Even the association between planets and metals was legitimized again.[145]

In the more practical pharmaceutical treatise, *Pharmacopée Universelle*, Lemery became more aware that the establishing of a science and the nature of its language were intimately related. His approach to the question of technical nomenclature became less confused and, by comparison with Boyle's, more systematic. The *Pharmacopée* was in fact prefaced by a pharmaceutical lexicon of more than 600 terms, explaining both their etymology and their meaning. For the first time since Agricola's attempt in 1556, a wide range of chemical names was analyzed etymologically, and their meanings were compared with their medical or chemical nature:

> The etymological explanations are neither as useless nor as indifferent as many imagine; they often provide an idea of the nature of each thing so that one is aware of what it might be even before seeing it. In fact those who gave the names, particularly the Greeks, tried to en-

---

142"Il n'est pas difficile de voir que tout ce que nous venons de rapporter des influences [planets-metals] est très-mal fondé puisqu'il n'y a personne qui ait vu d'assez près les planettes pour sçavoir si elles sont de la même nature que les métaux . . . Mais il n'y a rien qui confirme leur opinion, et nous reconnaissons tous les jours que les facultés qu'ils attribuent aux Planettes et aux métaux sont fausses. Les métaux, à la verité, nous servent dans la Médicine, et nous en retirons de bons remèdes, mais leurs effets se peuvent mieux expliquer par des causes prochaines, que par celle des Astres," *ibid.*, 40. The English translation is taken from the English edition, *A Course of Chymistry* (London, 1698).

143*Ibid.*

144"Ars sine Arte, cujus principium mentiri, medium laborare et finis mendicare," *ibid.*, 46.

145For instance, *Jupiter-etain, ibid.*, 75–76.

close in each of these names the most precise explanation of the things they aimed to describe.[146]

The etymological approach outlined by Lemery did not spring from a humanistic tradition as had that of Agricola, who considered the entire question of scientific language from a philological and humanistic angle, but from a more practical need to clarify the relationship between names and meanings. The lexicon proposed by Lemery, even if it listed many alchemical terms, was relatively clear, and with the aid of the descriptions, it was not difficult to infer the object from the name.[147] However, to avoid the obscurity of alchemical nomenclature, Lemery was forced to specify the meaning of each word rather than invent new names.

The creation of a new nomenclature would have implied the establishment of a renewed scientific tradition, but Lemery, like Boyle before him, could not escape the scientific background that supported his view of chemistry. By the end of the seventeenth century, *chemistry* was still a meeting place of different cultural traditions, an incoherent mixture of alchemy, metallurgy, medicine, and pharmacy. Those naturalists, such as Lemery and Boyle, who contributed most to its earliest progress were, at the same time, those most aware of its theoretical and experimental limits. This awareness is also evident in one of the clearest manuals of chemistry of the eighteenth century,[148] published by Hermann Boerhaave in 1732. This book still belonged to the chemical context of the seventeenth century. Boerhaave's views on mechanism, in particular, were shaped by the debate initiated by Lemery and later entered by Hartsoeker and Homberg, both of whom tried to find a compromise between the experimental investigations they performed in the laboratory and a mechanistic view of matter.

Boerhaave became more and more conscious of the objective experimental difficulty in applying any general theory of matter to chemical phenomena, and he suggested abandoning the attempts to discover the nature of primitive constituents and principles. Chemistry was still a long way from a sound scientific structure and from determining the number and nature of simple substances; in the laboratory, he said,

---

[146]Lemery, *Pharmacopée Universelle, contenant toutes les compositions de Pharmacie qui sont en usage dans la Médicine*, 3rd. ed. (Paris, 1734), 11.

[147]"Lexicon Pharmaceutique, où l'on donne l'Etimologie de plusieurs termes dont on se sert en Pharmacie," in Lemery (1734), 12–59.

[148]Hermann Boerhaave, *Elementa Chemiae* (Leyden, 1732), 2 vols. I have used the French translation, which includes an interesting introduction (*Élémens de Chymie* (Paris, 1754), 6 vols.). and the English translation *Elements of Chemistry* (London, 1735).

"chemical operations do not produce anything so simple" as an element.[149] After composing and decomposing chemical substances, it was extremely difficult, and often impossible, to get back to the primitive compound or element:[150]

> It is therefore necessary to prescribe to our Art certain fixed limits which we must not transgress if we want to avoid mistakes and discover the truth.[151]

This theoretical prudence has to be interpreted as an early sign of a scientific awareness of the limitations of chemistry, beyond which there was only abstract speculation. In this connection, Boerhaave offered a definition of chemistry that reflected his cautious empiricism:

> Chemistry is an art, that teaches us how to perform certain physical operations, by which *bodies that are discernible by the senses, or that may be rendered so,* and that are capable of being contained in vessels, may by suitable instruments be so changed, that particular determined effects may be thence produced, and that the causes of these understood by the effects themselves, to the manifold improvement of the various arts.[152] [my italics]

By focussing his attention on the immediate data of the senses and on the experimental investigation of compounds, Boerhaave was gradually liberating chemistry from its speculative and metaphysical nature, but he was unable to formulate an alternative theory of matter. Undoubtedly the notion of *mixt body* that now became an accepted concept in chemical laboratories left both the mechanistic and the metaphysical approach to the study of matter behind and became a cornerstone of a new and important phase of eighteenth-century chemistry, although its systematic use did not lead to an effective means of classification. Indeed, like his predecessors, Boerhaave was unable to construct a new chemical nomenclature or to express effectively the infinitely diverse form of an increasingly complex matter. However, the awareness of this inner complexity was an important step towards a new definition of the components.

In the last analysis, the attempts to define chemistry from a theoretical point of view encountered too many insuperable practical and experimental problems. Even those naturalists who relied upon a flex-

[149]Boerhaave (1754), vol. I, 153.
[150]*Ibid.*, 154.
[151]*Ibid.*
[152]Boerhaave (1735), 19.

ible form of *experimental philosophy* and a vague and empirical adaptation of mechanism could not yet overcome the enormous difficulties involved in relating the chemical phenomena collected in the laboratory to a general theory of matter or even to a sound definition of chemistry. These difficulties left the problem of chemical nomenclature open to virtually any kind of solution. Since it was declared impossible to isolate and precisely define a chemical element or substance, it was not yet feasible to contemplate the idea of systematizing the nomenclature of chemistry other than in a very general way. Chemistry was a science dealing with compounds and mixts that were frequently known by their therapeutic or chemical effects rather than by their inner nature. Nonetheless, the awareness of the difference between the methods of alchemy and of chemistry and the vigorous criticism that these chemists voiced against the old nomenclature were decisive steps towards a modern definition of chemistry. This awareness was demonstrated by the increasing publication of dictionaries and lexicons of chemistry aiming to provide the *empirical* translation of the speculative language of the past.[153]

The Phlogistic Nomenclature

The most prominent figure in chemistry in the first half of the eighteenth century was undoubtedly the German physician Georg Ernst Stahl. Yet important as he was in providing an entirely new insight into chemical knowledge, there are few critical studies giving us a clear picture of the complexity and general architecture of his chemical theory.[154] The dearth of secondary literature is partly due to the difficulty and, sometimes, the obscurity of Stahl's style. In many of his writings, he switches from German to Latin for no apparent reason other than his preference for Latin for certain arguments in relation

[153]See, for instance, J.C. Sommerhoff, *Lexicon Pharmaceutico-Chymicum* (Nuremberg, 1701), and J. Harris, *Lexicon technicum* (London, 1704), which are both *descriptive* dictionaries of scientific terms of a nature completely different from Ruland's *Lexicon Alchimiae.*

[154]On Stahl, see Pierre Duhem, *La chimie est-elle une science française?* (Paris, 1916), 58–85; Metzger, *Newton, Stahl, Boerhaave et la doctrine chimique*; Irene Strube, "Die Phlogistonlehre Georg Ernst Stahls (1659–1734) in Ihrer historischen Bedeutung," in *NTM,* I (1960), 27–51; Sandra Tugnoli Pattaro, *La teoria del flogisto. Alle origini della rivoluzione chimica* (Bologna, 1983), 85–146.

to mineralogy and metallurgy, whereas the literature and nomenclature were mostly vernacular.

A second difficulty that may have discouraged scholars from undertaking a detailed study of Stahl's chemistry is presented by his prolific output and his many digressions across the domains of medicine and philosophy. Moreover, the repetitive nature of his vitalistic arguments and his tortuous mode of expression do not facilitate an understanding of his ideas.

And, last but not least, Stahl's phlogiston theory has been regarded as a "wrong" interpretation of chemical phenomena, an obstacle to the correct and quantitative approach shown by Lavoisier. This picture has led many historians to consider Stahl's chemistry conservative or retrogressive, and therefore not worthy of study.

But the fact is that the phlogiston theory, and more generally Stahl's conception of matter led to a better understanding of important processes such as combustion and calcination, and that Lavoisier himself regarded Stahl's theory as the primary source of his own early work and of his revolutionary intuition concerning calcination.

It does not fall within the field of this study to discuss Stahl's chemical philosophy in general terms, but I shall try to explain the historical reasons for Stahl's obscure nomenclature.

Stahl's phlogiston theory grew out of the development of a number of observations on the composition of matter by Johann Joachim Becher. The source is of no little significance, since Becher's chemical works contained a mixture of chemical experiments, theological pronouncements, biblical references, and alchemical beliefs in the purification and transmutation of metals.[155] From this mixed bag of notions, Stahl attempted to develop his own views on the chemical composition of matter and to eliminate all metaphysical and alchemical reasoning. To this end, he chose to comment on one of the "clearest" of Becher's works, the *Physica subterranea.* Here Becher based chemistry on three earthly principles. The first earth, called *terra vitriscibili,* was the principle of all metals and stones, *prima metallorum,* or *lapidum principis,* and it was considered the cause of their colors, forms, and physical properties. It was also present in crystals, spaths, and quartz.[156] The second

[155]On Becher, see Partington, *A History* (1969), vol. 2, 637–652.

[156]"De primo metallorum et lapidum principio, quod lapis fusilis seu terra lapidea est, impropre sal dicta," from J.J. Becher, *Physica subterranea, profundam subterraneorum genesis . . .*, Stahl (ed.) (Leipzig, 1738), 61–66.

earth, "unproperly called sulphur," was *terra pinguis*,[157] which was supposed to be the principle of the inflammability, acidity, and unctuosity of bodies.[158] The third earth, which was called *terra fluida* and had formerly been called *mercurius*, was the principle of volatility, opacity, and malleability.[159]

It is evident from this scheme that Becher's classification of principles owed a great deal to Paracelsus. However, his emphasis on the concrete qualities of matter, rather than on a philosophical interpretation of it, marked an important difference between the two systems. Moreover, Becher underlined the importance of the notion of mixt bodies, and he proposed the following hierarchy of matter, which remained the standard classification for almost a century:

> *Principles—Mixed bodies* (composed by the combination of principles)—*Compounds* (formed by the combination of mixed bodies)—*Super compounds* (formed by the combination of compounds and mixed bodies).

Such a division, to which was added the central chemical role played by *terra pinguis*, influenced Stahl in his formulation of the phlogiston theory. The central features of Stahl's system were in fact the notion of mixture and the pervasive role of sulphur, which was the principle of inflammability and assumed to be present in all bodies.

From the vantage point of this new shape of chemical matter, Stahl criticized those who reduced chemistry either to medicine or to mechanism, as in neither case were they supported by a sufficient experimental background. The theoretical assumptions made by both philosophies in seeking to explain the structure of matter with a few principles were contradicted by the complexity of chemical phenomena.

With his acute sense of the inherent causes of scientific change, Pierre Duhem pointed out in an interesting essay on the notion of the *mixt* the historical necessity of the concept of *physical combination*, which more than anything else helped chemistry to emancipate itself from all external and metaphysical explanations.[160] That most eigh-

---

[157]"De secundo mineralium principio, quod terra pinguis est, improprie Sulphur dicta," *ibid.*, 66.

[158]*Ibid.*, 66–75.

[159]"De tertio mineralium principium, quod fluida terra est, improprie Mercurius dicta," *ibid.*, 76–84.

[160]Pierre Duhem, *Le mixte et la combinaison chimique: Essai sur l'évolution d'une idée* (1902), 2nd ed. (Paris, 1985).

teenth-century chemists considered this an invaluable achievement is clear from a note by the British physician Peter Shaw, who translated Stahl's *Fundamenta Chemiae* into English and described the fundamental importance of the chemical concept of *mixture* in the following words:

> An adequate *Notion* as to the business of *Mixture* and *Composition*, being of the utmost importance in the Theory and Practice of *Chemistry*, 'tis well worth attempting to clear it up a little farther [. . .].
>
> By the word *Mixt* is understood certain *Corpuscules* of such a degree of smallness, with regard to our senses, as not to be cognizable by them, unless in a numerous parcel [. . .].
>
> That as the number of *Mixts*, is but small, confined perhaps to *Silver*, *Gold* and some few other almost simple or homogeneous bodies; their distinction in most cases may commodiously be dropp'd, and *Compounds* substituted for them.[161]

It is obvious from this that a strict determination of the nature of elements, as well as of their names, was not needed. The question of nomenclature was therefore irrelevant to the conception of matter and, more generally, to the chemistry outlined by Stahl. Since matter was complex in structure and made up of mixts and compounds, it would have been illogical to establish a causal correspondence between name and substance. In Stahl's system chemical substances were in fact elusive by definition, and their nature was determined by establishing a hierarchy of *combinations of mixts* rather than by a static definition of *principles.*

Therefore it is not surprising that Stahl's nomenclature was always extremely unsystematic and often obscure. This was due to the unimportance in the chemistry of mixts of the question of their naming.

Stahl's general definition of chemistry helps us to understand this attitude better:

> Chemistry is the art of dissolving natural mixt bodies by various means.
>
> The objects of chemistry are inanimate bodies, either mixed or compositions of mixts. All mixts and compositions that require simples, are usually called elements [. . .].
>
> The subjects of chemistry are mixed bodies; the principles of mixture (*mixtionis*) are earth, water and ether; from these [principles] are engendered the principles *principiata,* or concrete, which come

[161]Stahl, *Philosophical Principles of Universal Chemistry,* translated by Peter Shaw (London, 1730), 7.

together in the composition of bodies and are called salt, sulphur and mercury.[162]

The principles of mixed bodies fulfilled the general function of the Aristotelian four elements and were regarded by Stahl as *principia*, whereas the chemical principles (sulphur, salt, and mercury) were derived by the action of fire.[163] This distinction liberated chemistry from the question of the ultimate composition of matter, which was now recognized as a combination of different kinds of chemical compound.

This being his general conception, Stahl turned his attention to the effect of fire on chemical phenomena. He studied in particular the action of the principle of inflammability: Becher's *terra pinguis*. Stahl regarded this principle as the main cause of combustion and acidity. The most important theoretical function of this principle was to explain the two essential chemical processes *calcination* and *combustion*, which he saw as attributable to the same cause. In these processes Stahl assumed a loss of the inflammable principle, which he called *phlogiston*.[164] In his famous treatise on sulphur, Stahl claimed that *phlogiston* was present in all bodies but that its presence could not be perceived immediately by the senses except from certain secondary qualities, such as color and smell and inner disposition:

> From all these combined circumstances, I have judged that no more fit name could be given to this material than that of inflammable matter or principle. Indeed, as up to the present time none has been able to find or recognize any portion of it *except in combination, and no one consequently can give a definition of it nor any name after some property which uniquely belongs to it, it seems to me nothing is more reasonable than to*

---

[162]"Chymia est ars corpora naturalia mixta mediantibus variis motibus dissolvendi [. . .].

Objectum chymiae sunt corpora inanimata mixta, et ex mixtis composita. Omnia mixta et composita requirunt simplicia, quae vocari solent Elementa [. . .].

Subjectum chymiae sunt corpora mixta: principia mixtionis sunt, terra, aqua, et aether; ex hisce emergunt principia principiata seu concreta, quae ad corporum compositionem concurrunt et vocantur propria Chymicorum nostra sententia duo tantum sunt, sal et sulphur. Mercurius spectat vel ad aetherem vel ad ipsam aquam.

Corporum varietas et compositio dependet a varia mixtura principiorum mixtionis terrae, aquae, et aetheris, et principiorum principiatorum cinnabris et sulphuris," from Stahl, *Fundamenta chymiae dogmatico-rationalis et experimentalis* (Nuremberg, 1732), Part I, 1–3; another definition of chemistry, similar to this, is provided in Part II, 64–65.

[163]*Ibid.*, 2 and 66–67.

[164]The term phlogiston was introduced by Hamerius Poppius in 1618, and it was used to denote the adjective inflammable. Stahl began to use it in relation to the definition of the properties of sulphur in 1679, but he gave the first systematic definition of it only in his *Züfallige gedanken und nützliche bedencken über den Stret, von dem sogennanten Sulphure . . .* (Halle, 1718).

> *name it after the general effects it produces even in its final combinations, that is why I give it the Greek name of "phlogiston," phlogistic or inflammable.*[165] [my italics]

Being related to chemical combinations rather than to chemical substances, the term *phlogiston* was not unambiguous, and Stahl introduced many other names to denote the same thing.[166]

This qualitative feature of *phlogiston*, far from being a sign of inaccuracy or of metaphysical commitment, allowed Stahl to reconcile the empirical data with a general theory of the composition of matter. For the first time in the history of chemistry, a theory was embodied in the facts it aimed to explain. Moreover, by focussing attention on some important and specific aspects of chemical combination, Stahl pointed the way to a fruitful research program.

The identification of the processes of combustion and calcination brought two different kinds of phenomena, formerly regarded as completely distinct, together in one body of knowledge. But it was the full understanding of the qualitative form of chemistry that was probably Stahl's most important contribution.

Stahl's theory of calcination, which implied the liberation of phlogiston from the calcined metal, was a natural observation to those who approached such a process qualitatively; the theory explained this phenomenon rationally, and by relating it to the process of combustion, it implicitly stated the crucial role of fire and heat, which eventually became the central concerns of eighteenth-century chemistry.

The simplicity and empirical plausibility of Stahl's theory account for its enormous success, while its qualitative nature explains why the problem of nomenclature was ignored. The emphasis on the qualities of chemical combinations left the question of a strict denomination of substances outside the horizons of chemistry.

Until the 1760s chemistry remained within the qualitative framework constructed by Stahl, although discrepancies between the laboratory experience and the elusive nature of phlogiston had already begun to arouse some perplexity by the end of the 1750s. Mathematicians such as d'Alembert questioned the role and even the existence of phlogiston in the article *Feu* of the *Encyclopédie*; on another scientific front, Buffon regarded phlogiston as a hypothesis unsupported by any empirical evidence. Nevertheless, skeptical judgments such as these

---

[165]Stahl, *Traité du Soufre* (Paris, 1766), 57. English translation by John Maxson Stillman, *The Story of Alchemy and Early Chemistry* (1924), reprinted (New York, 1960), 426.

[166]In the work by Tugnoli-Pattaro (1983), 112, eight different names and definitions of phlogiston used by Stahl are listed.

were still quite isolated and did not have a great impact on professional chemists. In this connection it should be remembered that all the chemical articles of Diderot's *Encyclopédie* reflected Stahl's hegemony. Gabriel-François Venel, the author of most of them, helped to create a wide interest in Stahl's theory in France.[167] With his own development of some of Stahl's principles, Venel distinguished the method of physics from that of chemistry, arguing the superiority of the latter. Venel remarked that physicists claimed to reduce the study of matter to mechanical laws and to the mere investigation of the mass of bodies. Chemists, by contrast, went further and by investigating the combination of mixed bodies and including all their secondary qualities tried to discover the inner constitution of matter.[168] The superiority of chemistry, argued Venel, lay in its closeness to the real nature of matter. Given the definition of chemistry supplied by Venel, it is obvious that the reform of its nomenclature could not be a priority. Venel's criticism of mathematics, which he regarded as an artificial medium between man and nature, implied a general indifference, if not an actual hostility, to any systematic attempt to reform the technical language of chemistry. As Venel remarked:

> Chemists are reproached, very unjustly, for taking pleasure in their obscurity; for this charge to be reasonable, it would be necessary to show them clear and certain principles; for otherwise they cannot be blamed for preferring obscurity to error.[169]

This attitude found even more radical expression in Rouelle's courses in chemistry.[170] These enjoyed extraordinary success and between 1742 and 1768 were attended by most of the learned community of Paris. Rouelle became *demonstrateur* of chemistry at the Jardin du Roi in 1742 and *maitre apothecaire* in 1750,[171] and his scientific training in-

[167]In Chapter One, I have already outlined the general contents of Venel's defintion of chemistry.

[168]Venel, article "*Miscibilité ou solubilité,*" in *Encyclopédie* (1765), vol. 10, 524.

[169]"On reproche aussi très-injustement aux chimistes de se plaire dans leur obscurité; pour que cette imputation fût raisonnable, il faudroit qu'on leur montraît des principes évidents et certains; car enfin ils ne seront pas blâmables tant qu'ils préféront l'obscurité à l'erreur," article "*Chymie,*" *ibid.*, 271–272.

[170]On Rouelle, see Rappaport (1960); *idem*, "Rouelle and Stahl and Phlogistic Revolution in France," in *Chymia*, 7 (1961), 73–102; Claude Secretain, "Un aspect de la chimie prélavoisienne—Le cours de G.-F. Rouelle," in *Mémoires de la Société Vaudoise des sciences naturelles*, 7, (1943), 219–444.

[171]In a hand-written register kept at the Bibliothèque Interuniversitaire de Pharmacie of Paris (sign. ms. 22), 191–192, the immatriculation of Rouelle as *Maître Apothecaire* is reported on May 15, 1750.

volved the pharmaceutical preparation of remedies. Rouelle's courses were never published, but many manuscript copies have survived. The impossibility of obtaining Rouelle's chemical system in printed form has left much room for speculation concerning his influence, which has been sometimes exaggerated, sometimes understated. The fact that Lavoisier, Macquer, Bayen, Turgot, Diderot, Rousseau, and d'Holbach attended Rouelle's courses has been taken as evidence of the scientific importance of the lectures. Undoubtedly Rouelle was a gifted orator, who helped to spread a wider interest in chemical matters; however, with the exception of his classification of salts, the content of his courses was not particularly original. The remainder consisted of a paraphrase of Stahl's works with the addition of some of the chemical investigations carried out by Homberg and Hellot. A short survey of the *Cours* will reveal the traditional nature of his approach.

Rouelle defined chemistry, according to one version dated 1759, as "an art which teaches us to separate by the means of instruments, several bodies; to combine them to the end to be reacquainted with their properties and render them useful to several arts."[172]

In 1767 Rouelle replaced this definition with a more sophisticated one:

> Chemistry is a physical art which by means of certain operations and instruments teaches us to separate from bodies many substances entering into their composition and to recompose them with each other [. . .] to produce the original bodies or form new ones.[173]

The course continued with a short description of chemical principles, based on the notion of mixt bodies and chemical combination outlined by Stahl. This was followed by a survey of the chemical elements fire, air, water, and earth. The main part of the course was devoted to the chemical analysis of the vegetable kingdom, undoubtedly the field that Rouelle knew best. In these lectures Rouelle provided detailed descriptions of the distillations and extractions of plants. However, these operations were already a consolidated body of knowledge at the beginning of the eighteenth century, when Bourdelin, Boulduc, Homberg, and Geoffroy carried out a systematic chemical analysis of the

---

[172]From the *Introduction to chymistry by Mr. Rouelle*, fol.1 (ms. kept at BIP., sign. ms. 17). The manuscript is an English version of the first lessons of Rouelle's course of chemistry.

[173]From the *Cours de chymie recueilli des leçons de M. Rouelle Apothicaire, Demonstrateur en chymie au Jardin du Roy, de l'Academie Royale des sciences de Paris, par A.L. de Jussieu, 1767*, fol.1 (manuscript kept at the BMHN., sign. ms. 1202). This version of Rouelle's lectures by the famous botanist A.L. de Jussieu is one of the last compiled.

vegetable kingdom.[174] Furthermore, this chemical analysis of the vegetable kingdom was complicated both by the inaccuracy of the instruments and by the complexity of organic chemistry. Retrospectively, Lavoisier expressed reservations concerning Rouelle's emphasis on vegetable chemistry:

> One began with the vegetable kingdom, that is, with what is the most difficult and complicated, with the part of the science which was least advanced and which is still the least advanced today.[175]

Lavoisier also spoke of the lack of clarity in Rouelle's presentation of his ideas.[176]

The importance of Rouelle in the history of eighteenth-century chemistry lies rather in his published memoirs on the classification of salts, read at the *Académie des Sciences* in 1744 and 1755, than in his courses.[177]

In these two memoirs, Rouelle proposed a new method of classification, introducing the term *base* to denote a generic alkali. The reaction acid + *base* resulted in a salt that could be neutral, acid, or basic according to the proportions of the reagents.

The main innovation in this classification was in the defining of a substance through a chemical reaction. As the salt was a substance derived from known reagents, its name had to be deduced from the reagents themselves and not from its therapeutic or other external properties. Unfortunately, most constituents of salts were still unknown, and Rouelle used the nomenclature of traditional salts without any significant changes.[178] But this method of binomial classification of salts did eventually influence the important investigations carried out by Macquer and Bergman in the late 1760s.

As I have tried to show, Stahlian chemistry was not concerned with the theme of chemical nomenclature, which lay outside its conceptual framework. The notions of *mixt* and *phlogiston* became the

---

[174]On these sources, see Frederic Lawrence Holmes, *Eighteenth-Century Chemistry as an Investigative Enterprise* (Berkeley, 1989), 61–83.

[175]A draft of Lavoisier's manuscript, dated 1792, was recently published by Bernadette Bensaude-Vincent, in *BJHS*, 23 (1990), 456–460.

[176]"Le célèbre professeur réunissait à beaucoup de méthode dans la manière de présenter ses idées beaucoup d'obscurité dans la manière de les énoncer," *ibid.*, 457.

[177]Rouelle, "Mémoire sur les sels neutres, dans lequel on propose une division méthodique," in *MARS* (1744), 353–365; *idem*, "Mémoire sur les sels neutres," in *MARS* (1754), 572–588.

[178]See Crosland (1978), 124.

deus ex machina of all chemical combination, and their enormous influence was due to the qualitative structure of early eighteenth-century chemistry. Another reason for the popularity of the phlogiston theory was that it explained calcination and combustion, giving chemistry a theoretical autonomy.

Only by considering these elements in their full historical context is it possible to understand the conceptual break with the chemical scenario that was implied by Lavoisier's attempt to overthrow phlogiston. From this perspective it is quite instructive to realize that Stahl's chemical works were considered by Lavoisier the most important achievement of modern chemistry and that Lavoisier's earliest interest in chemistry was stimulated by reading and commenting on a manuscript by Stahl on sulphur, which he purchased from Hellot's library in April 1766.[179] What Lavoisier found in the works of Stahl was a coherent *chemical theory* supported by *empirical evidence*, two elements that had otherwise been lacking in earlier attempts to give a foundation to the *science* of chemistry.

An attentive philosopher, such as Immanuel Kant, was not far from the truth in considering Stahl's phlogiston as revolutionary in its impact on chemistry as Galileo's discoveries had been on physics.[180]

---

[179]In 1766 Lavoisier bought a part of Hellot's library. Among the items there was a Latin manuscript by Stahl on the nature of sulphur dated 1718. The manuscript, entitled *Cogitationes et utiles reflexiones super contentione de sic dicto Sulphure et quidem tam de Sulphure communi combustibili aut volatili quam de incombustibili aut fixo, Hallae 1718*, is a volume in 4° of 316 leaves (613 pp.) with 37 marginal notes in Lavoisier's hand. Since Lavoisier did not seem to be very confident concerning several topics in the manuscript, these notes were presumably written in 1766. Interestingly Lavoisier focused his attention on the part of the manuscript concerning calcination and combustion! The manuscript is now kept at the BI (sign. ms. 1530). A critical edition of the manuscript is in preparation.

On Hellot's library, see Doru Todéricu, "La Bibliothèque d'un savant chimiste et technologue Parisien du XVIIIe siècle. Livres et manuscrits de Jean Hellot (1685–1766)," *Physis*, 18 (1976), 198–216.

[180]"When Galileo caused balls, the weights of which he had himself previously determined, to roll down an inclined plane; when Torricelli made the air carry a weight which he had calculated beforehand to be equal to that of a definite volume of water; or in more recent times, when Stahl changed metals into oxides, and oxides back into metals, by withdrawing something and then restoring it, a light broke upon all students of nature. They learned that reason has insight only into that which it produces after a plan of its own, and that it must not allow itself to be kept, as it were, in nature's leading-strings, but must itself show the way with principles of judgment based upon fixed laws, constraining nature to give answer to questions of reason's own determining," Immanuel Kant, *Kritik der reinen Vernuft* (1787), translated into English by Norman Kemp Smith under the title, *Critique of Pure Reason* (1929), (London, 1973), 20.

New Things, New Names

> If the seventeenth-century chemist with a knowledge of less than a dozen neutral salts was able to express himself without using a systematic nomenclature, this was clearly not possible when, about 1770–80, the number of possible neutral salts had increased to several hundred.[181]

A similar statement could in fact be made about the change that had taken place by the 1750s, when a wide-ranging debate on the classification of salts raised, for the first time in the history of chemistry, an explicit call for a general reform of chemical nomenclature.

The works of Macquer represented the first sign of an emerging awareness of this need. At the very beginning of his scientific career, Macquer had realized that the old definitions of chemistry were utterly inadequate to express its present experimental status and to promote its improvement.[182] We have already seen in Chapter One that Macquer made an effort to draw a definitive distinction, both historical and epistemological, between chemistry and alchemy.[183] Even in the preface to his first chemical work, *Élémens de chymie théorique*,[184] published in 1749, Macquer firmly criticized alchemy, claiming that because of it:

> . . . chymistry became an occult and mysterious Science; its expressions were all tropes and figures, its phrases metaphorical, and its axioms so many enigmas: in short, an obscure unintelligible jargon is the justest character of the Alchymistic Language. Thus, by endeavouring to conceal their secrets, those gentlemen rendered their Art useless to mankind, and brought it into deserved contempt.[185]

By 1766, Macquer's position on this controversial topic was even

---

[181]Crosland (1978), 130.
[182]For a complete and detailed biography of Macquer, see the PhD thesis of William C. Ahlers, *Un chimiste du XVIIIe siècle: Pierre Joseph Macquer (1718–1784). Aspects de sa vie et de son oeuvre*, Thèse présentée en vue de l'obtention du Doctorat de Troisième Cycle, École Pratique des Haute Études, VIe Section (Paris, 1969). On Macquer's role in the reform of chemical nomenclature, see W.A. Smeaton, "The Contributions of P.-J. Macquer, T.O. Bergman and L.B. Guyton De Morveau to the Reform of Chemical Nomenclature," in *Annals of Science*, 10 (1954), 87–103, and Crosland (1978), 136–138.
[183]See pp. 19–20.
[184]Macquer, *Élémens de chymie théorique* (1749), 2nd ed. (Paris, 1753).
[185]*Ibid.*, xi. I have taken the English translation of this passage from Macquer, *Elements of the Theory and Practice of Chymistry* (London, 1758), vol. 1, viii–ix.

more radical, and he remarked that "the alchemical mania was a leprosy that disfigured chemistry and obstructed its progress."[186] The view held during the seventeenth and the first half of the eighteenth century of an evolutionary process by which alchemy slowly led to chemistry changed radically with Macquer, who was certainly among the first to argue that the methods of alchemy and chemistry were totally different.[187] With this in mind, he claimed that, unlike alchemy, chemistry began when "the Genius of truth Philosophy (*physique*) prevailed in Chymistry as well as in the other sciences."[188] This argument, which represented a total rejection of Venel's definition of chemistry, was pursued by Macquer to its logical conclusion:

> Chymistry may now, in some degree, be compared to Geometry: each of these Sciences takes in a most ample field of enquiry, which every day enlarges very considerably; from each are derived several Arts, not only useful but even necessary to Society; each hath its Axioms and its undeniable principles, either demonstrated from internal evidence, or founded on constant experience; so that the one, as well as the other, may be reduced to certain fundamental truths, on which all the rest are built. *These fundamental truths connected together, and laid down with order and precision, form what we call the Elements of a science.*[189] [my italics]

This passage reveals a completely new approach to chemistry, and the analogy between the elements of chemistry and Euclid's elements of geometry was audacious and innovative.[190] To Macquer, it was of paramount importance to liberate chemistry from any residue of alchemical thought and to introduce at the same time a quantitative approach to the investigation of the composition of matter. The emphasis put on both linguistic and experimental precision in science

---

[186]Macquer (1766), vol. 1, xxxii.

[187]We saw earlier in this chapter that from Agricola onwards the criticism of Alchemy's use of metaphorical language became a common feature of metallurgical texts.

[188]Macquer (1753), xi. English translation (1758), ix.

[189]Macquer (1753), xiii. English translation (1758), ix–x.

[190]See the previous subchapter and note the similarity of Macquer's position to that of Spinoza. It is interesting to observe that what Boyle in the seventeenth century considered impossible to demonstrate experimentally became an innovative research program for eighteenth-century chemical investigations with Macquer. Undoubtedly the positions of Spinoza and Macquer on this matter were inspired by different aims, but it is extremely significant that they were both convinced that without the introduction of the Cartesian *esprit de géometrie* into chemistry, this science would have never achieved the dignity of other natural sciences.

was a significant step towards a definition of chemistry based on ponderal and quantitative argumentation.

Another differentiating feature of Macquer's approach was in the disposition of the contents of his book. Reversing the traditional hierarchy of Rouelle's courses, Macquer began with the study of the mineral kingdom. This reorganization had both a pedagogic and a scientific purpose. Macquer declared that to understand his work the reader did not need any previous chemical knowledge because he began from the most simple and certain truths and ended with the most complex.[191] After a chapter on chemical principles, the *Élémens* began with the analysis of salts, acids, and metals, these being the most thoroughly known substances in chemistry.[192] Incidentally, it should be noted that the importance attached by Macquer to the study of salts eventually led to a rapid development of this research program.

More generally, the *Élémens* was divided into mineral chemistry, vegetable chemistry and fermentation, chemical analysis, chemical affinities, and a brief survey of the most common furnaces and vessels used in the laboratory.[193]

Unlike most of his predecessors, Macquer did not give priority to the systematic description of chemical operations:

> In this Part I say nothing of many manual operations, or the several ways of performing Chymical processes; reserving these particulars for my Treatise of Practical Chymistry, to which this must be considered as an Introduction.[194]

The distinction between theoretical and practical chemistry made here was an important one. Before Macquer, the technological, experimental, and theoretical strands of works on chemistry were often interwoven almost at random. Macquer correctly appreciated the different levels of chemical research (theory—experiment—technology)

---

[191]"The general Plan on which I proceed is to suppose my Reader an absolute Novice in Chymistry; to lead him from the most simple truths, and such as imply the lowest degree of knowledge, to such as are more complex, and require a greater acquaintance with Nature," Macquer (1758), xi.

[192]Macquer (1753), 19ff. A detailed analysis of the *Élémens* can be found in Ahlers (1969), 80–85.

[193]*Ibid.*, xxi–xxii.

[194]*Ibid.*, xxii. In 1751 Macquer published the *Élémens de chymie Pratique, contenant la description des Opération fondamentales de la chymie, avec des explications et des remarques sur chaque opération*, 2 vols. (Paris, 1751). In this work Macquer made a traditional survey of the eighteenth-century chemical laboratory.

and dealt with them separately. These distinctions actually implied different degrees of chemical knowledge and different approaches.

In 1766 Macquer published his *Dictionnaire de chymie*,[195] which soon became the main reference book of eighteenth-century European chemistry. The theoretical assumptions made in the *Élémens* found widespread practical application in the *Dictionnaire*. Macquer underlined the need to reform the chemical nomenclature of salts and made some significant modifications to the traditional one. For instance, he gave a common identity to metallic salts derived from the same acid, by giving them the same generic name accompanied by a specific one derived from the metal that combined with the acid, explaining:

> It would be proper to give the same denomination *vitriol* to all vitriolic salts with a metallic base, and to name, for example, *Vitriol d'or*, the vitriolic salt composed of vitriolic acid and gold; *vitriolic d'argent, ou de lune*, the salt resulting from the union of the same acid with silver, and so on. Perhaps it would even be correct to include under the general name *vitriol* all vitriolic salts whatsoever.[196]

However, as Smeaton has pointed out, "these admirable suggestions were not, unfortunately, applied by their author other than in his *Dictionnaire de chymie*"[197] and Macquer relied upon the traditional lexicon in his later works.

A complete reform of the nomenclature of salts was a goal that required not only new nomenclative rules but also the tremendous mental effort involved in overcoming the consolidated tradition of considering salts on the basis of their secondary qualities. Despite these objective difficulties and the limited impact it had on his own work, Macquer's attempt was so persuasive that it was immediately successful. Antoine Baumé adopted the reform,[198] and Torbern Bergman was inspired to go further.[199]

---

[195] *Dictionnaire de chymie, contenant la théorie et la pratique de cette science, son application à la physique, à l'histoire naturelle, à la médicine et à l'economie animale* . . ., 2 vols. (Paris, 1766).

[196] *Ibid.*, article "*Vitriols*," vol. 2, 673–674. I have used the English translation by Smeaton (1954), 88.

[197] Smeaton (1954), 89; see also Crosland (1978), 137–138.

[198] Antoine Baumé, *Chymie Expérimentale et Raisonné* (Paris, 1773), vol. 1, 287 and 315.

[199] On Bergman's chemical contributions, see Hugo Olsson, "An Introductory Biography," in *Torbern Bergman's Foreign Correspondence* (Stockholm, 1965), xi–xviii; and my "T.O. Bergman and the Definition of Chemistry," in *Lychnos* (1988). For Bergman's contribution to mineralogical nomenclature, see the earlier part of this chapter.

In 1767 Bergman commenced work on a systematic treatise on salts,[200] and a year later, he wrote to Macquer, sending him the results of his preliminary observations on their crystalline configuration.[201] Although Bergman never published the work he had in mind, the study of salts became the main object of his investigations.

In stressing the necessity of a new systematic definition of salts based on their configuration, Bergman developed Rouelle's and Macquer's ideas considerably.

We have a surviving documentation of this reform in a manuscript dating from the late 1760s. To begin with, Bergman writes:

> I think that according to their structure salts could be divided into:
>
> > Simple salts, i.e. those containing only one saline matter like acids and alkali; double salts, that are composed of two substances, both of saline origin; medium or terrestrial salts, in which one of the two substances is made of earth; medium or metallic salts, in which one of the substances is a metal; triple salts, that are composed of three substances.
> >
> > As far as denominations are concerned, I think salts need a reform. The old names are absurd and intolerable and the most reasonable ones are based upon properties that either are wrong, uncertain or common to several substances. The number of salts is growing steadily, that is why it is necessary to establish a few constant rules. Here is my project, which I will try to improve and which is quite similar to the method proposed in the *Dictionnaire de Chymie* [by Macquer].
> >
> > As far as simple salts are concerned, I have nothing to add. It is necessary to say *alkali de tartare* instead of *sel de tartre* and *acide de vitriol concentré* instead of *huile de vitriol*.
> >
> > As far as compound salts are concerned, they should be classified according to their principles, mentioning the base before anything else. Thus, instead of saying *tartre vitriolé*, it would be better to say *alkali fixe vegetable vitriolé* and instead of *sel admirable de Glauber*, a name that suggests some quackery, I say *alkali fixe vitriolé* etc.
> >
> > This system makes it possible to: 1) have a general uniform

[200]On January 20, 1769, Bergman announced to Macquer that he was preparing a "*histoire systematique des sels.*" This unpublished letter is kept in UUB, sign. Okat. Waller—Waller samling.

[201]"J'ai l'honneur de vous offrir quelques observations sur les sels, principalement sur la figure des cristaux, à la verité très defecteuses, mais il faut bien de loisir pour en avoir un suite complette," Unpublished letter by Bergman to Macquer, dated 6.10.1769, now kept in BN, sign. ms. fr. 12305.

system; 2) understand the composition from the name and 3) easily denominate new salts.[202]

Inspired by the recommendations made by Macquer on the classification of salts, Bergman began to reform the nomenclature and to eradicate the alchemical linguistic heritage. In this context Bergman appreciated the importance of giving substances a name derived from their internal composition.

Macquer was very impressed by the project of his younger colleague, and in 1770 he wrote to Bergman making the following comments on his reform:

> I have read with the greatest satisfaction your results and observations and experiments on neutral salts that you had the kindness to communicate to me. . . . I very much share your opinion on the nomenclature of neutral, or as you named them very well, double salts and I am delighted that a man of your merit has the courage to change all the bad names; if all the good chemists could agree like this, it would be possible to reform an infinite number of improper, obscure and false names that cannot but lead to error.[203]

The increasing importance of a reform of the nomenclature of salts made more acute the need for a new definition of chemistry as a whole and for the abandonment of the obsolete and unscientific earlier terminology.

Bergman published a first draft of his reform in 1775 when, commenting on a work by H.T. Scheffer, he added several synoptic tables on the classification of salts and a new system for naming them.[204] The work was written in Swedish, which accounts for its limited impact on the European chemical community, but it is interesting to observe that some of the linguistic proposals contained in it marked the first systematic change to many traditional names. Bergman divided simple salts according to the following scheme (See also Fig. 12):

---

[202]The manuscript in two versions is entitled *Observations esparses sur les sels* and is kept in UUB, sign. D 1459 g. There are two other manuscripts on the classification on salts that contain the lectures given by Bergman on neutral and double salts in late 1772 and in the spring of 1773. The titles of these ms. are *Prof. Bergmans föreläsningar om nautral salterna, Höst Termin, 1772,* and *Publique föreläsningar om medel salterna, Vår Termin 1773,* and they are both kept in KB.

[203]Letter by Macquer to Bergman dated 18.5.1770, from Bergman (1965), 238–239.

[204]Henrik Theophilus Scheffer, *Chemiske Föreläsningar, Rörande SALTER, JORDARTER, VATTEN* . . . . (Uppsala, 1775).

- Simple salts
  - acids
    - mineral
      - vitriolic
      - nitrous
      - arsenical
      - saltium
      - fluoric
    - vegetable
      - tartaric
      - *acetosell* (oxalic)
      - acetic
    - animal
      - uric
      - formic
    - common to all kingdoms
      - aerial
  - alkali
    - fixed
      - vegetable
      - mineral
    - volatile
  - neutral salts
    - mineral acid with
      - vegetable alkali
      - mineral alkali
      - volatile alkali
    - vegetable with all alkali
    - animal with all alkali

To classify double and triple salts, Bergman used a binomial nomenclature based on the Linnaean model and on the mineralogical classification.[205] The difficulty of using the Swedish language was overcome by using Latin names where Swedish was inadequate. Latin was able to deal more effectively with the problem of the final suffixes of names and, more generally, the classification of substances in a precise order. It was probably for this reason that Bergman decided, after his first attempt, to write his major works in Latin and to tackle the reform of chemical nomenclature within the framework of the Latin syntax.

In 1775 Bergman published his famous *Dissertatio de attractionibus electivis*, which was an attempt to give chemistry a physical precision of

[205] *Ibid.*, 5; see also 94 and 104–105 where Bergman presented his classification of middle salts.

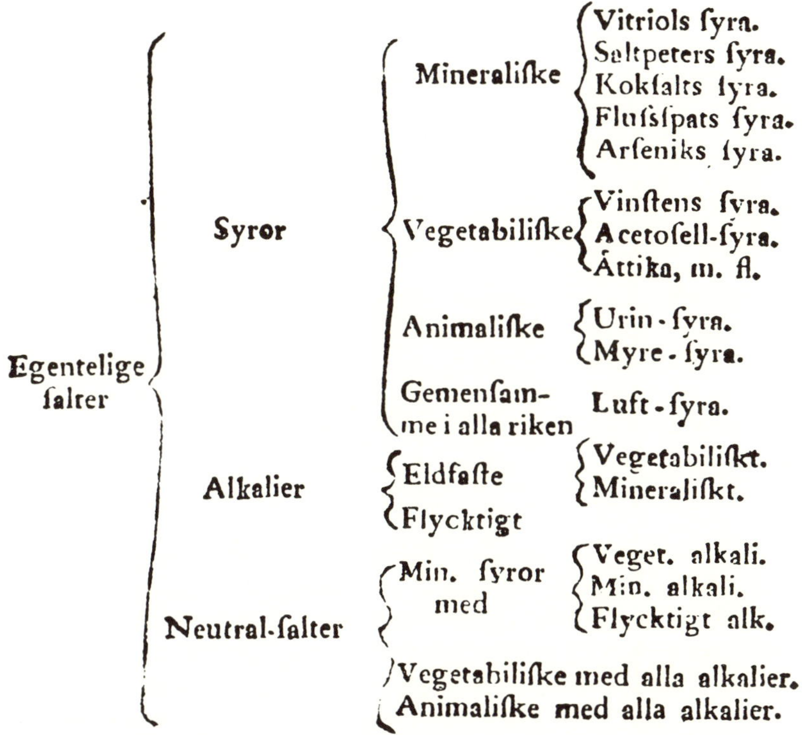

Figure 12 Bergman's classification of simple salts. From H.T. Scheffer, *Chemiske föreläsningar rörande salter, jordarter, vatten . . .* (Uppsala, 1775).

expression. Back in the early years of his scientific career, the Swedish chemist had written a physical dissertation on universal attraction that was based on the ideas of Newton.[206] The dissertation of 1775 was an analytical development of the *Table des affinités* presented by Geoffroy at the Parisian Academy of Sciences in 1718 and of the later works on this theme by Macquer, Gellert, and Baumé.[207] All of these works attempted to apply the Newtonian law of universal attraction to chemistry. How-

[206]Bergman, *Dissertatio physico-mathematica de attractione universali* (Uppsala, 1758).

[207]Bergman's *Dissertatio* was published in the *Nova Acta Societatis Regiae Scientiarum Upsaliensis,* vol. 2 (1775), 161–250; Geoffroy, "Table des différentes rapports observés entre différentes substances," in *MARS* [1718] (1719), 202–212; Macquer, *Élémens de chymie théorique* (1749); Gellert, *Afangsgründe zur Metallurgischen Chymie* (Leipzig, 1755); Baumé, *Manuel de chymie* (Paris, 1763).

ever, the difficulty of determining the intensity of the adhesive forces of microscopic bodies and particles brought the project to a standstill. Establishing the strength of such forces numerically was in fact inherently impossible. Aware of this problem, Bergman declared:

> In this dissertation, I shall endeavour to determine the order of attractions according to their respective force; but a more accurate measure of each, which might be expressed in numbers, and which would throw great light on the whole of this doctrine, is yet a desideratum.[208]

The attempts made by Guyton and Kirwan to determine quantitatively the force of cohesion of mercury with other metals confirmed only the enormous difficulty of applying the Newtonian method successfully to chemistry and the numerical discrepancies to which it gave rise.[209] On the other hand, Newtonian and quantitative patterns began to play an important role in emancipating chemistry from its qualitative structure. Although a role for mathematics in an eighteenth-century laboratory was hardly conceivable, as a rhetorical device it helped chemists to persuade the scientific community of the precision of their investigations and of the final liberation of chemistry from both alchemy and medicine. Here Newton's natural philosophy offered chemists a successful theoretical basis for a modern and scientific definition of chemistry. To understand the intellectual change in the perception of chemical phenomena and nomenclature, it is useful to analyze *Opuscula physica et chemica*, published by Bergman in 1779. The foreword (*De indagando vero*) to this work[210] was a philosophical introduction to the study of natural science. To his own question about the role of the scientist in relation to nature, Bergman replied that "the labour of the natural philosopher is chiefly employed in endeavouring to discover the means and the methods used by nature in her questions (that is, causes and their connections) and from thence to form what is called a theory."[211] It was necessary first to ascertain the accuracy of

---

[208]Bergman, *A Dissertation on Elective Attractions* (1775) (London, 1785), 4.

[209]On the role of mathematics in eighteenth-century chemistry, see Michelle Sadoun-Goupil, "Les tentatives de mathématisation de la chimie au XVIIIe siècle. Echecs et oppositions," in *Sciences et techniques en perspective*, Année 1981–1982, vol. 1, 1–20; Anders Lundgren, "The Changing Role of Numbers in 18th Century Chemistry," in Frängsmyr, Heilbron, and Rider (eds.), *The Quantifying Spirit in the Eighteenth Century* (Berkeley-Los Angeles, 1990), 245–266; Michelle Goupil, *De flou au clair? Histoire de l'affinité chimique* (Paris, 1991), 147–189.

[210]Bergman, "Of the investigation of the truth," in *idem*, *Physical and Chemical Essays*, trans. into English by Edmund Cullen, vol. 1 (London, 1784). I have here included with some small changes parts of my essay on Bergman (1988).

[211]Bergman (1784), xxi.

experience and of the chain of facts before eventually establishing the connections in some order.

Unlike Descartes, who started with the assumption of a world constructed *a priori* to explain the dynamics of empirical phenomena, Bergman claimed that it was better to follow the opposite method, i.e., that of Newton, for "he first solicitously collects the facts; these he examines with accuracy and compares with acuteness: hence he deduces the laws of nature, and, from effects well established, he infers their causes."[212] By choosing this approach, however, Bergman surely did not want to support an extreme empiricism, such as Priestley's natural history of chemical phenomena, but rather to give rules for tracking down the principles of a science and explaining them.[213]

Thus, it was not by superficial empirical description, nor, worse, by false analogy and metaphysical extensions that it was possible to investigate the nature of chemical substances. It was by discovering the most hidden properties of bodies through analysis and by not being misled by immediate inferences. Behind the appeal for an accurate and methodical observation of phenomena lay the awareness that a *true* reconstruction of nature was possible only with the aid of an effective theory.

The scientific method to which Bergman referred was *analysis*:

> In the investigation of causes, we must begin by phenomena, sufficiently varied, and well observed, and proceed in order, from proximate causes to more remote.[214]

This passage reflected the tone and content of the famous *Query 31* of Newton's *Opticks*, where the analytical method was defined as follows:

> As in Mathematics so in Natural philosophy, the Investigation of difficult Things by the Method of Analysis ought ever to precede the Method of Composition.
>
> The Analysis consists of making Experiments and Observations, and in drawing general Conclusions from them by Induction [. . .] By this way of Analysis we may proceed from Compounds to Ingredients and from Motions to the Forces producing them: and in general, from Effects to their Causes, and from particular Causes to more general ones [. . .].[215]

---

[212]*Ibid.*, xxi–xxii.

[213]*Ibid.*, xxv.

[214]*Ibid.*, xxxii.

[215]Newton, *Opticks, or a Treatise of the Reflections, Refractions, Inflections and Colours of Light* (1717), reprinted (New York, 1952), 404.

As applied by Bergman, Newton's analytical method offered a general means of approaching the science of chemistry. If it was applied more literally, analysis became a laboratory tool with which all substances were broken down into their primitive constituents and classified according to their chemical nature. The two meanings of the term *analysis* became the point of departure of Bergman's chemical investigations.

The feasibility of a methodological renewal of chemistry was directly connected in Bergman's philosophy with the urgent need to reform chemical nomenclature:

> Finally, I am giving denominations to things, as agreeable to truth as possible. I am not ignorant that words, like money, possess an ideal virtue, and that great danger of confusion may be apprehended from a change of names; in the mean time it cannot be denied that chemistry, like other sciences, was formerly filled with improper names. In different branches of knowledge we see those matters long since reformed; why then should chemistry, which examines the real nature of things, still adopt vague names, which suggest false ideas, and favour strongly ignorance and imposition.[216]

If the analytical method led chemical investigation to a closer observation of phenomena, chemical nomenclature should accordingly be based upon the *real nature of things* and their internal composition. In reality, the achieving of this aim was complicated by the fact that chemists were able to detect the constituents of only a few chemical compounds, while the chemical nature of most acids and organic substances was known only by their effects.

To make his position clearer, Bergman published in 1779 an elementary textbook of chemistry for beginners in this science.[217] This work contained a series of interesting reflections on chemistry and its role in relation to other sciences.

Bergman distinguished among three types of natural science:

1. Natural history, which studies the exterior features of bodies, that is, those parts that are already primarily known to the scientist, but need to be adequately classified according to order, genus, species, etc. Such sciences included botany, zoology, and, to a certain ex-

---

[216]Bergman (1784), xxxvii.

[217]Bergman, *Anledning til föreläsningar öfver chemiens beskaffenhet och nytta, samt naturliga kroppars almännaste skiljaktigheter* (Uppsala-Åbo, 1779), translated into English under the title *An Essay on the Usefulness of Chemistry, and its Applications to the Various Occasions of Life* (London, 1783).

tent, mineralogy, where the method was that of cataloguing facts. Accordingly, the naturalist "observes, by the help of characteristics previously ascertained, what genus, species and variety they [objects] belong to."[218]

2. Natural philosophy or physics, which, unlike natural history, investigates the essential and substantial properties of the bodies, and neglects their secondary qualities in order to concentrate exclusively on mechanical properties.
3. Finally, chemistry, "which, not content with investigating the properties of bodies, traces out the causes of these properties, by examining their ingredients and composition [. . .]. Chemistry is therefore that science which examines the constituent parts of bodies with reference to their nature, proportions and method of combination."[219]

This definition implied a clear awareness of the limits and objectives of chemistry. If Macquer appealed quite simply for an application of physics to chemistry, without specifying what was involved in the project, Bergman went further and tackled the identity of chemistry by distinguishing it from physics. This distinction, however, was made from a position very different from the attempts to distinguish the two sciences made by Venel and Rouelle.

Bergman claimed that the elements in physics were mechanical particles that could only be distinguished on the basis of their mass, whereas in chemistry the elements were principles of different natures with specific and varied characteristics.[220] In the first case, only mechanical qualities were taken into account, whereas in chemical investigation, all the other qualities, such as flavor, color, and form, played an integral part. On the other hand, Bergman distanced himself from those, like Venel and Rouelle, who attempted to transform chemistry into a purely qualitative and descriptive science. With this twofold aim in mind, Bergman defined chemistry as follows:

> Chemistry, like mathematics, may also be distinguished into vulgar or elementary chemistry [*chemia vulgaris*], which concerns itself only with the grosser and more palpable elements of bodies; and transcendental chemistry [*chemia sublimior*] which, not neglecting the grosser and more palpable elements, finds means, by particular contrivances of its own, to collect and examine those elements of fine texture, which

218 Bergman (1783), 4.
219 *Ibid.*
220 *Ibid.*, 5–6.

> otherwise, in consequence of the degree of subtlety which renders them imperceptible to our senses, would fly off and escape our notice.[221]

In this definition the shifting meaning of *chemistry* was a compromise between qualitative and physical chemistry. Bergman realized that an application of mathematics to chemistry was impossible in practice but that, at the same time, a quantitative approach was now needed. Significantly the *chemia sublimior* did not neglect the "grosser and palpable elements," but rather included them in a more general and abstract theory of matter. Another function of the *chemia sublimior* was to introduce quantitative investigation of the proportions of the constituents of a mixt into the chemical laboratory. With this, the role of the mixt in chemistry lost its importance, and its place was taken by the investigative concept of *equiponderance*. This essential concept did not arise from the mere introduction of the balance in the laboratory, where it was already a common instrument during the sixteenth century, but from the elaboration of a theory of chemical reactions and combinations receptive to it. "In fact, a man who considers nature with even so little attention, cannot surely help being aware, that everything that we see happens in conformity to certain laws; and that the smallest circumstances have their causes as well as the greatest. In the system of nature every phenomenon is connected with every other."[222] This deep awareness of the unity of chemical endeavor and its analogies with Newtonian philosophy was applied by Bergman in his laboratory investigations. The arrival of theory in the laboratory was the truly innovative element that contributed to the creation of a new definition of chemistry.

Chemistry was not becoming an autonomous science because of the usefulness of its applications (it was still far from being an industrial reality), nor because of the increasing number of chairs in the European universities,[223] but mainly because of its theoretical emancipation from the experimental tradition that confined it to pharmaceutical and medical laboratories.

While working towards a new definition of chemistry, Bergman paid constant attention to the reform of the old nomenclature that gave chemistry a most antiquated image. It was now clear that the new quantitative ambitions of chemistry could not be expressed in the language of alchemy. Bergman developed his ideas on the nomenclature

---

[221] *Ibid.*, 16–17.

[222] *Ibid.*, 18–19.

[223] The growth in importance of chemistry in the universities occurred between 1770 and 1780 when this renewed definition of chemistry was already established.

of salts and acids as early as 1779,[224] but it was only in 1784 that he published a long and systematic memoir on the nomenclature of chemical and mineralogical substances.[225]

The aim of the memoir was to propose a systematic reform of the whole nomenclature of inorganic chemistry. Acids, alkalis, earths, and their combinations were classified and given innovative names. In his long essay Bergman named the acids as follows:

Vitriolicum	*Tartareum*
Nitrosum	*Benzoinum*
Muriaticum	*Citrinum*
Regalinum	*Succineum*
Fluoratum	*Galacticum*
Arsenical	*Formical*
Boracinum	*Sebaceum*
Saccharinum	*Phosphoreum*
Oxalinum	*Aëreum*[226]

The alkalis consisted of three species, namely, the *Potassinum*, the *Natrum* (soda), and the *Ammoniacum*.[227] The name of a double salt was derived from the genus of the acid contained in it, plus an adjectival form derived from the alkali or earth or metal, such as:

Vitriolicum potassinatum	instead of *tartari vitriolati*
Vitriolicum calcareatum	instead of *gipsi*
Muriaticum barytatum	instead of *baritae muriati*, etc.

The double salts formed by the combination of an alkali, an earth or a metal were considered by Bergman to belong to the genus *alkali* and they were given the following names:

Potassinum argillatum
Potassinum silicatum
Potassinum argentatum
Ammoniacum argentatum
Ammoniacum cuprum

---

[224]Bergman (1783), 81–83.
[225]Bergman, *Meditationes* . . . (1784).
[226]*Ibid.*, 123. On Bergman's reform, see Smeaton (1954), 99–110, and Crosland (1978), 148–152.
[227]Bergman (1784).

These names were formed with the name of the alkali followed by the adjectival form of the earth or of the metal.[228] We saw earlier in this chapter the indebtedness of the classification used by Bergman to Linnaeus' binomial nomenclature.

The reform of 1784 did not have the impact that Bergman hoped for. The fact that the memoir was published in 1784 has led some historians to conclude that the Swedish chemist was influenced by the famous essay on nomenclature published by Guyton two years earlier.[229] Smeaton, for instance, remarked that "Bergman had been influenced by Guyton's memoir of 1782"[230] on the strength of the chronological sequence of the two papers. However, new historical evidence now makes it possible to reverse this statement and show that Guyton was directly inspired by Bergman. In a letter from Guyton to Kirwan, dated January 6, 1785, the French chemist quoted a passage from Scheffer's work on salts, written in 1775.[231] As we previously saw, annotations added by Bergman to this edition outlined the new nomenclature of salts in its almost definitive form. It is therefore reasonable to believe that Guyton received the book from Bergman soon after its publication and that he then became interested in a general reform of chemical nomenclature. Moreover, Bergman had been studying this problem since 1769, whereas we have no evidence of Guyton's interest in nomenclature before 1771.

Notwithstanding this question of priority, it remains a fact that Bergman's nomenclature was neglected by the European chemical community, with the sole exception of Black's lectures in chemistry.[232]

For this general indifference, there are at least four reasons. The first is that Bergman devoted his *Meditationes* mainly to the study of minerals (*Fossilium*), and most of his memoir alternated between min-

---

[228]*Ibid.*, 123–124.

[229]Guyton de Morveau, "Mémoire sur les dénominations chymiques, la necessité d'en perfectionner le système et les règles pour y parvenir," in *Observations sur la physique*, 19 (1782), 370–382.

[230]Smeaton (1954), 100.

[231]I would like to thank Madame Michelle Goupil who kindly gave me a copy of this letter. From the correspondence between Guyton and Kirwan, it clearly emerges that Guyton either could understand Swedish or was helped by his future wife Claudine Picardet, who translated several memoirs by Scheele and Bergman from Swedish into French.

[232]Joseph Black, *Lectures on the Elements of Chemistry* (Edinburgh, 1803), vol. 1, 487–488. Black declared that during his chemical lectures he had always used Bergman's nomenclature, but since he never published them, their influence was restricted to the audience of Edinburgh University.

eralogical and chemical matters without any clear boundaries, as has already been mentioned, whereas the French chemists were very strict in distinguishing the two sciences and their objects. Second, Bergman manifested only incidentally an interest in a systematic reform of the nomenclature of pneumatic chemistry, which after 1770 became the main topic of chemical debate. Third, his memoir was written in Latin, a language that was no longer the medium of learned scholars.

And last but not least, Bergman's conception of the role played by language in science was too conventional to invest it with a wider theoretical function than the one of expressing the contents of empirical observations as clearly as possible. Unlike Linnaeus, Bergman did not conceive of a *real* correspondence between names and things. He said simply that names should not be absurd and should possibly express something of the nature of the substances to which they referred. Bergman's Newtonian position could not allow him to accept any metaphysical or philosophical method unless it was based on the evidence of experience.

The Nomenclature Established

Louis-Bernard Guyton de Morveau[233] was a "pupil" of Macquer[234] and a friend and correspondent of Bergman. These biographical circumstances are important when considering the relationship between Guyton's chemical works and those of his predecessors. Guyton shared in fact with Macquer and Bergman a general conception of chemistry as a branch of physics, and it is no coincidence that he agreed with them on the necessity of a general reform of chemical nomenclature.

---

[233]On Guyton de Morveau's life and works, the best biography is still the study by Georges Bouchard, *Guyton Morveau, chimiste et conventionnel* (Paris, 1938). The best study of Guyton's contribution to chemistry is the essay by W.A. Smeaton, "L.B. Guyton de Morveau (1737–1816), in *Ambix*, VI (1957), 18–34.

[234]In a manuscript written in 1784 and quoted by Ahlers (1969), 89–90, Guyton remarked that he learned his first notions of chemistry by reading Macquer's chemical works, rather than by following the courses given by Rouelle. The text of the manuscript is worth quoting:

> "Ceux qui avoient suivi les cours de M.M. Rouelle m'ont souvent répété qu'ils étoient les seul instituteurs de la bonne chymie; c'étoient des disciples, qui remplis de la parole de leurs maîtres, en parloient avec l'enthousiasme de la reconnoissance; mais J'étoies le disciple des livres de M. Macquer [. . .]. C'est lui qui a fait aimer cette science, qui en a répandu le goût, qui *l'a dégagé du langage obscur des adeptes*; qui l'a veuyée du mépris qu'ils avoient attiré sur elle . . ." (my italics).

Guyton, however, tried harder than Macquer and Bergman to apply quantitative patterns to chemical investigation. In his first chemical work, which was published in 1762, Guyton suggested that the determination of the quantity of phlogiston contained in each substance could lead to a *mechanical* explanation of chemical affinities.[235] Even if the elusive nature of phlogiston made the failure of this program inevitable, the wish to quantify the proportions of chemical substances led Guyton to tackle the vexing problem of the gain in weight of calcined metals from a completely different angle. Suddenly a common operation, such as calcination, acquired an entirely new perspective. It was in fact the emphasis of the notion of *weight* that made chemists aware of the active and pervasive role played by air.

While trying to quantify the proportion of phlogiston in different substances, Guyton observed that calcined metals gained weight instead of losing phlogiston as Stahl believed. In Guyton's conception of chemistry, a discrepancy between a measurable variable, such as weight, and the theory which pretended to explain it was not acceptable. To this contradiction Guyton proposed an ingenious explanation: contrary to what Stahl thought, phlogiston was lighter than any other substance, and this quality explained why the calcined metal was heavier than the original metal. However, the more Guyton tried to demonstrate the presence of phlogiston *physically*, the more "volatile" the inflammable principle became. But despite this evident contradiction, the phlogiston theory still offered such a natural explanation of calcination that only a greater mental effort and a greater number of experimental discrepancies could cast doubt upon its existence.

In 1777, Guyton, in collaboration with Maret and Durande, published in Dijon a collection of lectures on chemistry that he had begun delivering the previous year.[236] At the beginning of the first volume Guyton summarized the authors' intentions:

> Having defined this science, and having developed its definition in order to show the extent of its subject matter and purposes, we shall provide an idea of the language which is appropriate to it by the brief explanation of the most essential words, ranged in a methodical order. That is to say, 1) the terms which belong to the theory; 2) the generic qualities of bodies; 3) the names of the operations and results; 4) finally, those of the instruments. We shall show then that all

[235]Guyton, *Digressions Académiques ou Essais sur quelques sujets de physique, de chymie et d'histoire naturelle* (Paris, 1762), 6–7.

[236]Guyton, Maret, Durande, *Élémens de chymie, théorique et pratique . . .*, 3 vols. (Dijon, 1777–1778).

> chemical theory lies in these two words: *attraction* and *equiponderance*; that the practice lies in these two other words: *dissolution* and *crystallization*.[237]

If we compare this text with Rouelle's definition of chemistry two years before, we may perceive the radical shift away from the earlier interpretations of the phlogiston theory. The Newtonian notion of attraction, together with the physical concept of *equiponderance*, which reversed the qualitative structure of Stahl's chemistry, were now placed by Guyton at the center of the system.

An immediate consequence of the introduction of equiponderance in chemistry was the new role acquired by the balance, which now became an indispensable instrument in chemical analysis. The scientific meaning of the notion of attraction was not as clear as the concept of *equiponderance*, and the hope of quantifying the affinities between substances had still to compromise with the old qualitative concept of *sympathy*. Nevertheless, the explicit reference to Newton's definition of attraction freed the concept of affinity from its alchemical roots and promised more intimate relations between the method of physics and that of chemistry:

> This great law of physics, which makes the stars move, became in the laboratory of Uppsala [with Bergman] as well as in ours, the key to all operations [. . .].
>
> This conformity seems to finally to promise the reconciliation of opinions for a long time divided on such a capital issue, and it inspires in us the confidence to believe that the physicist and the chemist will stop either basing themselves exclusively on the rules of mathematics and calculation or attaching themselves to those of sympathies as sterile as they are imaginary.[238]

The conscious attempt to obtain a numerical determination of the intensity of the force of adhesion between mercury and other metals was far from successful.[239] In reality, mathematics was no more than a rhetorical device, and it was only with the publication of Berthollet's *Essai de statique chimique* in 1803 that a real attempt was made to apply numerical expressions to the study of chemical attractions.

The importance of Guyton's proposal lay mainly in his abandonment of the qualitative approach of Stahlian chemistry. The historical significance of this shift becomes evident when we consider it in con-

---

[237] *Ibid.*, vol. 1, 1–2.

[238] *Ibid.*, vol. 2, vi–vii.

[239] *Ibid.*, vol. 1, 63.

nection with the chemical nomenclature. In Guyton's view this problem was of the greatest urgency, since the growing number of new chemical objects required an immediate linguistic response:

> A part of the words which compose chemical language belong to the theory; these are the signs of ideas we have seized from the march of nature in the effects we observe. Another part indicate the qualities of bodies; many [words] are adopted to mean the operations and their products; finally, there are some proper and most familiar names for instruments.[240]

The care taken by Guyton to distinguish the various levels of chemical names, from the most *natural* and abstract theoretical names down to the more conventional and familiar names of technological instruments, reveals that scientific language was becoming an essential part of the reform of chemistry as a whole. The theoretical terms listed by Guyton were in fact entirely new additions to the chemical lexicon. Some of them, such as *pesanteur, absolue, specifique, gravitation, attraction, adhésion, équiponderance, équiponderable,* etc.,[241] would have been more at home in a physics treatise had not the conceptual horizon of chemistry changed so radically. It is significant that Guyton saw an intrinsic value only in these terms. Knowledge of the qualities of bodies was still too imperfect for a systematic and total reform.[242]

Despite these reservations, Guyton began in 1776 and 1777 to change some of the old terms in his contribution to the supplementary volumes of Diderot's *Encyclopédie.*[243] In this preliminary work, Guyton started to introduce ideas on nomenclature that eventually came to form general rules for the coining of new names.[244] In the article *Hépar,*[245] for example, the introduction of this neologism was explained as follows:

> This word is derived from Latin and means *liver.* This is how we have named the product of the combination of sulphur, antimony and arsenic with alkali. We even speak quite often of *liver of sulphur, liver of antimony;* but both of these designations are quite incorrect because

[240] *Ibid.*, 16–17.
[241] *Ibid.*, 19–20.
[242] *Ibid.*, 21–29.
[243] Diderot (ed.), *Supplément à l'Encyclopédie,* 4 vols. (Amsterdam, 1776–1777). According to Smeaton (1957), 25, Guyton wrote the following articles: *affinité, air, air fix, alkali phlogistiqué, calcination, causticité, causticum, combustion, crystallisation, crystallographie, dissolution, équiponderance, hépar, phlogistique.*
[244] Crosland (1978), 153–154.
[245] Diderot (ed.) (1777), vol. 3, 347.

they have been applied to these mixtures only because of a purely reddish colour.[246]

Guyton changed the word *foie* to *Hépar* because the latter had the advantage of not evoking any false impressions of the nature of the substance to which it referred and motivated his choice of a Latin and "insignificant" word by pointing out that the chemist:

> . . . should at least prefer a name which is further removed from common use, because it is much better if the technical terms of a science express nothing which is known and recall no idea than indicate false resemblances which lead beginners astray and always astonish the most educated people.[247]

These few ideas were developed and modified in the celebrated memoir of 1782, which was entirely devoted to chemical nomenclature. For the first time in the history of chemistry, a scientific publication was devoted to the task of solving this problem. Although largely indebted to Bergman's contributions, Guyton's remarks on this subject were presented in a different conceptual frame, which regarded the reform of scientific language as a reform of the science itself. In this connection it is important to remember Guyton's own opinion of the matter:

> The denominations of those things which form the object of a science or an art [. . .] compose what we call its own language. The state of perfection of the language indicates the state of perfection of science; its progress can be certain and rapid only in so far as its ideas are represented by precisely determined signs, correct in their meaning, simple in their expression, convenient in their use, easy to remember in which are retained as far as possible without error, the analogy which brings them together to the system by which they are defined and the etymology by which their meaning can be guessed.[248]

By directly identifying the state of a science with the state of its language, Guyton stressed the scientific implications of a reform of chemical nomenclature. Since the development of chemistry now depended on such a reform, it was no longer a mere question of overcoming the obscurity of alchemy or providing an unambiguous lexicon, but of creating a nomenclature that was completely new. This emphasis on the importance of scientific language was based on Condillac's *Logique*,

---

[246]*Ibid.*

[247]*Ibid.* Crosland's translation (1978), 154.

[248]Guyton (1782), 371.

a philosophical textbook published in 1780. Condillac stated that "reasoning is perfected only to the extent that languages are themselves perfected; and [that] the art of reasoning, reduced to its greatest simplicity, can only be a well made language."[249] This philosophical position led to a rigid linguistic realism that projected onto words the function of expressing objects in their true and unique nature. In this respect, Guyton's linguistic assumptions were different, since his belief in the importance of language did not yet imply a "realistic" correspondence between the names and the things. This cautious approach became clearer when Guyton tried to apply his linguistic philosophy to the concrete reform of chemical language. The old names, Guyton remarked, had been formed by relying upon: 1. the apparent physical consistency of substances (*oil of vitriol*); 2. false analogies (*butter of antimony*); 3. uncertain and variable colors (*liver of arsenic, saffron of Mars*); 4. exterior forms of crystals (*cubic niter*); 5. medical properties (*sedative salt*); and 6. methods of preparation (*precipitated and sublimated*).

> To this we can add terms such as *arcane* and *magistère* which were introduced by the alchemists, and the signs of the planets which they considered as synonymous with those of certain metals; we can also add the products which are designated only by the names of their inventors, such as *teinture de Stahl, de Ludovic, de Paracelse*, etc. Now let us ask whether it is possible to find a way in this chaos, and whether the understanding of such a nomenclature is not more difficult than the understanding of the science itself.[250]

Guyton was in no doubt about the answers to these questions.

There were also two new factors that contributed to the pressure for a reform of the old names. First, some chemical substances, namely, the salts, needed new names capable of indicating the different states of saturation or of combination. In one word, a *systematics* of names was now required. Second, the multiplicity of chemical objects needed a methodical distribution.[251]

The first step of Guyton's reform was to give some general rules for naming, such as:

1. "A phrase cannot be regarded as a name; chemical entities and their products must have their own names, which refer to them all

[249]Condillac, *La logique* (1780), translated into English by W.R. Albury (New York, 1980), 303. The influence of Condillac on the reform of chemical nomenclature undertaken by Guyton and Lavoisier is analyzed in detail in the next chapter.
[250]Guyton (1782), 372. Smeaton's translation (1954), 88.
[251]Guyton (1782), 371.

the time, without any necessity for recourse to circumlocutions."[252] To implement this principle, descriptive definitions, such as *liqueur alkaline saturée de la matière colorante du bleu de Prusse*, had to be replaced by the more sober name *alkali dephlogistiqué*.

2. Denominations had to conform to the nature of the thing they referred to as closely as possible. To this principle Guyton added three corollaries:
   2.1 A simple substance should preferably have a simple name.
   2.2 The denomination of a chemical compound was clear and exact only in so far as it recalled its constituent parts by names in conformity with their nature.
   2.3 The names of discoverers of substances should be excluded from the general nomenclature.[253]
3. When there was uncertainty concerning the character that ought in principle to determine the name, a name that expressed nothing was to be preferred to one that might express a false idea.
4. Designations that had their roots in the most generally known language were to be preferred, so that the word could easily be found from the sense and the sense from the word.
5. Names should be chosen with due regard to the genius of the language for which they were formed [. . .]. This rule was not the least important.[254]

By comparison with Bergman's nomenclature, the only innovations introduced by Guyton were principles three and five. Guyton's emphasis on the superiority of a vernacular nomenclature to Bergman's Latin one was particularly well received. Despite this successful intuition, Guyton's ideas were in certain respects less advanced than Bergman's. Bergman had established a Latin nomenclature intended to be universal in its etymological structure, whereas Guyton's reform did not imply any linguistic hierarchy and mixed Greek, Latin, Swedish, and French words indiscriminately as long as they conformed to his principles. As a result, his reform seemed to rely more on subjective choice than on objective and universal rules. Moreover, Guyton's table of chemical nomenclature presented a system of French names derived from Latin but lacking the advantages of classical syntax. Thus it turned out to be impossible in French to incorporate different suffixes

---

[252]*Ibid.*, 373. With few exceptions, I have used Crosland's English translation (1978), 157–159.
[253]Crosland (1978), 158.
[254]*Ibid.*, 158–159.

TABLEAU DE NOMENCLATURE CHYMIQUE,

Contenant les principales dénominations analogiques, & des exemples de formation des noms compoſés.

RÈGNES.	ACIDES.	*Les Sels formés de ces Acides prennent les noms génériques de*	BASES *ou ſubſtances qui s'uniſſent aux Acides.*	EXEMPLES *pour la claſſe des Vitriols.*	EXEMPLES *pris de diverſes claſſes.*
Des trois Règnes.	Méphitique *ou* Air fixe. . .	Méphites.	Phlogiſtique.	Soufre vitriolique *ou* ſoufre commun.	Soufre méphitique *ou* Plombagine.
	Vitriolique	Vitriols.	Alumine *ou* Terre de l'argille. .	Vitriol alumineux *ou* Alun.	Nitre alumineux.
	Nitreux	Nitres.	Calce *ou* Terre calcaire. . .	Vitriol calcaire *ou* Sélénite.	Muriate calcaire.
Minéral.	Muriatique *ou* du ſel marin.	Muriates.	Magnéſie.	Vitriol magnéſien *ou* Sel d'epſom.	Acète de magnéſie.
	Régalin.	Régaltes.	Barote *ou* Terre du Spath peſant.	Vitriol barotique *ou* Spath peſant.	Tartre barotique.
	Arſenical	Arſeniates.	Potaſſe *ou* Alkali fixe végétal .	Vitriol de potaſſe *ou* Tartre vitriolé.	Arſeniate de potaſſe.
	Boracin *ou* ſel ſédatif . .	Boraxs.	Soude *ou* Alkali fixe minéral .	Vitriol de Soude *ou* Sel de Glauber.	Borax de Soude *ou* Borax commun.
	Fluorique *ou* du ſpath fluor.	Fluors.	Ammoniac *ou* Alkali volatil .	Vitriol ammoniacal.	Fluor ammoniacal.
			Or	Vitriol d'or.	Régalte d'or.
	Acéteux *ou* Vinaigre . .	Acètes.	Argent.	Vitriol d'argent.	Oxalte d'argent.
	Tartareux *ou* du Tartre. .	Tartres.	Platine	Vitriol de platine.	Saccharte de platine.
	Oxalin *ou* de l'Oſeille . .	Oxaltes.	Mercure	Vitriol de mercure.	Citrate de mercure.
Végétal.	Saccharin *ou* du Sucre . .	Sacchartes.	Cuivre	Vitriol de cuivre *ou* Vitriol de Chypre.	Lignite de cuivre.
	Citronien *ou* du Citron . .	Citrates.	Plomb	Vitriol de plomb.	Phoſphate de plomb.
	Lignique *ou* du Bois. . .	Lignites.	Etain . . . . . . , . .	Vitriol d'étain.	Formiate d'étain.
			Fer	Vitriol de fer *ou* Couperoſe verte.	Sébate martial.
	Phoſphorique	Phoſphates.	Antimoine (*au lieu de* Régule d')	Vitriol antimonial.	Muriate antimonial *ou* Beurre d'antimoine.
Animal.	Formicin *ou* des Fourmis .	Formiates.	Biſmuth	Vitriol de biſmuth.	Galacte de biſmuth.
	Sébacé *ou* du Suif . . .	Sébates.	Zinc	Vitriol de zinc *ou* Couperoſe blanche.	Borax de zinc.
	Galactique *ou* du Lait . .	Galactes.	Arſenic	Vitriol d'arſenic.	Muriate d'arſenic.
			Cobalt	Vitriol de cobalt.	Saccharte de cobalt.
			Nickel	Vitriol de Nickel.	Formiate de Nickel.
			Manganèſe	Vitriol de manganèſe.	Oxalte de manganèſe.
			Eſprit de-vin	Ether vitriolique.	Ether lignique *ou* Ether de Goettling, &c. &c. &c.

N. B. Lorſque les acides particuliers déjà entrevus dans la molybdène, l'étain, &c. ſeront plus connus, on en formera les noms d'acide *molybdique* & *molybdes*, d'acide *ſtannique* & *ſtannes*, &c. Il en ſera de même des nouvelles baſes. Le nouveau demi-métal trouvé par M. Bergmann dans les fers caſſans, pourra être nommé *ſydérotète*, caché dans le fer.

Les ſoufres & les éthers deviennent eux-mêmes noms de genres, & ſe diſtinguent par l'épithète de l'acide.

Les noms de ces baſes, ou leurs adjectifs, ajoutés aux ſubſtantifs qui indiquent les genres des acides, forment les dénominations exactes, comme on le voit dans les exemples ſuivans.

Les dix-huit acides, les vingt-quatre baſes & les produits de leur union, forment ainſi quatre cents ſoixante-quatorze dénominations claires & méthodiques, indépendamment des *hépars* ou compoſés à trois parties, dont les noms viennent encore dans ce ſyſtême, comme *hépar de ſoude*, *hépar ammoniacal*, *pyrite d'argent*, &c. &c.

FIGURE 13 Guyton's table of chemical nomenclature. From L.B. Guyton de Morveau, "Mémoire sur les dénominations chymiques," in *Observations sur la physique*, 19 (1782).

capable of denoting different states of saturation of the same substance, whereas in Latin Bergman successfully used, for example, the suffix *-atum* to denote the acid part of salts.[255] These difficulties were only overcome a few years later in 1787 when a systematics of *chemical* suffixes was introduced into French.

Apart from these technical problems, Guyton's nomenclature was presented efficiently and systematically. In the table of chemical nomenclature (Fig. 13), five simple earths (*quartz, alumine, calce, magnésie, barote*), the three alkalis (*potasse, soude, ammoniac*), 18 acids, and 24 bases (earths, alkalis, metals, phlogiston, and alcohol) were listed.[256] This methodical table offered the reader a display of more than 500 chemical substances and compounds on a single page.

Thanks to this brilliant synthesis, Guyton's reform of the nomenclature was an immediate success. In a letter to Guyton, dated July 24, 1782, Macquer declared: "Your new nomenclature is excellent, and, for myself, I am ready to adopt it."[257]

In 1786 Guyton published the first chemical volume of the *Encyclopédie Méthodique*, on which he had been working since the late 1770s. In the *Advertissement* Guyton declared with great satisfaction that his nomenclature had already been adopted by Buffon, Bergman, Fontana, Macquer, Kirwan, Landriani, Fourcroy, Crell, and Leonhardi.[258] If, on the one hand, it is difficult to evaluate this widespread approval, on the other hand, it is a fact that Guyton's nomenclature had stimulated great interest in the issue of chemical language and that it was the first step towards the general reform undertaken by Lavoisier and his colleagues in 1787.

---

[255] *Ibid.*, 161.
[256] Guyton (1782), 382.
[257] *Encyclopédie Méthodique—Chimie, Métallurgie, Pharmacie*, vol. 1, Part 1 (Paris, 1786), v.
[258] *Ibid.*, v–vi.

CHAPTER FOUR

"L'Èpoque de la Nomenclature Chimique"

"Les sciences, semblables a des plantes, demandent un air doux et beaucoup de soins."

Antoine Laurent Lavoisier (1760), Arch. Acad. (Paris), Dossier Lavoisier, 1710.

"L'extrême exactitude qu'exigent les matières scientifiques, oblige à la plus grande sévérité dans la choix des expressions."

Madame Lavoisier, Preface to R. Kirwan, *Essai sur le Phlogistique* (Paris, 1788), VII.

"In der [. . .] Chemie die Bedeutung eines Körpers und also auch ein Name nicht mehr bedingt durch seine blosse Zusammensetzung, sondern vielmehr durch seine Stellung in der *Reihe*, der er angehört. Finden wir also, dass ein Körper einer solchen Reihe angehört, so wird sein alter Name ein Hindernis des Verständnisses und muss durch einen *Reihennamen* ersetzt werden. . . ."

Friedrich Engels, *Dialektik der Natur*, in *Marx/Engels—GESAMTAUSGABE* (Moscow, 1935), 652–653.

From One Revolution to Another

The early years of Lavoisier's scientific career have been the subject of much conjecture, of adventurous hypothesis, and of painstaking investigation.[1] Most of the known historical references to the education of the young Lavoisier constitute a regrettably poor source of information. With the exception of the years between 1766 and 1773, which have been partly illuminated by Meldrum and by Guerlac and his pupils, our knowledge of Lavoisier's early evolution as a scientist still relies on the work of Edouard Grimaux, who is probably the only historian to have had the opportunity and the patience to study all the surviving manuscripts and letters of the French chemist.

In 1955 Maurice Daumas drew attention to some autobiographical notes, probably written in 1792, in which Lavoisier briefly recalls his first approach to science. Despite the fact that Lavoisier relates several otherwise unknown episodes of his early career, the importance of this manuscript, recently reprinted by Bernadette Bensaude-Vincent,[2] has been completely overlooked.

As we know, Lavoisier entered the Collège des Quatre Nations in 1754, and it is likely that he left there in 1761 to study law. While at the Collège, Lavoisier attended the courses in mathematics and astronomy of La Caille. The year 1763 was a busy one for Lavoisier: he began working with Guettard and studying botany with Bernard de Jussieu while at the same time earning the degree of Bachelor of Law. While it has been

---

[1]Little is known of Lavoisier's early career in science. In addition to the biographies by E. Grimaux, *Lavoisier* (Paris, 1888), M. Daumas, *Lavoisier théoricien et expérimentateur* (Paris, 1955), and Guerlac (*DSB*, 7, 66–91), it is possible to find some original information on the period 1763–1773 in the following works: Andrew Norman Meldrum, *The Eighteenth Century Revolution in Science—The First Phase* (London-New York-Calcutta, 1930); H. Guerlac, "A Note on Lavoisier's Scientific Education," *ISIS*, 47 (1956), 211–216; J.B. Gough, "Lavoisier's Early Career in Science: An Examination of Some New Evidence," *BJHS*, 4 (1968), 52–57; Robert Siegfried, "Lavoisier's view of the Gaseous State and Its Early Application to Pneumatic Chemistry," *ISIS*, 63 (1972), 69–78; J.B. Gough, "The Origin of Lavoisier's Theory of the Gaseous State," in H. Woolf (ed.), *The Analytic Spirit* (Ithaca and London, 1981), 15–39; F. Abbri (1984), 65–119.

[2]Daumas (1955), 93; Lavoisier, *Sur la manière d'einseigner la chimie*; Bensaude-Vincent (ed.) (1990), 456–457.

assumed that Lavoisier attended Rouelle's chemistry courses on the rue Jacob in 1762 and 1763, there is no actual evidence of this.

In an autobiographical note, Lavoisier adds some new and interesting facts:

> When I began for the first time to attend a course in chemistry I was surprised to see how much obscurity surrounded the first approaches to the science, even though the professor whom I had chosen passed for the clearest and most accessible to beginners, and even though he took infinite pains to make himself understood. I had taken a useful course in physics, I had followed the experiments of the Abbé Nollet, I had also studied elementary mathematics with some success in the works of the Abbé La Caille and had attended his lessons for a year.[3]

These experiences, as we shall see, deeply influenced Lavoisier's early approach to chemistry. What is more important is that Lavoisier mentions a fact that is at variance with the generally accepted account of Grimaux, namely, that his first chemistry teacher was not Rouelle, but a little-known apothecary, Charles-Louis de la Planche, who was later recalled by Lavoisier as the most lucid professor of chemistry of the day.[4]

Lavoisier relates that during his first course La Planche attempted to bring some order to the obscure lexicon of alchemy by explaining the chemical terms and their meanings, although he does not seem to have been very successful. There are no works or manuscripts providing detailed confirmation of Lavoisier's comments on La Planche, but two printed prospectuses advertising La Planche's courses in chemistry for the years 1750 and 1764 are preserved in the Bibliothèque Nationale in

---

[3]"Lorsque j'ay commencé pour la première fois à suivre un cours de chimie, quoique le proffesseur que j'avais choisi passât pour le plus clair et le plus à portée des commençans, quoiqu'il prît infiniment de peine pour se faire entendre, je fus surpris de voir de combien d'obscurité les abords de la science se trouvaient environnés. J'avais fait un bon cours de philosophie, j'avais suivi les expériences de l'Abbé Nollet, j'avais étudié avec quelque fruit la methématique élémentaire dans les ouvrages de l'Abbé de la Caille et j'avais suivi pendant un an ses leçons. J'étais accoutumé à cette rigueur de raisonnement que les mathématiciens mettent dans leurs ouvrages jamais ils ne prennent une proposition que celle qui la précède n'ait été découverte," quoted in Bensaude-Vincent (1990), 457. The English translation is by Albury (1972), 112.

[4]"M. de la Planche était le plus clair des proffesseurs qui enseignaient alors la chimie," *ibid.*, 457. There is no biography on La Planche. Nevertheless from fol. 12 recto of the 38th *Registre des Marchands Apothicaires de la Ville de Paris* kept at the BIP, we know that La Planche was an active member of the Guild of Parisian Apothecaries, and his signature appears for the first time on May 12, 1750 and for the last (fol. 50 recto) in January 1767.

Paris.[5] The plan for the 1764 course is the more interesting, as it approximately coincides with Lavoisier's initiation into chemistry.

The course began with the enumeration and explanation of the chemical principles and continued with a demonstration of the existence of phlogiston or the inflammable principle.[6] It is interesting that La Planche promised electrical experiments to demonstrate the universal nature of this principle. The investigation of electrical phenomena in a chemistry course was an original idea at that time and was not copied by other chemists until the late 1770s.

The rest of the course was less original, consisting of a description of chemical instruments and an analysis of the vegetable, animal, and mineral kingdoms. It followed the same pattern as Rouelle's courses, which, Lavoisier pointed out, began like all eighteenth-century chemistry textbooks with the principles of the science. This arrangement had the disadvantage of not enabling the pupil to familiarize gradually himself with the objects of chemistry.[7]

Although his prospectus placed the vegetable kingdom first, La Planche seems to have devoted much of the course to the analysis of mineral substances. From his syllabus we infer that he intended, among other things, to discuss the nature of gypsum (*gipse*); to make a detailed analysis of acids and salts; to carry out a chemical analysis of antimony; to examine the calcination of lead and its transformation into minium and also the calcination of lead combined with niter to produce nitric litharge (*litarge nitreuse*), etc.[8] To all who are acquainted with Lavoisier's first chemical research into gypsum and calcination, the contents of La Planche's course are revealing and confirm Lavoisier's statement that it was important.

After taking La Planche's course, Lavoisier turned to the chemistry courses that Rouelle gave for three years in his private laboratory on the rue Jacob. It is difficult to establish the exact length of Lavoisier's attendance, but it seems likely that he took La Planche's course in 1761 after

[5][Laurent-Charles de La Planche], *Plan d'un COURS d'Operations de Chimie & PHARMACIE, suivant les principes de BECHER, de STAHL, de NEWTON & de BOERHAAVE*, 4pp. n.n. (Paris, 1750) (BN, Te 147.61); *idem, Plan, D'UN COURS DE CHYMIE, suivant les principes de BECHER, de BOERHAAVE & de STAHL, dans la vue d'enrichir la Médicine, d'éclaircir la physique & de perfectionner les Arts*, 8 pp. (Paris, 1764) (BN, R.5805).

The course beginning in September 1750 was given by La Planche in his private laboratory in *rue du Roulle à l'Alambic*. The course beginning in 1764 was given in a laboratory in *rue de la Monnoye*. I have not been able to find evidence on the continuity of La Planche's courses during the years 1751–1763.

[6]La Planche (1764), 1.

[7]Bensaude-Vincent (1990), 457.

[8]La Planche (1764), 5–7.

leaving the Collège Mazarin, and then went to Rouelle between 1762 and the spring of 1764, when he began to accompany Bernard de Jussieu and Jean-Etienne Guettard on their botanical and geological excursions.

By comparing what Lavoisier tells us himself with La Planche's prospectus, we can draw the conclusion that Lavoisier's early interest in mineralogy, in the calcination of metals, and in the chemical analysis of gypsum was stimulated by La Planche rather than by Rouelle. This is also consistent with the fact that Rouelle seems to have been far better acquainted with plant chemistry than with mineralogy. With due deference to Guerlac,[9] it should be noted that most of the manuscript copies of Rouelle's chemistry courses are devoted to the vegetable kingdom and that Lavoisier himself referred to Rouelle's keen interest in this field. Moreover, Lavoisier declared that what he learned of the composition of neutral salts and the preparation of mineral acids came from La Planche and not from Rouelle.[10]

Further important, and hitherto unnoticed, evidence of Lavoisier's chemical background prior to 1766 is provided by a draft of a chemistry course written by Lavoisier and another person, whose identity remains uncertain.[11] This manuscript rearranges matters in a distinctive way and introduces new approaches to various chemical questions.

Lavoisier begins his course by defining chemistry as the "art which teaches us to separate the different substances of which bodies are composed, to examine them, recognize them and combine them either to recompose the bodies already analyzed or to form new ones."[12] This Stahlian definition is followed by a description of the utility of chemistry in the arts of dyeing, glassmaking, pharmacy, etc.

The second part of the course consists of an examination of the principles of chemistry and their composition. As in Stahl's system, matter is regarded as a hierarchic combination of mixts rather than as made up of single elements.

The most interesting and original part of Lavoisier's course was the section devoted to the investigation of the nature of fire and electricity. Significantly, Lavoisier examines the properties of fire from a physical angle. He claims that it is easier to understand fire from its effects than

---

[9]"Rouelle, Lavoisier's teacher of chemistry, was one of the most enthusiastic supporters of the new mineralogical emphasis," Guerlac (1956), 215.

[10]Bensaude-Vincent (1990), 457.

[11]I have provided a full description of this manuscript in my forthcoming *A New Course in Chemistry: Lavoisier's First Chemical Paper.* The manuscript is kept at Arch. Acad., Dossier Lavoisier 380; hereafter Dossier Lavoisier 380

[12]Dossier Lavoisier 380, fol. 2 recto.

by investigating its intrinsic quality. Accordingly he assumes that there exists in nature an entity (*être*) that produces fire, heat, and light.[13] Yet our notions of fire, heat, and light are primarily based on our physical perception of them, rather than on accurate and objective data concerning their nature. The perception of heat, for instance, differs from person to person, and "a cellar which seems warm in winter, seems cold in summer."[14] The difficulty of relating the effects of heat to anything other than human sensation led physicists (*physiciens*) to invent instruments, such as the pyrometer and the thermometer, for more objective and accurate measurement of these phenomena. The pyrometer measured the expansion and contraction caused in bodies by heat, the thermometer the heat of air and other substances.

It seems likely that Lavoisier copied the descriptions of these two instruments from Peter van Musschenbroek, whose works are quoted in the text and were also to be found in Lavoisier's own library.[15] It is interesting to note that Lavoisier applied these notions to a course in chemistry and that he provides a *physical* description of fire and the instruments that detect heat quantitatively, whereas he does not say a word about fire as a chemical principle in the manner of the chemistry textbooks of the period.

Lavoisier regarded fire as the cause of expansion and contraction of bodies and as an extremely subtle fluid that surrounded the whole earth with particles considerably smaller than those of which other natural bodies were composed. He thought that the matter of fire (*mattiere de feu*) did not penetrate, unless excited, the parts that made up mixt bodies but only those that composed the aggregates. Accordingly Lavoisier distinguished between the operations of *heating* a body, which meant that the fire affected only the aggregated parts and not the mixt, and *burning* a body, i.e., when fire also affected the parts making up the mixt.[16]

In contrast, Stahl saw fire both as a principle of the composition of

---

[13]"Il est plus aisé de donner une idée des effets du feu que de faire connoitre la nature meme de cet element. Nous sçavons quil existe dans la nature un être qui suivant les differens modifications qu'il prend produit le feu, la chaleur, la lumiere . . .," *ibid.*, fol. 5 recto.

[14]*Ibid.*

[15]The catalogue of Lavoisier's library lists the following works by Musschenbroek: *Tentamina experimentorum naturalium captorum in Academia del Cimento* . . . (Lugduni Batavorum, 1731); *Essai de physique* . . ., 2 vols. (Leyden, 1739); *Dissertatio experimentalis de magnete* (Vienna, 1754); *Cours de physique expérimentale*, 3 vols. (Paris, 1769). The catalogue of Lavoisier's library is in preparation.

[16]Dossier Lavoisier 380, fol. 10 verso.

a body and as a cause of heat. In the first sense, fire denoted the principle of inflammability, which, as we saw in Chapter Three, was named *phlogiston* by Stahl and assumed to be inherent in all natural bodies. The distinction made by Stahl between these two notions of fire was taken for granted in France after 1750, and Macquer, who regarded himself as a supporter of the physical approach to chemistry, presented it as established fact in the article *"Feu"* in his *Dictionnaire de chymie* (1766).[17]

It is important to underline that Lavoisier mentions only the physical properties of fire and heat in his manuscript and that he does not consider Stahl's chemical principle. But the importance of Lavoisier's first chemical work goes beyond this significant difference. After a brief discussion of Newton's theory of light, Lavoisier turns to an aspect of natural philosophy which he examines in the context of chemical problems for the first time, namely electricity.

The ideas proposed by Lavoisier on this matter are not particularly original, but they testify that he was well acquainted with the works of Musschenbroek, Franklin, and Nollet at an early stage of his scientific career.[18] Concerning the argument between Franklin and Nollet on the nature of electricity, Lavoisier at first seemed inclined to favor Nollet, who regarded the effects of electricity as a movement of effluence and affluence,[19] but immediately afterwards, in the chapter devoted to electricity, Lavoisier examined in detail the extraordinary analogy between the phenomena of lightning and electricity and Franklin's experiment with the kite flier.[20] The inclusion of this experiment in what was supposed to be a chemistry course was defended by Lavoisier on the grounds that phenomena related to electricity were useful in "explaining a great number of operations of Nature."[21]

Although Lavoisier did not embellish the chemistry course with original experiments or prescient examinations of the gaseous state of matter, we may say that this manuscript is a document demonstrating the originality of his approach to scientific problems in general and chemical phenomena in particular. No earlier contemporary chemical texts included a study of electricity or laid so much emphasis on a

---

[17]Macquer, *Dictionnaire de chymie* (1766), vol. 1, 498, and vol. 2, 199 and ff.

[18]Dossier Lavoisier 380, fol. 14 verso. Lavoisier explicitly mentioned Franklin's *Experiments and Observations on Electricity,* probably in its French translation of 1752 and Nollet's *Leçons de physique expérimentale,* 6 vols., of which he owned the 1754 edition.

[19]On Nollet's theory, see Heilbron, *Electricity in the 17th and 18th Centuries* (Berkeley-Los Angeles-London, 1979), 282–289.

[20]On Franklin's theory and the experiment with the kite flier, see Heilbron (1979), 346–352.

[21]Dossier Lavoisier 380, fol. 18 verso.

physical approach to chemical problems, and in the chemistry courses given by Rouelle, La Planche, Macquer, and Baumé in Paris in the 1760s, there is no sign of chemistry being conceived in the light of natural philosophy.

It is nevertheless obvious that despite its originality Lavoisier's course still drew on the earlier tradition and that, being a beginner in the science, Lavoisier still relied largely on Stahl's theory of the composition of matter. However, the introduction of a purely physical approach to the investigation of what were considered the qualitative features of matter makes the discontinuity with his predecessors quite striking and unquestionable. This early manuscript also reveals that even when Lavoisier was still a firm Stahlian and had not yet discovered the active role of gases in chemistry, his ideas on the disposition of chemical matter were far from the qualitative thinking of the Stahlian chemists. Furthermore, some of the ideas contained in the course, such as the physical nature of heat and fire and the importance of accurate pyrometers and thermometers, remained constant features in Lavoisier's chemical investigations. Although he did not develop quite the same interest in electricity, Lavoisier studied the chemical phenomena associated with it in 1767, in 1782 when he collaborated with Volta, in 1786, and in 1792 when, while planning a new textbook on chemistry, he started to outline a course in electricity.[22]

In 1988 Arthur Donovan lamented the lack of historical evidence explicitly connecting Nollet's advocacy of experimental physics and Lavoisier's program for the reform of chemistry, but found the circumstantial evidence and the similarity of doctrines striking and added that Lavoisier had already decided that the way to make chemistry a science was to refashion it according to the methods and goals of experimental physics.[23] Although the picture revealed by a detailed analysis of Lavoisier's early chemical ideas is of too complex a nature to be traced back solely to Nollet's experimental physics, Donovan's statement is not far from providing the correct historical source of Lavoisier's interest in chemistry.

The year 1766 is usually regarded as the starting point of Lavoisier's interest in pneumatic chemistry,[24] but before examining the accuracy of this assumption, it is necessary to mention an important episode (see also Chapter Three) that has been overlooked by most historians. On February 15, 1766, the chemist Jean Hellot died and

[22]See LC, vol. I, 77; Daumas (1955), 58 and 64; Bensaude-Vincent (1990), 459.
[23]Donovan (1988), 222.
[24]See Gough (1968) and Siegfried (1972).

soon afterwards his library was sold by auction. Although there are no known official reports of the auction, we fortunately have the indirect information that Trudaine de Montigny, an honorary member of the Académie des Sciences, purchased Hellot's writings and possibly a part of his library.[25] It is certainly no mere chance that the catalogue of Lavoisier's own library lists more than ten items that had formerly belonged to Hellot.[26] Since Trudaine and Lavoisier were close friends and scientific colleagues, we may assume that Lavoisier was involved in the purchase of Hellot's library. This pedantic prelude is necessary in order to determine with a reasonable degree of exactitude the date when Lavoisier purchased Hellot's books and manuscripts. They include an important Latin manuscript by Stahl, dating from 1718, running to over 600 pages and dealing with the nature of sulphur. As well as being of inestimable scientific importance as the only known extended version in Latin of Stahl's major chemical work on the phlogiston theory,[27] this document is crucial testimony because the young Lavoisier annotated and studied the text thoroughly. It was this manuscript that led Lavoisier to turn his attention to the experimental paths of combustion and calcination. If, as seems likely, Lavoisier purchased Stahl's manuscript as early as 1766, we may assume that its pages constitute the main source of his early experiments on the calcination of metals. There are in fact no references in Lavoisier's earlier manuscripts to any of the central themes touched on by Stahl in his treatise on sulphur (calcination and combustion), whereas after 1766 these are mentioned frequently. In these circumstances, the studies by Meldrum, Guerlac, Gough, and Siegfried on the origins of Lavoisier's experiments of 1772 and 1773 assume a new significance. It is likely in fact that Lavoisier became interested in these operations long before 1772 and certainly before becoming acquainted with the works of Hales, Black, and Priestley on pneumatic chemistry.[28] This does not mean

---

[25]Arthur Birembaut, "L'enseignement de la minéralogie . . .," in Taton (ed.) (1986), 381.

[26]On Lavoisier's own library, see my "The Library of Antoine Laurent Lavoisier 1743–1794," in E. Canone (ed.), *Filosofi, scienziati e i libri. Da Cusano a Leopardi,* forthcoming.

[27]G.E. Stahl, *Züfallige gedanke und nützliche bedencken über den Stret, von dem sogenannten Sulphure* . . . (Halle, 1718). This work was translated into French by d'Holbach in 1766(!). The Latin manuscript version, of which I gave a complete bibliographical description in note 179 of Chapter Three, is an extended version of the *Züfallige* . . . although it deals with the same matter and has the same arrangement. A critical edition of the manuscript is in preparation.

[28]Guerlac (1961), 11–35, convincingly demonstrated that Lavoisier could not have read the works of the British chemist by 1772.

that when studying Stahl's manuscript Lavoisier immediately saw the active role played by air in the calcination of metals and understood the crucial inferences which might be drawn from this observation, but I believe that this text helped Lavoisier to decide the experimental priorities that eventually led to his new theory of calcination.

Although the background to Lavoisier's chemical ideas remains obscure because of the lack of documentation,[29] the new evidence presented here fills some of the gaps and suggests that even before 1766 Lavoisier saw chemistry differently from his predecessors and contemporaries and that in 1766 his ideas found fertile ground as he developed Stahl's observations on calcination and combustion.

Against this background, some of the statements contained in the famous sealed note of 1772 acquire a different and more consistent meaning. On November 1, 1772, Lavoisier remarked:

> About eight days ago I discovered that sulphur, in burning, far from losing weight, on the contrary gains in weight [. . .].
>
> This discovery, which I have established by experiments that I look upon as decisive, has led me to think that what is observed in the combustion of sulphur and phosphorus may well take place with regard to all substances that gain weight by combustion and calcination: and I am persuaded that the increase in weight of metallic calces is due to the same cause [. . .].
>
> *This discovery* [appears] *to me one of the most interesting of those that have been made since the time of Stahl* . . . [my italics].[30]

The importance of the experimental background to Lavoisier's observations on calcination has already been thoroughly illuminated by Meldrum and Guerlac, but a further comment on Lavoisier's reference to Stahl may be added to their shrewd analysis. The combustion and calcination of sulphur were operations that Stahl described in his

---

[29]The archives of the Académie des Sciences, the BI, and the Olin library, still possess large quantities of manuscripts and letters by Lavoisier that have never been published and not even mentioned. The manuscripts published by Fric, Daumas, Guerlac, Gough, Siegfried, Perrin, Holmes, and Bensaude-Vincent represent quantitatively only a minute part of this immense collection. Significantly most of Lavoisier's manuscripts published in the last 40 years are glosses of the investigations by Grimaux, Berthelot, and, above all, by Daumas' work. Hundreds of manuscripts, of which many deal with chemistry, still lie forgotten as Lavoisier left them 200 years ago. In contrast with the publication of the correspondence that is now reaching its final stages, Lavoisier's papers and manuscripts are still a virgin field, and I believe that a new edition of Lavoisier's collected works would illuminate many of the still obscure passages of his scientific career.

[30]The original version of the sealed note is printed in Guerlac (1961), 227–228. The English translation is by Meldrum (1930), 3.

work, and he also recognized that calces were usually heavier than the original metal. It is therefore significant that Lavoisier relates his observation of the *constant* increase in weight of metallic calces to Stahl's work. In doing so Lavoisier seems to be aware that his discovery explains the consequences of a chemical operation that was at the center of Stahl's chemical system but which the German physician overlooked. It is therefore not surprising that Lavoisier saw his discovery as important in relation to Stahl's theory and not to some more recent ones. In reality it seems quite clear that Lavoisier regarded Stahl as the main authority on chemical matters and that in making an observation that did not fit into Stahl's system he realized that something new and important was being introduced into the chemical scenario. At this point it is not of great interest whether Lavoisier was still a Stahlian or not when he wrote these lines, nor whether his chemical knowledge of calcination continued the Stahlian tradition or already revealed another level of understanding. In reality a literal reading of Lavoisier's own words suggests that this question has been relevant to modern historians of science more as a possible source of support for their belief in cumulative progress or in revolutionary change than in shedding light on the chemical sources of Lavoisier's discovery. The scientific debt to Stahl for this discovery was fully acknowledged by Lavoisier himself, but the recognition of this influence emphasized rather than diminished the discontinuity introduced by Lavoisier's idea on the increase in weight shown by calces. This conflict between tradition and innovation emerges very clearly in another celebrated paper written by Lavoisier during the same period.

On August 8, 1772, Lavoisier wrote a *memorandum*[31] in which he outlined some preliminary reflections on the experiment of the *miroir ardent.* The final part of this manuscript was entirely devoted to the examination of a new kind of air (*air fixe*), and in it Lavoisier remarked that although this air entered into the composition of the greater part of the minerals, no chemist had yet included the air in the definition either of metals or of any mineral bodies.[32] In the same year, in his *Systême sur les Elemens,*[33] Lavoisier pointed out even more clearly the discrepancies between the experimental results achieved in pneumatic chemistry and

[31]The text of this manuscript was printed for the first time by Guerlac (1961), 208–214.

[32]"Il paroit Constant que [la pluspart] l'air entre dans la Composition de la plus part des Mineraux, *des Metaux meme Et en tres grande abundance.* aucune chimiste Cependant n'a fait encore entrer l'air dans la definition ni des Metaux ne d'aucun corps mineral," *ibid.*, 214.

[33]This manuscript was published for the first time by René Fric in 1952 and was reprinted by Guerlac (1961), 215–222.

the lexicon that claimed to represent them. A large number of experiments appeared to prove that air "plays a large part in the composition of minerals [. . .]. However, these experiments seem to be unknown, they have not been linked at all to chemical theory and, in spite of the striking facts they had (and still have) before their eyes, modern chemists keep the same old definitions which they have given to minerals and have not admitted air at all."[34]

If Lavoisier acknowledged the fact that a large range of experiments in pneumatic chemistry had been carried out before him by other chemists who had recognized the active role of air, he also pointed out the weakness of this experimental tradition in being unconnected to a general theory. In both papers Lavoisier also lamented the total lack of *definitions* of the minerals combined with air. This early observation is the first reference to the need for a general reform of chemical nomenclature to take account of the pneumatic discoveries. Lavoisier seems almost surprised to find "modern chemists" using the old terminology to denote the combination of inorganic substances with air. At this stage he probably did not realize that to observe the importance of the role of air in chemical combination, to formulate a general theory on the basis of these observations, and to reform the old nomenclature accordingly—which is what he was thinking of doing—was not simply an empirical consequence of the discovery of fixed air but a major conceptual achievement.

But by February 20, 1773, Lavoisier had become fully aware of the difference, and on that date he remarked:

> However numerous may be the experiments of Messrs. Black, Mag-Bride (*sic*), Jaquin, Cranz, Prisley (*sic*) and de Smeth, in this direction, nevertheless, they come far short of the number necessary for a complete body of doctrine. It is established that fixed air shows properties very different from those of common air [. . .]. These differences will be exhibited to their full extent when I shall give the history of all that has been done on the air that is liberated from substances and that combines with them. The importance of the end in view prompted me to undertake all this work, which seemed to me destined to bring about a revolution in physics and in chemistry. I have felt bound to

[34]"Une foule dexperinces paroissent prouver que lair entre pour beaucoup dans la composition des mineraux [. . .]. Cependant ces experinces Semblent ignorées elles n'ont point ete liee a la theorie chimique et les chimistes modernes malgre les faits frappans quils avoient Sous les yeux [ont tojours] on laisse subsister les [memes] anciennes deffinitions quils avoient donne des mineraux ils ny ont point fait entrer l'air," *ibid.*, 215.

look upon all that has been done before me merely as suggestive; I have proposed to repeat it all with new safeguards, in order to link our knowledge of the air that goes into combination or that is liberated from substances, with acquired knowledge, and to form a theory.[35]

The main importance of this famous and much-studied passage lies rather in the theoretical and methodological commitments made by Lavoisier with such confidence, commitments that may fairly be described as revolutionary. No other textbook on pneumatic chemistry published before 1773 even contemplates the possibility of forming a general theory capable of encompassing the new discoveries with regard to the airs. The question of redefining the mineral kingdom to accommodate these new discoveries was not an issue. Indeed, outside Lavoisier's mind, there was nothing in 1773 to indicate the existence of the general field of investigation eventually known as *pneumatic chemistry*.

In describing all that had been done before him on the study of airs as "separate pieces of a great chain,"[36] and as merely "suggestive," Lavoisier was not being particularly presumptuous but rather showing an awareness that earlier observations concerning air were made only incidentally and that no other chemist foresaw the importance of the investigative field in its generality and entirety.

By 1773 Lavoisier had already stated three priorities in his vision of the future of chemistry: 1. to examine the chemical role played by air; 2. to formulate a theory capable of explaining phenomena which were not explained in Stahl's theory; 3. to devise a new nomenclature for pneumatic chemistry.

To understand the conceptual shift represented by this project, it is interesting to compare Lavoisier's paper with those of Black and Priestley on fixed air. Although Joseph Black published only three papers on pneumatic chemistry during his scientific career,[37] his part in laying the foundations for the remarkable advances in this field was central. There is in fact no doubt that Black's experiments on *magnesia alba* and quicklime, performed in 1753, were the first clear and non-equivocal signs of the discovery of a new kind of air (CO_2), called "fixed air" by Black, and of the active part played by air in chemical

[35]Quoted by Guerlac (1961), 229–230; English translation by Meldrum (1930), 9.

[36]Quoted by Guerlac (1961), 230; Meldrum (1930), 9.

[37]On Joseph Black, see William Ramsay, *Life and Letters of Joseph Black, M.D.* (London, 1918); J.G. Crowther, *Scientists of the Industrial Revolution* (London, 1962), 9–92; J.R. Partington, *A History of Chemistry*, vol. 3 (London, 1962), 130–143; A.D. Donovan, *Philosophical Chemistry in the Scottish Enlightenment* (Edinburgh, 1975), 165–277; Guerlac, *Essays and Papers in the History of Modern Science* (Baltimore and London, 1977), 283–303.

reactions. After his discovery, Black remarked that this kind of air, which was dispersed through the atmosphere, had the property of fixing itself in other bodies and taking part in their combination. "To this I have given the name of *fixed air*, and perhaps very improperly; but I thought it better to use a word already familiar in philosophy than to invent a new name, before being more fully acquainted with the nature and properties of this substance," he said, and as late as 1788, he still supported this cautious approach, declaring: "Chemistry is not yet a science. We are very far from the knowledge of first principles. We should avoid anything that has the pretension of a full system," and "in Philosophical Enquiries, the principal or only object is *New Discoveries*; but in attempting improvements of Arts, the object in view is to make the art more profitable."[38] Black's opinions on the introduction of new names and on the role of chemical theory are strikingly different from Lavoisier's and reveal an empirical and pragmatic attitude that was typical of British philosophy of science.

Although Black initially planned to continue his investigations into pneumatic chemistry, it is significant that he abandoned them for other scientific research and that it was not until 1790 that he officially expressed an opinion on the chemistry of the airs. By that time, however, the chemical revolution was practically complete. It is true, on the other hand, that Black's courses included some pneumatic chemistry, but most of his lectures were devoted to the nature of heat and the classification of salts. As Guerlac has pointed out, Black's work did not influence the pneumatic investigations carried out by Lavoisier, who outlined his theory independently of Black's experimental contribution.

The fate of Priestley's work in the discussion of aerial chemistry is more interesting. Priestley's interest in pneumatic phenomena is in fact almost contemporary with Lavoisier's, and it lasted until his death. In 1774 Priestley published his first major work on the subject,[39] summarizing all his scattered publications and experiments during the period between 1772 and 1774. It is not the purpose of this work to go into Priestley's experimental endeavors,[40] but it is desirable to empha-

---

[38]First quotation in Donovan (1975), 211; second in Ramsay (1918), 128; third from Black, *Lectures on the Elements of Chemistry* (Edinburgh, 1803), vol. 1, 547.

[39]Joseph Priestley, *Experiments and Observations on Different Kinds of Air* (London, 1774). On the origin and sources of this work, see Guerlac, "Joseph Priestley's First Papers on Gases," in Guerlac (1977), 304–313.

[40]On Priestley's contribution to chemistry, see J.R. Partington (1961–1970), vol. 3, 245–276; Robert E. Schofield (ed.), *A Scientific Autobiography of Joseph Prestley (1733–1804)* (Cambridge, Mass., 1966); *idem, Mechanism and Materialism: British Natural Philosophy in the Age of Reason* (Princeton, 1970); Abbri (1984), 145–151.

size the difference between his approach to chemical problems and that of Lavoisier.

After introducing his work with a quotation from Bacon, Priestley announced to the reader what he might expect from this work:

> As to myself, I find it absolutely impossible to produce a work on this subject that shall be any thing like *complete*. [. . .] But, paradoxical as it may seem, this will ever be the case in the progress of natural science, so long as the works of God are, like himself, infinite and inexhaustible. In completing one discovery we never fail to get an imperfect knowledge of others, of which we could have no idea before.[41]

The *structural* impossibility of achieving a complete knowledge of natural phenomena forced Priestley to adopt a sequential style in the narration and disposition of his discoveries. Following the classical leitmotif of Bacon's natural history, Priestley decided to confront his readers with a detailed *history* of his experiments, so that "other adventurers in experimental philosophy"[42] might be inspired to collect new facts. The emphasis on detailed descriptions of experiments is not a mere introductory appeal for a more empirical philosophy of science: Priestley attached great methodological importance to these descriptions, and he regarded "patient industry" and passive attention as the main qualities of the natural philosopher. Furthermore, the patient industry of collecting facts was guided by divine providence, which only could "strike out and diffuse" the knowledge of nature that was otherwise inaccessible to ordinary human understanding.[43] The attitude of some naturalists in drawing abstract conclusions or general theories from *any* number of empirical observations was therefore incompatible with the natural limits placed on our understanding by God. Likewise, with respect to the specific field of the different kinds of air recently discovered, the investigations were "by no means sufficiently advanced for anything that would be tolerably complete of the kind" (for a theory).[44] More bluntly, he says: "*Speculation* is a cheap commodity. *New and important facts* are most wanted, and therefore of most value."[45] Priestley continued to abstain from speculative assumptions for the rest of his life and even in 1779, when most European chemists were immersed in the theoretical discussion of the controver-

[41]Priestley, *Experiments* . . ., vol. 1, vii.

[42]*Ibid.*, x.

[43]*Ibid.*, vol. 2 (London, 1775), ix.

[44]*Ibid.*, vol. 3 (London, 1777), x.

[45]*Ibid.*, xxx.

sial nature of gases, Priestley underlined as strongly as ever the necessity of suspending judgment:

> If we would content ourselves with the bare knowledge of new *facts*, and suspend our judgment with respect to their *causes*, till, by their analogy, we were led to the discovery of more facts, of a familiar nature, we should be in a much surer way [sic] to the attainment of real knowledge.[46]

It would be possible to find numerous similar quotations, but this may suffice to indicate the difference in the attitude to scientific method expressed by Lavoisier and Priestley with regard to pneumatic chemistry.[47] In contrast to Priestley, Lavoisier observed in his first manuscript on the chemistry of gases that all the scattered experiments performed by other scientists on the nature of air needed a theory and that a redefinition of chemical objects was accordingly required. Priestley never felt this urgency; on the contrary, he always preferred the authority of *pure facts* to the futility of speculation and theory. This early difference between Priestley and Lavoisier is confirmed by comparing their positions on the question of nomenclature. As we have already seen, Lavoisier, as soon as he recognized the important role played by gases in chemical combination, asserted the *necessity* of introducing them "into the definition" of the substances with which they were combined, namely, metals and minerals. On the other hand, Priestley not only felt no need to promote a general reform of traditional nomenclature but even explicitly warned against the introduction of a single new name. On the specific case of *fixed air*, a name given by Black, and on the naming of gases in general, Priestley remarked:

> The term *mephitic* is equally applicable to what is called fixed air, to that which is inflammable, and to many other kinds: since they are equally noxious, when breathed by animals. *Rather, however, than either introduce new terms, or change the signification of the old ones, I shall use the term fixed air, in the sense in which it is now commonly used, and distinguish the other kinds by their properties, or some other periphrasis.* I shall be under the necessity, however, of giving a name to those kinds of air,

---

[46]Priestley, *Experiments and Observations Relating to Various Branches of Natural Philosophy*, vol. 1 (London, 1779), ix–x.

[47]For a detailed analysis of Priestley's philosophy in relation to his scientific methodology, see J.G. McEvoy, J.E. McGuire, "God and Nature: Priestley's Way of Rational Dissent," in *Historical Studies in Physical Sciences*, VI (1975), 325–404; McEvoy, "Continuity and Discontinuity in the Chemical Revolution," in Donovan (ed.) (1988), 195–213; for a more general picture, see also R. Anderson and C. Lawrence (eds.), *Science, Medicine, and Dissent: Joseph Priestley* (London, 1988).

to which no name has been given by others, as *nitrous, acid* and *alkaline.*[48] [my italics]

Significantly Priestley preferred to keep the same term for three or more kinds of air and confine himself to periphrasis to distinguish them. This attitude, the opposite of Lavoisier's, reduced the scientific value of defining and naming the substances that were newly discovered. In the last resort, Priestley considered the process of discovery far more important than the interpretation of *what* was discovered. Being a consistent Baconian, Priestley put the knowledge of things before their naming, and he saw no possible connection between the two operations. He did not foresee that the question of nomenclature, which had already arisen in mineralogy, was going to become one of the most important issues in chemistry and that the whole conception of the chemistry of gases could be changed by the change of a few names.

Lavoisier recognized that Priestley was a gifted experimental scientist, but he immediately pinpointed the weakness of his work, labelling it "a train of experiments, not much interrupted by any reasoning" and "an assemblage of facts."[49] This fitted the image Priestley had of himself when he recognized that his "knowledge of chymistry [was] very imperfect"[50] and that he was a "bad theorist."[51] These two statements must not be interpreted as false modesty: in reality they reveal Priestley's lack of understanding of the methods, the objects, and the characteristics of chemistry. His approach to chemistry was exactly the same as his approach to his investigations in the fields of electricity, light, optics, and

[48]Priestley, *Experiments* . . . (1774), vol. 1, 24.

[49]"un assemblage de faits interrompu par aucun raisonnement," Lavoisier, *Opuscules Physiques et chymiques* (Paris, 1774), 108–109. The entire passage taken from the English translation by Thomas Henry (London, 1776), 120–121, is: "Nothing now remains to complete the object which I proposed to myself in this first part, but give an account of a numerous series of experiments, communicated last year to the Royal Society of London by Dr. Priestley. This work may be regarded as the most elaborate, and most interesting of any which appeared since that of Dr. Hales, on the fixation and separation of Air. No modern work has seemed to me more adapted to evince how many new roads Philosophy and Chemistry still point out to travel over.

Dr. Priestley's work, being, as it were, a train of experiments, not much interrupted by any reasoning, an assemblage of facts, mostly new, whether considered by themselves, or by the circumstances which accompany them, it may imagined that is little capable of abridgment."

[50]Schofield (1966), 192; see also p. 51.

[51]When presenting his new observations to Fabbroni in 1779, Priestley declared: "The *facts* appear to me rather extraordinary. You must help me to explain them, for I am a very bad theorist," from Schofield (1966), 171.

natural philosophy. In fact Priestley conceived chemical phenomena as an integral part of natural philosophy rather than a research field of their own, and he did not think that facts and experiments could have any restricted and *specific* meanings within different scientific disciplines.[52]

The strikingly different approaches of Lavoisier and Priestley in their early works when describing the *same* experiments were manifestations of two different conceptions of science that rested on two distinct cultural traditions. The insistence on the urgency of formulating a general theory on the basis of a restricted range of known phenomena revealed a typically Cartesian approach to science, whereas the priority given to an unsystematic catalogue of observations and experiments was a classical Baconian theme.[53] Lavoisier's faith in a rationalistic explanation may be contrasted with Priestley's belief in factual evidence. We shall return to this philosophical conflict later.

In 1774 Lavoisier published his first systematic work,[54] but instead of publicizing his revolutionary ideas of 1772 and 1773, he preferred to approach the chemistry of gases in a more circumspect manner and devoted almost 200 pages to the examination of all previous studies of the role of air. In the second part of his work, in which he added his own original experiments, Lavoisier preferred not to make any general pronouncement on the nature of elastic fluids. However, to quote Meldrum:

> . . . when rightly read the *Opuscules* reveals the slow progress of knowledge. In producing it Lavoisier gained much; he broke ground that was new to him: he asserted himself on an important question that was controversy: he felt the difficulties that he must overcome in order to solve the great problems that he had taken up.[55]

Between 1775 and 1781, Lavoisier collected more experimental evidence to support his first theoretical draft of 1772 and at the same time

---

[52]As Abbri (1984), 180, correctly put it: "Priestley aveva una buona conoscenza della chimica del suo tempo, ottenne dalle sue ricerche risultati di grandissimo rilievo scientifico, ma le indagini chimiche costituivano per lui un capitolo della storia della filosofia naturale, simile a quelli dell'elettricità e dell'ottica."

[53]It has been argued by McEvoy and Albury that Lavoisier's epistemology was Lockean and Condillacian and Priestley's Boylian and mechanistic. As will later become clear, this distinction touches only superficially the philosophical difference between the two scientists. Although disguised in vague appeals to experience and sensations, Descartes' influence on eighteenth-century French science and philosophy was in my view stronger than Locke's empiricism.

[54]Lavoisier, *Opuscules physiques et chymiques.* On the contents, the impact, and the diffusion of this work, see Meldrum (1930), 14–34, and the more detailed study by Abbri (1984), 109–167.

[55]Meldrum (1930), 33.

became confident enough to assert that metallic calces gained weight because they absorbed air and that they could not lose phlogiston as everybody believed. The discovery of oxygen and the process that led to the enumeration of its properties was undoubtedly the experimental turning point for Lavoisier's theory of calcination.

Around this initial belief, which Meldrum fairly described as a "fixed idea,"[56] Lavoisier began to construct and gradually develop a more coherent general theory that enabled him to explain far more than the single process of calcination. The experimental path of this interlocutory period has already been the object of a detailed study by Abbri (1984), and I shall therefore content myself with examining it in relation to the general questions of the definition of chemistry and its nomenclature.

On September 5, 1777, Lavoisier presented his *Considérations générales sur la nature des acides* at the Academy of Sciences,[57] including in it his first direct attack on the phlogiston theory and laying the foundations of an alternative. In this paper Lavoisier decided to pursue his initial intuition to its logical conclusion and claimed that the kind of air that was now known as the purest air or *eminently respirable air* (oxygen), and that was called *dephlogisticated air* by Priestley, was the universal principle of acidity:

> On the basis of these truths, which I already consider firmly established, I shall henceforth refer to dephlogisticated air or eminently respirable air in its state of combination and fixity, by the name of *acidifying principle* or, if one prefers a Greek word with the same meaning, *oxygenic principle*. This designation will avoid periphrasis, will introduce more rigour in my mode of expression, and avoid the ambiguities which would constantly arise if I used the word air.[58]

The first appearance of the word "oxygen" on the chemical scene signified a substantial break with the traditional lexicon, and it was immediately clear to most chemists that Lavoisier attached more than

[56]*Ibid.*, 6.

[57]The Memoir was read in front of the Academy on November 23, 1777, and was published in *MARS* 1778 (1781), 535–547, and was reprinted in LO, vol. 2, 248–260.

[58]"D'après ces vérités, que je regarde déjà comme très-solidement établies, je désignerai dorénavant l'air déphlogistiqué ou air éminement respirable dans l'état de combinaison et de fixité, par le nom de *principe acidifiant*, ou, si l'on aime mieux la même signification sous un mot grec, par celui de *principe oxygine*, cette dénomination sauvera les périphrases, mettra plus de rigueur dans ma manière de m'exprimer, et évitera les équivoques dans lesquelles on serait exposé à tomber sans cesse, si je me servais du mot d'air," LO, vol. 2, 249.

its immediate meaning to this name and that both the gaseous state and acidity in their entirety were now being regarded from a completely different perspective. The gain in weight seen in a calcined substance was due to the fixation of oxygen. Oxygen was also the principle of acidity and played a vital part in the respiration of animals; as in any other form of combustion, animal respiration transformed oxygen into fixed air (CO_2)[59] and this combustion was made possible by the oxygen.[60] By the end of 1777, these were the properties and the semantic implications of the term *principe oxigine.*

In his memoir on combustion, Lavoisier insisted on the necessity of subsuming the pneumatic discoveries in a general theory and announced the method he intended to adopt:

> Dangerous though the spirit of systems is in physical science, it is equally to be feared lest piling up without any order too great a store of experiments may obscure instead of illuminating the science: lest one thereby make access difficult to those who present themselves at the threshold: lest, in a word, there be obtained as the reward of long and painful efforts nothing but disorder and confusion. *Facts, observations, experiments, are the materials of a great edifice. But in assembling them, we must not encumber our science. We must, on the contrary, devote ourselves to classifying them, to distinguishing which belong to each order, to each part of the whole to which they pertain.*[61] [my italics]

The reference in this passage is extraordinarily clear. The obsession of some scientists, such as Priestley, with merely collecting facts was hindering the progress of science as much as the *spirit of systems.* On the contrary, suggested Lavoisier, facts had to be classified and ordered within a general conceptual framework.

In 1778 Lavoisier planned to publish the second volume of his *Opuscules physiques et chymiques,*[62] the first having appeared four years earlier, and he returned to what became for him the fundamental

---

[59]On April 9, 1777, Lavoisier read to the Academy his *Mémoire sur la respiration des animaux et sur les changements qui arrivent à l'air en passant par les poumons,* now reprinted in LO, vol. 2, 174–183.

[60]On November 12, 1777, Lavoisier read to the Academy his *Mémoire sur la combustion en général,* now in LO, vol. 2, 225–233.

[61]*Ibid.*, translated by Gillispie (1960), 225.

[62]LO, vol. 5, 267–270. The second part of the *Opuscules*... was never published, and Lavoisier got no further than the writing of a short introduction and a general outline of the work. Grimaux (1888) claimed that the manuscript was written in 1778, but Daumas (1955) expressed some doubts and pointed out discrepancies in the text. It is unlikely, however, that Lavoisier worked on this text after 1780.

question of the method to be employed in chemical investigations. In the introduction to his work, he declared that he did not intend to provide the reader with a list of experiments. To do this it would have been enough "to copy [. . .] the laboratory records." Lavoisier, by contrast, preferred to follow "the method of geometers, a precise method which always leads to the evidence, if we can base our physics on sure and demonstrated data."[63] We do not know whether Lavoisier was satisfied with this arrangement at the time he wrote it, but it is significant that he abandoned the plan to publish the second volume. Instead, he began in 1780 to prepare an elementary textbook of chemistry,[64] and once again wondered which method to adopt. In his short introductory notes, Lavoisier pointed out the problems to be overcome:

> Difficulty of choosing the order which is appropriate for a treatise of chemistry—defects of all existing treatises. They are obliged, from the beginning, to draw upon knowledge which ought not to be set forth until much later in the body of their work, according to a plan which they have made for themselves. It is extremely difficult to remedy this defect.
>
> Reflections on the logic applied to sciences, necessity of well conceiving the whole before forming a treatise.[65]

His reflections in 1778 and these notes together leave no doubt about the Cartesian commitment to science felt by Lavoisier during this period. However, a purely Cartesian approach to chemistry, i.e., a direct application of the method of geometry, turned out to be impracticable and Lavoisier himself admitted:

> The order which I believed I ought to adopt is that of geometry, but modified and corrupted relative to the state of imperfection which chemistry is in—an imperfection which does not permit the laying down of certain demonstrated principles, as in geometry, and which consequently necessitates another procedure.[66]

Even if Lavoisier became increasingly prudent in assessing the possibility of making chemistry an exact science, he did not totally

---

[63]*Ibid.*, 267–268.

[64]Maurice Daumas (1955) (pp. 98–101) was the first to discover and date the preparatory study of an elementary textbook in chemistry, conceived by Lavoisier between the end of 1779 and the first half of 1780. Albury (1972) (pp. 128–134), provided some new insights and more accurate data on the dating of the manuscript.

[65]Quoted and translated into English by Albury (1972), 130.

[66]*Ibid.*

renounce the Cartesian ideal: he simply searched for a more appropriate and practical method of applying it successfully. The publication of Condillac's *Logique* in 1780 fulfilled this need and, as Albury has pointed out,[67] after studying this work, Lavoisier was able to produce a new and detailed answer to his methodological questions. In a second draft, probably written in 1781, of the introduction to his elementary textbook, Lavoisier finally declared:

> The first rule of procedure in an elementary treatise of science is that one can advance to the discovery of a truth which one does not know only in so far as it is found in the truths which are known. Thus every research is a problem to solve which supposes some data. Now in the physical sciences these data are the facts. Thus the first thing in physics and in chemistry is to assume nothing and to begin always with sure experiments. These experiments ought thereafter to be considered as the data of a problem, by means of which one must proceed to the search for the unknown.[68]

In this paraphrase of Condillac's logic, Lavoisier described the main features of the analytic method—a method that, he believed, was universally true and "applicable to all the sciences." The content of this method will be examined later in this chapter, but it may be observed that by committing himself to applying analysis to chemistry, Lavoisier emphasized even more strongly his debt to the Cartesian approach to science and that he was absolutely correct in concluding: "No author, no professor of chemistry has seen science from this point of view."[69]

The Overthrow of Phlogiston

The years between 1785 and 1789 marked a definitive shift in the history of modern chemistry. Before 1785 Lavoisier was virtually alone in advocating his oxygen theory. Even those who agreed with him on the urgency of a physical and quantitative approach to chemistry, such as Macquer and Guyton, were by no means inclined to abandon the phlogiston theory. On the other hand, between 1775 and 1785, there was little consensus on the nature of phlogiston, and each chemist interpreted it in the terms of the pneumatic system he supported.

---

[67] *Ibid.*

[68] *Ibid.*, 131.

[69] *Ibid.*, 132.

Moreover, the extraordinary experiment on the composition of water performed by Henry Cavendish in 1783 served to create more confusion and uncertainty among the supporters of Stahl.[70] The major chemists of Europe, such as Scheele, Kirwan, Priestley, Guyton, Bergman, Baumé, and Macquer, began to attach different meanings and different chemical roles to the notional phlogiston. Phlogiston was variously identified as inflammable air (hydrogen), as heat, as fire, and as a chemical principle or instrument, either heavy or light, solid or fluid. Each of these opinions was supported by experimental evidence, and all shared the quality of accommodating the new experiments to phlogiston rather than vice versa.[71] Lavoisier quickly drew attention to the contradictions entailed in this method and decided that the time had come for a direct frontal attack on phlogiston:

> This entity, introduced into chemistry by Stahl, far from having brought light to bear upon it, seems to have created an obscure and unintelligible science for those who have not made a highly specialized study of it; it is the *deus ex machina* of the metaphysicians: an entity which explains everything and which explains nothing, to which one ascribes in turn opposite qualities.[72]

In 1785 Lavoisier presented a monograph in which his criticism of phlogiston became more explicit and sweeping.[73] However, *Réflexions sur le phlogistique* was an attack on phlogiston but not, as most historians have assumed, on Stahl. On the contrary, Lavoisier regarded the German physician as the "patriarch of chemistry," whose discoveries and observations concerning the calcination and combustion of metals had led to a "kind of revolution in science."[74] These discoveries forced Stahl to assume the existence of an "inflammable principle" of metals capable of explaining combustion in terms of liberation of the fire fixed within them. "From the fact that metals are combustible it necessarily followed that those substances contained an inflammable prin-

---

[70]On the crucial importance of this experiment to phlogiston theory, see the illuminating account by Abbri (1984), 298–314.

[71]For a detailed description of the different phlogiston theories set forth in the early 1780s, see Abbri (1984), 354–369.

[72]Lavoisier, "Considérations générales sur la dissolution des Métaux dans les acides" (1782), in LO, vol. 2, 510, translated by C.E. Perrin, "The Triumph of the Antiphlogistians," in Woolf (ed.) (1981), 44.

[73]Lavoisier, "Réflexions sur le Phlogistique," read to the Academy on June 28, 1785. This memoir was published in *MARS* 1783 (1786), 505–538 and reprinted in LO, vol. 2, 623–655.

[74]LO, vol. 2, 624.

ciple."[75] Thus the appearance of phlogiston in chemistry was regarded by Lavoisier not only as a major achievement of chemical theory, but also as a historical necessity that enabled new generations of chemists to make new discoveries.[76] Lavoisier shrewdly regarded the recent past of his science with more objectivity and detachment than many historians of chemistry, who have seen Stahl's chemical system as regressive, or even as a series of errors. What Lavoisier actually intended to challenge with his memoir was not the notion of phlogiston put forward by Stahl but rather the phlogistic theories introduced by pneumatic chemists to explain the active role of air in chemistry. Stahl used phlogiston to explain the mechanism of combustion and calcination, but his disciples went further and

> . . . made phlogiston a vague principle, which is not strictly defined and which consequently provides all the explanations demanded of it. Sometimes it has weight, sometimes it has not; sometimes it is free fire, sometimes it is fire combined with earth; sometimes it passes through the pores of vessels, sometimes these are impenetrable to it. It explains at the same time causticity and non-causticity, transparency and opacity, colour and absence of colour. It is a veritable Proteus that changes its shape at every instant.[77]

With these critical arguments, it was easy for Lavoisier to point out the consistency of his own theory to the reader and to persuade him that he admitted neither theoretical contradictions nor observable discrepancies into his chemical system. Furthermore, the oxygen theory had the relevant advantage of being constructed on the basis of the pneumatic discoveries rather than being a hypothesis adopted *a priori*. No theoretical accommodation was made, and all the new discoveries appeared to fit perfectly the nature and the properties of Lavoisier's classification of gas. The consistency of the new theory was also supported by a method that, being based upon the principles of physics and geometry, provided chemistry with an aura of scientific prestige.

Confident in his awareness of these circumstances, Lavoisier was able to conclude his memoir in a peremptory manner:

---

[75] *Ibid.*

[76] "Ce fait constant [the identity between calcination and combustion], que Stahl paraît avoir reconnu le premier, et qui est aujourd'hui généralement avoué de tout le monde, le mettait dans la *necessité* d'amettre un principe inflammable dans les métaux" (my italics), *ibid.*, 625.

[77] *Ibid.*, 640.

> My only object in this memoir is to extend the theory of combustion that I announced in 1777; to show that Stahl's phlogiston is imaginary and its existence in the metals, sulphur, phosphorus, and all combustible bodies, a baseless supposition, and that all the facts of combustion and calcination are explained in a much simpler and much easier way without phlogiston than with it. I do not expect that my ideas will be adopted at once; the human mind inclines to one way of thinking and those who have looked at Nature from a certain point of view during a part of their lives adopt new ideas only with difficulty; it is for time, therefore, to confirm or reject the opinions that I advanced. Meanwhile I see with much satisfaction that young men, who are beginning to study the science without prejudice, and geometers and physicists, who bring fresh minds to bear on chemical facts, no longer believe in phlogiston in the sense that Stahl gave to it and consider the whole of this doctrine as a scaffolding that is more of a hindrance than help for extending the fabric of chemical science.[78]

In these few concise sentences, Lavoisier demolished the principle that nearly all chemists of Europe still considered the hypostasis of their science. He attacked the attitude of those who still clung to it as *conservative* when compared with the refreshing insights of a few young chemists,[79] physicists, and geometers,[80] who increasingly accepted the new ideas.

In actual fact, in 1785, there were still very few people—with the exception of Chaptal and some of Lavoisier's collaborators—who were convinced that phlogiston was a word devoid of scientific meaning, and none of those who were dared to say so officially.[81] It is therefore impos-

[78]*Ibid.*, 655; the translation of this passage is taken from D. McKie, *Antoine Lavoisier—Scientist, Economist, Social Reformer* (1952), 2nd edition (New York, 1990), 157–158.

[79]On June 29, 1784, Jean-Antoine Chaptal, a chemist from Montpellier, wrote to Lavoisier saying: "J'ai eu l'honneur de vous adresser les différens petits ouvrages que j'ai publiés. Vous aves dû y trouver un zélé partisan de vos principes, parce que l'expérience journalière m'en demontroit la vérité," LC, vol. IV, 21. Jean Baptiste Meusnier, with whom Lavoisier had collaborated on the experiments on the decomposition and composition of water since 1783, was also probably a supporter of his theory.

[80]Laplace, Monge, Vandermonde, and J.A.J. Cousin were among the first to support Lavoisier's views.

[81]In his essay on the antiphlogistians, Perrin (1981) tries his best to reverse the historical value of these facts and claims that this "scenario exaggerates Lavoisier's degree of isolation in the period prior to 1785" (p. 41). However, it is difficult to use any other adjective when it is an undeniable fact that Lavoisier was, before that date, the only chemist in Europe who systematically and repeatedly questioned the existence of phlogiston, who published several memoirs against it, and who set forth a theory alternative to it.

sible to underestimate the conceptual rupture introduced into chemistry by the publication of this work without distorting the historical value of Lavoisier's memoir. Lavoisier challenged a scientific theory that was shared by all the chemists of Europe; he put forward his own rival theory and explicitly demanded that the chemical community should plunge into the arena of the controversy and choose between the two. The fact that *Réflexions sur le Phlogistique* broke with chemical tradition is historically confirmed by the number of hostile reactions to its appearance. If, as some historians wish to believe, Lavoisier's theory did not cause a revolution in chemistry, it has to be explained why his contemporaries thought that it did, regarding it as a serious threat not only to their general conceptions but also to their experimental enterprises. The more we compare Lavoisier's writings, his method, and his intellectual consistency with those of his contemporaries, the more we are confronted with the discontinuity introduced into chemistry. It is, however, obvious, as it was obvious to Lavoisier himself, that the shift from phlogiston to oxygen was the conclusion of a historical process that implied constant reference to and comparison with the sources of a consolidated experimental tradition.

From the animated discussion provoked by *Réflexions sur le Phlogistique*, Lavoisier understood that the time was ripe to collect his ideas in a systematic work. Importantly, in 1785 and 1786, prestigious chemists, such as Fourcroy, Berthollet, and Van Marum, were converted to the new system. In the early spring of 1787, after his visit to Paris, Guyton de Morveau also became convinced of the consistency of Lavoisier's oxygen theory and agreed to collaborate with him in devising a new chemical nomenclature. This project, which also included Fourcroy, Berthollet, Hassenfratz, and Adet, resulted in the publication of the well-known *Méthode de nomenclature chimique* by the end of August 1787.[82] The work was a collection of essays, most of which had already been presented at the Académie in 1787. The role played by Guyton in the compilation of this book has been the subject of much conjecture. The most common view attributes the original idea of the work to Guyton, who was already engaged in reforming the nomenclature of chemistry in 1782. Henry Guerlac, for instance, claimed that the *Méthode* was "originally suggested" by Guyton;[83] Denis Duveen and Herbert Klickstein also believe that "its development was due primarily to the ef-

---

[82]On the *Méthode*, see D.I. Duveen and H. Klickstein, *A Bibliography of the Works of Antoine Laurent Lavoisier 1743–1794* (London, 1954), 119–154.
[83]Guerlac, *DSB*, vol. 7, 80.

forts" of the chemist from Dijon.[84] By contrast, George Bouchard,[85] William Smeaton, and Maurice Crosland[86] have been either more cautious or more inclined to give the credit for the idea to Lavoisier. The puzzling question of whether Guyton actually inspired and carried out most of the reform of chemical language is complicated by the fact that the only surviving testimonies to the origin of the *Méthode* are the *procès-verbaux* of the Académie and a letter from Lavoisier to Meusnier. From the *procès-verbaux*, we know that on April 18, 1787, Lavoisier read the first memoir on the necessity of reforming chemical nomenclature; that this was followed on May 2 by a presentation of the new names by Guyton; that on May 5 Hassenfratz and Adet presented the new symbols; that on June 13 Darcet reported on the nomenclature, and on the same day, the table of nomenclature was presented; that this was followed on June 27 by a report on the new symbols read by Fourcroy. The printed work was finally presented to the Academy by Lavoisier on August 29, 1787.[87]

The letter from Lavoisier to Meusnier, dated March 4, 1787, is more informative. In it Lavoisier confirmed that Guyton was by then in Paris to help him devise a new nomenclature, but that he had not yet been converted to the new theory. Lavoisier added that he was confident that if he could keep Guyton in Paris for two more months, he would be able to bring him round to his way of thinking.[88] The fact that Guyton had not yet accepted Lavoisier's theory when the plans for a new nomenclature had already been outlined makes it difficult to believe that Guyton was the major influence in the publication of the

---

[84]Duveen and Klickstein (1954), 119. Although slightly more prudent, J.R. Partington, *A History of Chemistry*, vol. 3 (London, 1961), 481, also seems to suggest that the role played by Guyton was the more important.

[85]"Sous l'influence de Lavoisier, de notables modifications avaient été apportées au projet primitif. La Méthode de nomenclature était plus complète, plus précise que le projet primitif de Guyton, dans la distinction entre les corps simples et les corps composés et la classification de ceux-ci d'après leurs propriétés générales. Elle créait de noms nouveaux pour les corps récemment découverts ou étudiés, en particulier les gaz oxygène et hydrogène," from G. Bouchard (1938), 159–160.

[86]According to Smeaton "L.B. Guyton de Morveau 1737–1816, *Ambix*, IV (1957), p. 26, the role of Guyton did not go beyond collaboration. More explicitly Crosland (1978), p. 174, claims that "the chairman [of the work] was Lavoisier rather than Guyton."

[87]This chronology is taken from Daumas (1955), 60–61.

[88]The letter, which will be published in LC, vol. V, says: "M. de Morveau est dans ce moment à Paris et nous profittons de cette circonstance pour travailler avec lui [sur] une nomenclature chimique. C'est peut etre ce qu'il y a maintenant de plus pressé pour l'avancement de la science. . . .

Si M. Morveau reste deux mois à Paris il sera converti."

Méthode, which, it is worth emphasizing, was the first systematic manifesto of the new chemistry. Furthermore, there are no known letters by Guyton prior to the summer of 1787 that support the hypothesis that he had ever envisaged a new nomenclature before that date. On the contrary, all the evidence indicates that the *Méthode* was a work mainly conceived and inspired by Lavoisier. In the second reprint of the first edition of his *Traité élémentaire de chimie* (1789), Lavoisier added a third volume containing a reprint of the *Méthode*.[89] This suggests that he regarded the *Méthode* mainly as his own work and that he was implicitly entitled to the copyright. Interestingly none of his colleagues ever took the step of reprinting the whole work, although Fourcroy included a part of it in the fifth volume of his *Élémens d'Histoire Naturelle et de Chimie* (1789) in its third edition.[90] The strongest evidence supporting this thesis, however, lies in the consistency of Lavoisier's intention to include gases in the definition of chemistry, which he first declared in 1773. The urgent preoccupation with defining the aerial substances chemically was, as we have seen, an original idea proposed by Lavoisier, who after inventing the notion of *oxygen* increasingly extended the compass of this one name to a larger number of chemical objects. Guyton's nomenclature of 1782 was by contrast focussed on inorganic chemistry rather than on pneumatic objects.

This crucial difference suggests quite strongly that Lavoisier intended the new nomenclature to play its part in redefining chemistry to take account of the new pneumatic discoveries and that it is unlikely that Guyton, who until March 1787 still clung to the idea of phlogiston, would have agreed to support such a plan. As I proceed in this chapter, other evidence will confirm the paternity of the new nomenclature. But let us now contemplate the editorial structure of the work. The *Méthode de nomenclature chimique*[91] is a collection of monographs and a dictionary of chemistry. The first memoir is by Lavoisier and bears the title *Mémoire sur la nécessité de réformer & de perfectionner la nomenclature chimique* (pp. 1–25); the second, *Mémoire sur le dévelop-*

---

[89]See Duveen and Klickstein (1954), 174.

[90]*Ibid.*, 130.

[91]The full title page of the work is: *Méthode de Nomenclature Chimique, Proposée par MM. Morveau, Lavoisier, Bertholet & de Fourcroy. On y a joint Un nouveau Systême de Caractères Chimiques, adaptés à cette Nomenclature, par MM. Hassenfratz & Adet. A Paris, Chez Cuchet, Libraire, rue & hôtel Serpente, M. DCC. LXXXVII., Sous le Privilège de l'Académie des Sciences.*

The fact that the name of Guyton came first on the title page has led many historians to the conclusion that he was the main author of the *Méthode*. In reality the order of the names follows the ages of the collaborators: Guyton (born in 1737), Lavoisier (1743), Berthollet (1748), and Fourcroy (1755).

pement des principes de la Nomenclature méthodique (pp. 26–74), is by Guyton; the third, *Mémoire Pour servir à l'explication du Tableau de Nomenclature* (pp. 75–100), is signed by Fourcroy. The three introductory memoirs are followed by a *Synonimie Ancienne & Nouvelle, par ordre alphabétique* and by a *Dictionnaire Pour la nouvelle Nomenclature chimique* (pp. 107–237). Baumé, Cadet, Darcet, and Sage, all supporters of the phlogiston theory, wrote a *Rapport Sur la Nouvelle Nomenclature* (pp. 238–252); two young pupils of Lavoisier, Hassenfratz and Adet, wrote two *Mémoires sur de nouveaux Caractères à employer en Chimie* (pp. 253–287), and the work concludes with a *Rapport sur les nouveaux Caractères chimiques* (pp. 288–312) signed by Lavoisier, Berthollet, and Fourcroy.

THE PHILOSOPHICAL ROOTS OF OXYGEN

The theoretical sources and the methodological purpose of the *Méthode* were outlined by Lavoisier in the first memoir of the work as follows: "After having profoundly meditated on the metaphysics of languages, and on the relation between the ideas and the words, we have dared to form a plan."[92] This is an immediate statement that the *Méthode* was a project carried out *after* reflection *a priori* on the gnoseological and scientific value of language and that, in other words, the technical problems related to the construction of a new nomenclature were subordinate to the assumption of certain principles. The linguistic principles set out by Lavoisier were intended to be the guidelines for the new chemical nomenclature, and although they appeared abstract and very far removed from the chemical laboratory, they were claimed to engender an effective technical language.

Lavoisier begins his memoir by explaining in detail his belief in the "metaphysic of language":

> Languages are intended, not only to express by signs, as is commonly supposed, the ideas and images of the mind; but are also analytical methods, by means of which, we advance from the known to the unknown, and to a certain degree in the manner of the mathematicians: let us try to demonstrate this idea.

---

[92] *Méthode de nomenclature chimique* (1787), 5. For the English translations, I have used St. John's translation of 1788, henceforth *Method* (1788). Hereafter I will give the pagination of both the French and the English edition of the *Méthode*.

> Algebra is the analytical method by excellence; it has been invented to facilitate the operations of the understanding, and to render reasoning more concise, and to contract into a few lines what would have required whole passages of discussion; in fine, to lead, in a more agreeable and laconic (*plus commode & plus sûre*) method, to the solution of the most complicated questions. Even a moment's reflection is sufficient to convince us that algebra is in fact a language: like all other languages it has its representative signs, its method, and its grammar, if I may use the expression: thus an analytical method is a language; a language is an analytical method; and these two expressions are, in certain respects, synonymous.[93]

These ideas, as Lavoisier immediately acknowledged, had first been presented by Étienne Bonnot de Condillac. Since the philosophy of language has rarely been studied from the point of view of its influence on eighteenth-century science, it may be in order briefly to recapitulate its genealogy and contents.[94]

The influence of Condillac's philosophy on the French Enlightenment had been enormous, and his philosophical doctrine, known as *sensisme*, can be found in most texts of the *philosophes*. A friend of Diderot and Rousseau, Condillac published his first work, the *Essai sur l'origine des connoissances humaines*, in 1746; in it he explained the principles of *sensisme*. It has been argued that Condillac's philosophy was not very original and that its success in France was as a vehicle for Locke's empiricism. However, it must be borne in mind that Locke's philosophy had already been known in France since his *Essay Concerning Human Understanding* was translated into French at the beginning of the eighteenth century. Undoubtedly Condillac's philosophy was influenced by Locke and its doctrine of the empirical origin of human understanding certainly derived from the works of the English philosopher. Had Condillac merely paraphrased Locke's empiricism, however, it would be difficult to understand the impact of his philosophy on the French Enlightenment. In reality the reason for its success is to be found in the *method* and the philosophy of

---

[93] *Méthode* (1787), 6–7; *Method* (1788), 4–5.

[94] On the influence of Condillac's thought on the French scientific Enlightenment, see the following studies: M.P. Crosland, "The Development of Chemistry in the Eighteenth Century," in *Studies on Voltaire and the Eighteenth Century*, XXIV (1963), 369–441; C.C. Gillispie, *The Edge of Objectivity* (Princeton, 1960), 202–259; F. Dagognet, *Tableaux et langages de la chimie* (Paris, 1969), 15–55; W.R. Albury (1972); J. Lambert, "Analyse lavoisienne et chimie condillacienne," in Jean Sgard (ed.), *Condillac et les problèmes du langage* (Geneva, 1982), 369–377. Although it is explicitly devoted to the examination of this problem, the study by G.W. Anderson (1984) is more a reevaluation and reassessment of Foucault's thesis than a useful historical investigation.

language he presented, rather than in the doctrine on which they were based.[95] Condillac's ideas on the origin and structure of language, which, as we shall see, were completely different from Locke's, inspired at least two generations of French philosophers and scientists. The publication of the *Essai* in 1746 aroused a fierce debate on the origin of language that engaged, among others, Turgot, d'Alembert, Diderot, Rousseau, Maupertuis, Euler, Condorcet, A. Smith, de Brosses, du Marsais, Batteux, Buffon, Herder, and Monboddo. Between 1746 and 1780, the question of language and its origin became the main topic of European philosophical and scientific discussion, to such an extent that in 1772 the Academy of Sciences in Berlin arranged a competition on the subject (eventually won by Herder). The enormous popularity of Condillac's ideas is also illustrated by the fact that its 137 pages make the article *Condillac* in the *Encyclopédie Méthodique* by the materialist philosopher Naigeon one of the longest of the entire work.[96] For a better understanding of the sources of this success, it is therefore necessary to examine Condillac's philosophy of language at some length.

According to Condillac, language has an intrinsic gnoseological value, and it is only by its use that we are able to acquire and coordinate all possible knowledge. To demonstrate the essential nature of this cognitive property, Condillac proposes to reconstruct the development of language since its first appearance in human history. Using the *theoretical* or *conjectural* approach that was very popular during the Enlightenment Condillac tried to reveal the origin of human understanding and language, compensating for the lack of historical records or direct testimony with demonstrations of the present morphological structure of language. In a few words, having collected as much information as possible on the structure of language, Condillac tried to visualize its history and genealogy.[97] Since no records were left of the first steps of human

---

[95]On Condillac's philosophy of language, see the definitive study by N. Rousseau, *Connaissance et langage chez Condillac* (Geneva, 1986). Good monographic studies on Condillac are: M. Dal Pra, *Condillac* (Milan, 1942) and I. Knight, *The Geometric Spirit: The Abbé Condillac and the French Enlightenment* (London, 1968). For a bibliography and a reconstruction of the diffusion of Condillac's works, see J. Sgard (ed.), *Corpus Condillac, 1714–1780* (Geneva-Paris, 1981).

[96][Naigeon (ed.)], *Encyclopédie Méthodique—Philosophie*, vol. 2 (Paris, 1792), 1–137. In addition to this, Pierre Joseph Lacretelle, editor of the volumes of the *Encyclopédie Méthodique* devoted to *Logique, Métaphysique et Morale*, reprinted most of Condillac's philosophical works in the first volume.

[97]The most revealing example of this conjectural reconstruction was given by Condillac in his *Traité des sensations* (1754), where to demonstrate the sensistic origin of human understanding it was assumed that a statue was provided with human senses. This image or method became so successful that it was also used by Buffon, Bonnet, and Diderot.

language, the *mental experiment* proposed by Condillac was the only possible path of historical reconstruction.

Condillac saw the history of language as interwoven with that of human needs and passions.[98] The need to convey, for instance, warning of imminent danger to other human beings gave rise to cries or analogous sounds. "The use of those signs insensibly enlarged and improved the operations of the mind,"[99] says Condillac, and allowed men to communicate their immediate needs and passions. However, the improvement of language was extremely slow, and "the organ of speech was so inflexible that it could not easily articulate any other than a few simple sounds."[100] Condillac therefore connected the development of human language with the morphology of the human body and assumed that the biological evolution of "the organ of speech" was accompanied by a corresponding development in the articulation of language. This connection implied that there was a causal interrelation between articulate languages and nature and consequently that human language could never have been entirely the product of either human conventions (as Locke believed) or imagination (as Descartes believed). In the sequel to his *Essai*, Condillac tried to provide evidence of this linguistic realism by showing that the earliest civilizations still displayed signs of an original and natural *language of action*. Thus, even with the appearance of spoken language, the word "succeeding the language of action, retained its character [. . .]. In order to take the place of the violent contortions of the body, the voice was raised and depressed by a very sensible interval."[101] The prosody of the Greeks and Romans, for instance, was often marked by exclamations; gestures stemmed from the primitive form of the language of action. This historical evidence lent weight to the thesis that the essence of language was tied to human morphology and character. Having demonstrated the *natural* origin of human language, Condillac considered the origin of words in the same light:

> Language was a long time without having any other words than the names which had been given to sensible objects, such as these, *tree, fruit, water, fire,* and others.[102]

---

[98]I have used here the English translation by Nugent, *Essay on the Origin of Human Knowledge being a Supplement to Mr. Locke's Essay on the Human Understanding* (London, 1756), a translation that is unfortunately full of omissions and not always accurate.

[99]Condillac (1756), 174.

[100]*Ibid.*

[101]*Ibid.*, 179–180.

[102]*Ibid.*, 238.

Once Condillac's hypothesis on the sensory origins of language was accepted, it necessarily followed that the first words ever uttered by human beings corresponded to their simple impressions and sensations. Only by overcoming this direct relationship between sensation and substantive were verbs, adjectives, and other abstract forms of the language introduced. However, even these more complex forms were initially "*derived from the first names that were given to sensible objects,*" and since "people never understood one another better, than when they gave names to sensible objects,"[103] Condillac claimed that this primitive form of communication was free from misunderstandings.

If the language of our ancestors was so efficient and easy to understand, one may wonder what led to all the ambiguities, metaphysical disputes, and meaningless terminology of modern language. Condillac's answer to this question reveals the originality of his thinking. He believed that linguistic errors first made an appearance when man began to derive reality from language rather than vice versa. Instead of following the empirical path imposed by the primitive language of action, man preferred to coin new words from his own imagination and erroneous ideas. Ceasing to verify the connection between words and "sensible objects," man began to believe that words had a power of their own and that their values were independent of the external world. This belief led to the never-ending disputes on semantics during the Middle Ages, the ultimate outcome of which was the total detachment of language from reality. For this problem Condillac had a simple remedy: "If error owes its origin to the defect of ideas, or to ideas not properly determined, truth must arise from determined ideas. Of this we have a proof in mathematics."[104] Mathematicians are able to compose and decompose their expressions and constantly verify the truth of each passage, "now in all sciences whatsoever, as well as in arithmetic, the only way to come at the truth is by" composing and decomposing.[105] Thus the only method of attaining the truth was, in Condillac's view, the analytic one. The success of this method was evident from its effectiveness in mathematics, but the real and major advantage of analysis as a method lay in the fact that it was a natural one. To demonstrate this property, Condillac invites his readers to conduct another mental experiment:

> If God were to create an adult person, with organs so perfect, that the very first moment of his existence he enjoyed the full use of reason,

---

[103]*Ibid.*, 253.
[104]*Ibid.*, 304.
[105]*Ibid.*, 305.

> this man would not meet with the same difficulties as we in the investigation of truth. He would invent no signs but in proportion as he experienced new sensations, and made new reflections. His first ideas he would combine according to the circumstances in which he found himself; he would determine each collection by particular names; and when he wanted to compare two complex notions, he might easily analyze them, because he would meet with no difficulty in reducing them to the simple ideas with which he himself had framed them. Thus as he invented words only after he had framed his ideas, these would be always exactly determined, so that his language would not be subject to the obscurity and ambiguity which prevails in ours.[106]

But since "simple ideas can never occasion any mistake,"[107] Condillac seems to envisage the possibility of constructing a perfect language based on analysis. At this point it is important to emphasize the main source of this idea, namely, Cartesian philosophy. In declaring that the only way to search for truth was to use the analytic method, by "ascending" to the origin of ideas, reducing them to their simplest parts, and finally reassembling them, Condillac was literally following the path indicated by Descartes in the sixth rule of his *Regulae ad directionem ingenii* (1629), where he stated:

> for the distinguishing of the simplest things from those that are complex, and in the arranging of them in order, we require to note, in each and every series of things in which we directly deduce truths, which thing is simplest, and then to note how all the others stand at greater or lesser or equal distance from it.[108]

The urge for a rational method that enables human understanding to classify reality in its true order undoubtedly makes Condillac more an heir to Descartes than to Locke. This apparently paradoxical statement becomes less surprising if we compare Condillac's philosophy of language with the one propounded by Locke. Locke examined the structure and properties of language in the third book of his *Essay Concerning Human Understanding* (1690). Like Thomas Hobbes before him, Locke believed that the relationship between words and ideas was totally arbitrary and conventional, whereas the relationship between ideas and things was natural. He considered ideas generally to be the impressions of external things on our senses. In short, ideas were nothing more than sensations. In contrast to this natural connection:

---

[106] *Ibid.*, 319–320.
[107] *Ibid.*, 321.
[108] Descartes, *Philosophical Writings*, Norman Kemp Smith (ed.) (London, 1952), 25.

> Words, by long and familiar use, as has been said, come to excite in men certain ideas, so constantly and readily, that they are apt to suppose a natural connection between them. But that they signify only Men's peculiar ideas, and by a *perfectly arbitrary imposition*, is evident, in that they often fail to excite in others (even that use the same language) the same ideas we take them to be the signs of. . . .[109]

Although Locke believed that human language could be perfected and, if appropriately used, adequately express ideas, this passage confirms that he projected the faculty of achieving the truth onto the *idea* or sensation alone and considered language a relative, conventionalized, and often deceptive means of expression. Locke strongly criticized the "abuse of words" and the structural imperfection of words, and he seemed to believe that the defects of human language were due to the imperfect state of human knowledge. In consequence of this structural imperfection:

> Where men in society have already established a language amongst them, the significations of words are very warily and sparingly to be altered: because, men being furnished already with names for their ideas, and common use having appropriated known names to certain ideas, an affected misapplication of them cannot but be very ridiculous. *He that has new notions, will, perhaps, venture sometimes on the coining of new terms to express them: but men think it a boldness, and it is uncertain whether common use will ever make them pass for current.*[110] [my italics]

The extreme caution and freely acknowledged reluctance with regard to the introduction of new words into the lexicon of a language was a characteristic that eventually became the hallmark of English empiricism. The distaste for new terms is particularly typical of English scientists. Robert Boyle, who was a friend of Locke, heavily criticized the abuses of chemical and alchemical language, but, as we saw in Chapter Three, he was loath to introduce new terms.[111] A century later Joseph Priestley displayed the same attitude. However, what should be

[109]J. Locke, *The Works*, 6th edition (London, 1759), vol. 1, 185–186.

[110]*Ibid.*, 218. The Scottish chemist and physician William Cullen supported Locke's ideas on language and his prudence in the introduction of new names and remarked: "If upon any occasion one single term occurs which does not give you a clear idea, rest not till by considering your notes, reflecting on what you hear, or inquiring among your fellow students, you become perfectly acquainted with its meaning," quoted by Crosland (1963), 414–415.

[111]This attitude can be traced back to Bacon's and Hobbes' works in which language was regarded as a subordinated and mostly deceptive instrument of the human mind.

noted here is the profound difference between Locke's position and that taken by Condillac in 1746. The French philosopher considered the creation of words a process simultaneous with the creation of sensations and ideas. With respect to *numbers*, for instance, "we must not deceive ourselves, by imagining that the ideas of numbers, separated from their signs, are something clear and determinate."[112]

To sum up, the *Essai* successfully connected three ideas that belonged to the Cartesian tradition of thought rather than to the Lockean: 1. the only method of reasoning is by analysis, and it proceeds from the simplest elements of our ideas or sensations to more complex and abstract combinations of them; 2. "the ideas are connected with the signs, and it is only by this means [. . .] that they are connected to each other";[113] 3. the effective means of connecting our ideas with signs is to be found in the method used by mathematicians.

The ideas on language outlined in the *Essai* were further developed in the *Logique* (1780), a work that constitutes a landmark in the history of French life sciences. Since it had a direct influence on Lavoisier's reasoning concerning chemistry, an accurate description of its contents is needed. The *Logique* was intended to be a manual that set out the basic principles of the art of reasoning according to *sensisme*.[114] The first part of the work covers "how Nature itself teaches us Analysis; and how the origin and generation both of ideas and of the faculties of the soul are explained according to this method."[115] Condillac thought that the first elements of any possible knowledge were the sensations and the faculties by which they were perceived, i.e., the human sensory organs. These faculties are determined and developed by human needs and the needs train men to develop their first perception of an external world. "But because these needs and faculties depend upon the organization of the animal's body, and vary with it, it follows that we understand by the nature of an animal the conformation of its organs: and in fact that is just what nature is in its principle."[116] This important passage, reminiscent of Descartes' thinking on the mind-body relation-

[112]Condillac (1756), 116.

[113]*Ibid.*, 7.

[114]The *Logique* was printed for the first time in Paris in 1780 under the title of *Logique, ou les premiers developpemens de l'art de penser. Ouvrage élémentaire.* The enormous success of this work is shown by its 28 editions between 1780 and 1834. The work was translated into most European languages. I have used here the translation made by W.R. Albury, Condillac, *La Logique—The Logic* (New York, 1980), who wrote an informative introduction and annotated the text.

[115]Condillac, *The Logic* . . . (1980), 45.

[116]*Ibid.*, 51.

ship, links the growth of human knowledge with the conformation of the human body more clearly than the *Essai.* This connection does not merely imply that Condillac externally supports a form of naturalism, but, more importantly, that he finds it possible to establish *objectively* the origins of human knowledge using naturalistic arguments. Nature provides us with our first notions of the external world and through our sensations of pain and pleasure, we are taught by her to judge what to avoid and what to look for. But men grow up, and nature ceases to warn us of the dangers, and our judgments begin to fail. Accordingly errors begin "when, by judging of matters which have little relation to our primary needs, we are unable to test our judgments in order to see whether they are true or false."[117] To avoid these errors, we have to return to the path indicated by nature during our childhood: we have to apply to secondary needs what we applied to our primary needs. The method to which Condillac refers is obviously that of analysis, which he sees as the only method of acquiring knowledge. It involves a few simple principles and mental operations:

> To analyse, then, is nothing other than observe the qualities of an object in a successive order so as to give them, in our mind, the simultaneous order in which they exist outside it. Therefore analysis, which people think is known only by philosophers, is actually known to everyone; and I have taught the reader nothing. I have simply made him note what he continually does.[118]

It is almost superfluous once again to comment on the influence of Descartes in Condillac's belief that with the help of analysis truth was accessible to everybody. If the theory of knowledge on which Condillac relied was Locke's empiricism, the philosophical method he adopted was Cartesian, and it was on the latter that he largely based his system. This becomes clear if we continue to examine Condillac's philosophy of method. Condillac's definition of analysis implies that to acquire knowledge one must proceed "from the known to the unknown."[119] A *logical* consequence of following this method is that all kinds of hypothetical, metaphysical, and arbitrary assumptions are eradicated, and we are left with simple and unmistakable sensory data on which to build our knowledge. This method should also be applied to politics and morals: these matters, Condillac admitted, are "conventions we have made: nevertheless we have not made them alone; *nature made them with us, nature dic-*

[117]*Ibid.*, 57.
[118]*Ibid.*, 63 and 73.
[119]*Ibid.*, 87.

tated them to us, and it was not in our power to make other conventions."[120] Thus Condillac's natural ontology also included matters that were formerly considered either the effect of God's design (Leibniz) or the fruit of human conventions (Locke). Condillac's political determinism does not fall within the scope of our study, but it is worth repeating that like Descartes' Condillac's naturalism implied an investigative principle applicable to all fields. Condillac projected onto nature what Descartes projected onto God, but both philosophers saw mechanism, analysis, and determinism as the paths to the truth.

Once he had based the *Logique* on these naturalistic principles, Condillac went on to examine "analysis considered in its means and its effects, or the art of reasoning reduced to a well-made language," in the second part of his work.[121] According to Condillac, language is more than a useful implement of analysis because men "can analyse only by means of a language."[122] Indeed, the only way of conceiving the breaking down of an object is to assign to each of its parts a different sign that enables us to distinguish them. But if it is true that men "think only with the aid of words,"[123] the function of language also surpasses the faculties of the senses in cognitive power. This judgment, which would have been unacceptable to Locke, followed from Condillac's belief that "man is born with the elements of the language of action, and these elements are the organs which the Author of our nature has given us. *Hence there is an innate language, even though there are no innate ideas*" [my italics].[124] The origin of both philosophical and scientific errors is in the failure to recognize the cognitive value of language and in using it as a tool subordinate to our intelligence.

The originality of Condillac's position lies in his theory of knowledge, in which he reversed the place assigned to logic by modern philosophers and asserted that logic preceded knowledge. It is also interesting to note that, although relying on empirical principles, Condillac admitted the existence of the *innate* elements of language as a preliminary condition of human understanding.

Before Condillac, philosophy of language had developed in two directions: Descartes' rationalism and Bacon's empiricism,[125] both para-

[120]*Ibid.*, 127.
[121]*Ibid.*, 189.
[122]*Ibid.*, 211.
[123]*Ibid.*
[124]*Ibid.*, 213.
[125]On the origin and early development of modern philosophy of language, see André Robinet, *Le langage à l'âge classique* (Paris, 1978).

doxically sharing the same opinion of the function of language in philosophy and science. Descartes made language subordinate in value to ideas and thought that a clear and distinct perception of things alone was enough to ensure the attainment of the truth; Bacon, for his part, considered language a mere convention that usually entailed errors and unreliable hypotheses, and before words he placed the power of facts, empirical observations, and experiments.[126] Malebranche and the Port Royal School, on the one hand, and Hobbes and Locke, on the other, developed the opinions of their masters without introducing any fundamentally new ideas on the gnoseological function of language. Thus, before Condillac, language was thought of mainly as a means of communication that if used properly could represent either our sensations or our ideas.

As the first to assign language a privileged position in a philosophical system, Condillac has an extremely important position in the history of philosophy, and as we have seen, most of his contemporaries were influenced by his ideas. Condillac's philosophy of language is neatly and precisely summarized in the formula: "the art of reasoning is reduced to a well-made language."[127] Once the ontological principle is stated, the remaining problem is to define the features of such a "well-made language." Languages are analytic methods and "were exact methods so long as people spoke only of things which related to their primary needs."[128] When the primary needs were satisfied, men began to use words with increasing freedom and did not feel the need to connect the signs to the sensations to which they were supposed to refer. It then became far easier to forget that language is an analytic method that follows certain rigid paths. Only in mathematics, physics, and, to a certain degree, chemistry were the use of analysis retained. However, "the languages of the sciences are not best made," and the positive results they achieved in their way of expression were due only to "analysis which forms languages and creates the arts and sciences,"[129] and not vice versa. After these general observations, Condillac provides his readers with some more concrete examples of what he means by analytic language. Mathematicians are among the few who put these

---

[126]On Bacon's philosophy of language, see Paolo Rossi, *Clavis Universalis*, 2nd edition (Bologna, 1983), 162–197. On English linguistics during the seventeenth century, see the excellent study by Lia Formigari, *Linguistica ed empirismo nel Seicento inglese* (Bari, 1970).

[127]Condillac (1980), 249.

[128]*Ibid.*, 229.

[129]*Ibid.*, 237 and 255.

principles into practice effectively and create a language accordingly. Euler and Lagrange, in particular, demonstrated[130] that the analytic method provided mathematics with a method of discovery that was far superior to the synthetic method that laid down the principles of mathematics in *a priori* definitions and axioms.

This distinction had already been made in the same terms by Descartes more than a century earlier in his reply to the *Secondes Objections* of the *Méditations metaphysiques*, in which he encouraged philosophers to use the analytic method rather than the synthetic one. Analysis showed the paths of reason and was easier to teach beginners, whereas synthesis gave a demonstration without showing the path that led to it. Furthermore, the progression of ideas entailed in the analytic method had the advantage of revealing the presence of errors more easily and reliably.[131]

At this point there can no longer be any doubt that the philosophy envisaged by Condillac was inspired by Descartes' wish to construct a general method of investigating nature, rather than by Locke's more modest desire to ascertain the limits to human understanding. Like Descartes, Condillac did not consider algebra an empty scheme of conceptual representation, but saw it as an effective method of reflecting the real structure of the external world; moreover algebra was a method that showed the paths followed by nature itself.[132] Reasoning was nothing more than: 1. the enunciating of the data or the state of a problem; 2. the disengaging "of the unknowns, or the actual reasoning itself."[133]

---

[130]Condillac (see pp. 279–281) referred to the *Éléments d'Algèbre* of Leonard Euler, published in Lyon in 1744 with additional commentaries by Lagrange. Condillac praised Euler and Lagrange for having preferred the analytic method in the treatment of algebraic problems. By contrast he criticized the French mathematician Clairaut, who followed the synthetic method.

[131]Descartes thought that even those who declared that they were using the synthetic method were, in reality, implicitly relying on the analytic one. "Les anciens Geometres avoient coutume de se servir seulement de cette synthese dans leurs écrits, non qu'ils ignorassent entierement l'analyse, mais, à mon avis, parce qu'ils en faisoient tant d'état, qu'ils la reservoient pour eux seuls, comme un secret d'importance.

Pour moy, j'ay suivy seulement la voye analytique dans mes Meditations, pource qu'elle me semble etre la plus vraye, et la plus propre pur enseigner," *Oeuvres de Descartes*, Adam and Tannery (eds.), vol. IX, (Paris, 1904), 122.

[132]It is evident that Condillac's ideas on mathematics were far from Newton's. The English naturalist in fact assigned to mathematics the limited role of explaining a restricted number of physical phenomena. For Condillac, as for Descartes, mathematics was the only method for deducing the whole of nature.

For a concise history of the notion of analysis, see Albury (1972), 31–74.

[133]Condillac (1980), 307.

Thus if the algebraic method is followed, "the self-evidence of the reasoning process consists solely in the identity which is displayed from one judgment to the next,"[134] just as in a mathematical equation. Condillac's imaginative program reached its ultimate conclusion when he stated that any kind of problem could be expressed in "algebraic signs" and solved like an ordinary equation. Although it was never explicitly stated, Condillac thought of building a universal grammar based on algebra, capable of guiding human reasoning in all circumstances. This project was outlined in the *Langue des Calculs*, on which the French philosopher was working until just before his death.[135] Here Condillac set out his concept of language very clearly, stating that the elements of language were innate and that algebra was a natural language to which "nothing arbitrary" was admitted.[136] To illustrate the natural origins of algebra, Condillac considered the origin of numeration. The morphology of man's hands led man to his first calculation, which assigned a unit to each finger. The act of counting was therefore not the product of human convention, nor the discovery of human genius, but something inherent in the human body.[137] Like algebra, this primitive act of reckoning took man from the known to the unknown, from the first simplest forms of numeration to more abstract and complex mathematical expressions. This belief implies that "we see the unknown in the known itself"[138] and that consequently the unknown is nothing else than the known.[139] From this it follows that any science reduced to an original truth (*vérité originaire*) could be transformed into a series of identical propositions as in an equation and provide us with every possible indication of how to achieve new and original results. Condillac believed that mathematical equations could also lead to truths other than numerical ones with the same rigor,[140] and that by following the analytic method systematically man could reach the ideal degree of accuracy in which

---

[134] *Ibid.*, 299.

[135] *Idem*, *Langue des calculs*, in *Oeuvres Philosophiques*, vol. 2, 418–529.

[136] *Ibid.*, 420.

[137] *Ibid.*, 421–423.

[138] *Ibid.*, 427.

[139] "C'est que l'inconnue se trouve dans le connu, et il n'y est que parce qu'il est la même chose. Nous ne pouvons donc passer de ce que nous savons à ce que nous ne savons pas, que parce que ce que nous ne savons pas est la même chose que ce que nous savons," *ibid.*, 431.

[140] "J'ai voulu encore faire voir que si, avec les phrases de nos langues vulgaires, on démontre *rigoureusement* des vérités mathématiques, on pourroit, avec les mêmes phrases, démontrer tout aussi *rigoureusement*, des vérités d'un autre genre," *ibid.*, 454.

"the metaphysical analysis and the mathematical analysis [. . .] are precisely the same thing."[141] If all these assumptions were accepted, it followed that "calculer c'est raisonner, et raisonner c'est calculer."[142]

As Jean Dhombres has rightly pointed out, although Condillac did not produce any original mathematical results and his acquaintance with mathematics was restricted to the "simple" textbooks of Euler, Lagrange, and Lacaille,[143] he was the main protagonist of the mathematicization of eighteenth-century science.[144] Condillac's faith in the absolute power of mathematics inspired a whole generation of naturalists to apply more accurate and precise methods in the biological and qualitative sciences. In addition to this important contribution, the linguistic scheme introduced by Condillac could easily be applied to the concrete project of elaborating a scientific nomenclature. Condillac's successful formula can be expressed as follows:

> Nature—elements of language—sensations—ideas—knowledge of Nature

Condillac's ideas on language and gnoseology found an enthusiastic supporter in Lavoisier, who, after reading the *Logique,* constantly tried to apply them to his chemical system. For Lavoisier it was a natural choice to base a new scientific nomenclature on the principles of the new philosophy.

We can now read the contents of the *Méthode de nomenclature chimique* in a new light. In his introductory comments on the work, Lavoisier immediately states:

> An analytical method is a language; a language is an analytical method; and these two expressions are, in a certain respect, synonymous.
>
> This is a fact that has been explained with infinite exactness and perspicuity in the Logic of the abbé de Condillac, a work which can never be too much studied by the youth that dedicate themselves to the sciences, and from which we cannot avoid borrowing a few ideas. He has explained in that work how the language of algebra could be translated into the vulgar tongue and reciprocally; how the progress of the understanding and judgment is the same in both; and how the art of reasoning is the art of analyzing.[145]

---

[141] *Ibid.,* 466.
[142] *Ibid.,* 469.
[143] Jean Dhombres, "La langue des Calculs de Condillac. Ou comment propager la lumière?," in *Science et Technique en perspective* (1985), vol. 2, 197–230.
[144] *Ibid.,* 230.
[145] *Méthode* (1787), 7–8; *Method* (1788), 5.

Lavoisier's equating of the art of reasoning with analysis could only be of benefit to a science like chemistry in which, by definition, the notion of analysis was fundamental. Incidentally it is important to remember that in the *Logique* Condillac remarked that the analytic method was already successfully applied in sciences other than algebra and that "some parts of physics and chemistry have been treated with the same precision by a small number of excellent minds born for observing well."[146]

Lavoisier was well aware that the notion of chemical *analysis* needed an epistemological generalization in order to become a central principle of the science. For this reason he intended to transform its meaning, using it to signify a method of chemical investigation rather than an ordinary laboratory operation and making analysis the foundation of the chemical method.[147] Lavoisier was so firmly convinced of the power of such a semantic extension of the analytic method that he linked the reform of the entire language of chemistry to it:

> This method, which is necessary to be introduced into the study and teaching of the science of chymistry, is closely connected with the reformation of the nomenclature: a language perfectly well composed, a language agreeable to the successive and *natural order of ideas*, must occasion a necessary and immediate revolution in the method of teaching: it will not permit the professors of chymistry to deviate from the line of *nature*; and they must either reject the nomenclature or irresistibly pursue the delineated course.[148] [my italics]

The analytic method was not, as many historians have claimed, merely a means of teaching; it was also an irreplaceable tool "in the study" of chemistry. In introducing the analytic method into chemistry on such a broad front, Lavoisier was aware that he was the first to

---

[146]Condillac (1980), 239. It is impossible to know whether Condillac was referring to Lavoisier's work, but it is interesting to see that when he wrote this passage he had changed his opinion about the possibilities of chemistry. In his *Traité des systêmes* (1749), Condillac was skeptical about a fruitful application of analysis to chemistry and remarked: "En vain, par exemple, le chymiste se flatter d'arriver, par l'analyse aux premiers élémens: rien ne lui prouve que ce qu'il prend pour un élément simple et homogène, ne soit pas un composé de principes hétergènes," Condillac, *Oeuvres*, vol. 1, 197–198.

[147]In her beautiful *La philosophie de la matière chez Lavoisier* (Paris, 1935), 12–22, Hélène Metzger distinguished Lavoisier's *analyse mentale* from his *analyse chimique*. The mental analysis was the logical operation that considered the chemical combination of different substances in terms of numeric proportions of different kinds of matter; the chemical analysis was the concrete and experimental examination of substances.

[148]*Méthode* (1787), 12; *Method* (1788), 8–9.

approach chemistry in that way, and as early as 1780–1781, he justifiably remarked that "no author, no professor of chemistry has seen the science from this point of view."[149] From Lavoisier's point of view, this was certainly one of his most important and original contributions to chemistry. Briefly, he was suggesting that a chemist should: 1. collect the facts; 2. break them down into their simple parts; 3. reassemble them as they were originally. Since, as Condillac had said, no analysis was possible without the use of signs, the act of naming came about between phases two and three. If we compare this methodological scheme with the one proposed by Priestley, it is clear that the English chemist explicitly believed that only the first phase was allowed in experimental sciences and that any activity beyond the collecting of facts was a dangerous concession to metaphysics.

Having demonstrated the essential role of the analytic method in chemistry, Lavoisier distinguished three fundamental elements of a science: 1. the facts that distinguish the domain of each science; 2. the ideas that recall those same facts; 3. the words that express them. Following Condillac's suggestion Lavoisier effectively expressed this approach in the following formula: "the word should give birth to the idea, and the idea should represent the fact." Seen from this perspective, the goal of a chemical nomenclature was a representation of the facts and the ideas in their *true essence* without adding to or subtracting from their *real nature.* The chemical language anticipated by Lavoisier in his introduction was meant to be a "faithful mirror" of chemical reality, rather than merely a convenient and more effective means of expression:

> The perfection of the chymical nomenclature [. . .] consists in rendering the ideas and facts in their exact truth, without suppressing any thing, or making any additions whatsoever; *it should be nothing less than a faithful mirror.*[150] [my italics]

Lavoisier therefore criticized the obscurities of the alchemical lexicon on two different levels. Like Bergman and Guyton before him, Lavoisier pointed out that the terms used in alchemy and chemistry were too often unintelligble and enigmatic, but he also pointed out that this obscurity was due to the lack of an appropriate general method of investigation. He blamed early chemists for being "*systematiques,*" i.e., for

---

[149]"Aucun auteur, aucun professeur de chimie n'a vu la chimie sous ce point de vue," quoted by Albury (1972), 277.

[150]*Méthode* (1787), 13; *Method* (1788), 9–10.

constructing the science more from their own imagination rather than by following paths indicated by nature.

For this, Lavoisier suggested a radical and immediate remedy:

> It is now time to remove the variety of impediments which retard the progress of chymistry; to introduce into the science a true taste for analyzing: and we have sufficiently demonstrated that this reformation should be commenced by bringing the language to a state of perfection. We are very far, without doubt, from knowing the entire extent of all parts of chymistry: it must be therefore expected that a new nomenclature, although formed with all possible attention, must be far from absolute perfection; but provided that it has been undertaken upon good principles, provided that it is rather a *method of naming* than a *nomenclature*, it must naturally adapt itself to future discoveries, and indicate before-hand, the place and name of such new substances as may be found out, and we can never require more than some local and particular reformations.[151]

What impresses a modern reader about this passage is its confidence. Lavoisier asserts that an entirely new nomenclature will last for ever, with only occasional minor amendments. It is striking to see that, so far, he has been proved right and that the language of inorganic chemistry still uses the words and the *method of naming* that he outlined in 1787. This success, almost unique in the history of science, is due to two crucial innovations made by Lavoisier:

1. Lavoisier was more concerned to present the right *method* of reforming technical language than the details of the reform itself. This difference implied the superiority of the method of nomenclature to the previous attempts to reform the lexicon of chemistry. A method was in fact a conceptual tool that enabled scientists to achieve the truth in any situation, whereas the changes to chemical nomenclature made by Bergman and Guyton depended on more relative and circumstantial factors, such as the need to name new substances.
2. Lavoisier's second innovation was the intuition, which in 1787 was a conceptual breakthrough, that all new substances could be named automatically by applying the *permanent* principles set out in the *Méthode.* Interestingly Lavoisier, who was an accurate experimental scientist, shared with Condillac the philosophical belief that the analytic method and its language were a *method of discovery.*

[151] *Méthode* (1787), 16–17; *Method* (1788), 11–12.

Lavoisier concluded his introduction by outlining the specific applications of his method to chemistry. By consistently following the concept of analysis, he immediately made it clear that he regarded as elements those substances "which we cannot decompose; all such as we obtain in the last result from chymical analysis."[152] This definition, which has been interpreted by historians as a simple extension of Boyle's definition of an element, was in reality extremely innovative: with it Lavoisier recognized that *complexity* was an intrinsic property of matter; that the research program which investigated the nature of the elements or principles of chemistry was now obsolete; that the study of matter had been reduced to the investigation of chemical combination and the determination and isolation of chemical reagents.

As Ferdinando Abbri has remarked,[153] the mental shift provoked by Lavoisier's definition of simple substances implied a completely new definition of chemistry, which was now to be regarded as the science of studying chemical *changes* rather than investigating the chemical *composition* of matter. Thus, the notion of analysis proposed by Lavoisier was, in its essential and general sense, very far from the chemical laboratory. It was, on the contrary, a method of *conceiving* all chemical operations *a priori*. Many eighteenth-century chemists before Lavoisier employed the notion of analysis in chemical experiments, but none of them regarded it as the ruling principle of chemical science.

Following the analytic definition of simple substances, Lavoisier delineated the method of classification used in the *Méthode* and remarked:

> In respect to the bodies composed of two simple substances, as their number is very considerable, it appeared indispensibly necessary to arrange them in different classes. According to *the natural order of ideas*, the title of class and genus is that which reminds us of the properties common to a great number of individuals; but the title of species is that which indicates the particular properties of certain individuals. *This natural logic* belongs to all the sciences, and we have attempted to apply it to chymistry.[154] [my italics]

This classification system rested on Linnaeus' binomial nomencla-

[152]*Méthode* (1787), 17; *Method* (1788), 12.

[153]"La chimica di Lavoisier non ha più la pretesa di essere una "fisica" delle origini e della struttura del cosmo e della terra, ma è divenuta una scienza fisico-quantitativa del mutamento chimico," Abbri (1984), 385.

[154]*Méthode* (1787), 19; *Method* (1788), 13–14.

ture and had already been applied to chemistry by Bergman and Guyton in 1775 and 1782. However, it was only Lavoisier who insisted, without hesitation, that the categories of genus and species had a *natural state* and that their combination formed a *natural logic*. This ontological belief in the power of language evokes Linnaeus' linguistic realism (see Chapter Two), and despite the substantial differences between the two scientists, the similarity of their conceptions is not as questionable as it may at first seem.

In the summers of 1763 and 1764, Lavoisier accompanied Bernard de Jussieu, professor of botany at the Jardin du Roi in Paris, "dans ses herborisations."[155] Jussieu, whom Linnaeus designated "Major General" in his *Florae Officiarii*,[156] was the first French botanist to adopt the sexual system and the binomial method of classification. Jean-Etienne Guettard, who suggested Jussieu's excursions to Lavoisier and whose influence on the young chemist was considerable, was primarily a botanist. He was also a devotee of the Linnaean system and he began to compile a Linnaean catalogue of the plants at the *Jardin du Roi* as early as 1758.[157] Contrary to what is commonly believed, most of Guettard's published works deal with botanical rather than geological topics, and he was elected to the *Académie des Sciences* in the section for botany. From these brief biographical particulars of Lavoisier's tutors, we can easily understand that Linnaean botany was an important strand in Lavoisier's later chemical thinking. Since there is no record of Lavoisier's early interest in botany, Guerlac has suggested that "Jussieu's influence was probably slight; botany held little attraction for Lavoisier."[158] But in a detailed analysis of Lavoisier's *Mémoire sur la Nature de l'Eau* (1770), Ferdinando Abbri has shown that the French chemist actually based much of his memoir on the botanical works of Guettard, Bonnet, and Duhamel, and that Jussieu's role was vital in stimulating Lavoisier's interest in classifica-

---

[155]Grimaux (1888), 4. See also R. Rappaport, "Lavoisier's Geological Activities 1763–1792," *Isis*, 58 (1967), 375–384.

[156]*Vita Caroli Linnaei*, Elis Malmeström and Arvid Uggla (eds.) (Stockholm, 1957), 153. In this curious compilation, Linnaeus made a list of the officers of *Flora*, of which he was obviously the general. Bernard de Jussieu was second, and Guettard eleventh in rank. Apart from its anecdotal interest, this list enables us to see the degree to which Linnaeus considered his colleagues faithful to his sexual system.

[157]*Catalogus Plantarum Horti Regiis Parisiensis. Anno Domini 1758*, Arch. Acad., Dossier Guettard, *Cahier* n. 6.

[158]Guerlac, "Lavoisier," in *DSB*, vol. 7, 67. James W. Llana, in his "A Contribution of Natural History to the Chemical Revolution," *Ambix*, 32 (1985), 71–91, recognized the importance of Linnaeus' classificatory system to the chemical nomenclature of 1787, but he did not emphasize the direct and historical connections between the two.

tion and nomenclature.[159] This latter influence is even more important for our purpose, since, as we saw earlier, the problem of nomenclature and classification arose at a very early stage of Lavoisier's chemical thinking. Thus it is likely that Jussieu and Guettard, both fervent linnaeans, considered the problem of nomenclature an important part of their scientific investigations and passed on this belief to their pupil. In this respect it is worth recalling again that in December 1787 Lavoisier was among the first to join the *Société Linnéenne de Paris* (see Chapter Three).

These being his early sources, we can now better understand Lavoisier's belief in the *natural* status of the classificatory categories of *genus* and *species*. Yet, the differences between Linnaeus' and Lavoisier's linguistic realism were in other respects considerable. Linnaeus based his philosophy of language on the metaphysical *credo* of the fixity of species. His logic was accordingly the *static* operation of attributing a name to each species. Lavoisier, by contrast, based his realism on the linguistic principle laid down by Condillac, which stated that scientific language was the result of a historical *method* rather than of a static *act* of naming.

Accordingly we find in Lavoisier's nomenclature names like oxygen, whose semantic power went far beyond the immediate representation of one individual substance and fundamentally affected the perception of all chemical matter.

The Principles in Action

The principles outlined by Lavoisier were applied in the memoirs by Guyton de Morveau and Fourcroy that followed. Guyton divided Lavoisier's combinatorial system of matter into five classes: 1. the principles (*principes*) or simple substances; 2. the acidifiable bases or radicals of the acids; 3. the metals; 4. the earths; 5. the alkalis.

The principles of the first class were "la *lumière*, la *matière de la chaleur*, l'air appellé *déphlogistiqué*, puis *air vital*, le *gaz inflammable* & l'*air phlogistiqué*,"[160] and their nature was extremely controversial.

Guyton admitted that light and heat seemed to be two substances that produced the same effects, and that the experimental knowledge

---

[159]"Affermare che Guettard (geologo) e Rouelle (chimico) hanno avuto influenza su Lavoisier perché si occupò di geologia e di chimica, ma che un'influenza non sussite nel caso di de Jussieu perchè non si occupò della classificazione delle piante significa ridurre la ricchezza tematica delle scienze sperimentali e naturali del Settecento e considerarle sulla base delle codificazioni attuali," Abbri (1984), 84–85.

[160]*Méthode* (1787), 30.

of their inner properties was still too limited to establish their mutual relations and differences. Despite this fact, Guyton remarked that the traditional name of *matière de la chaleur* to designate heat was unclear and inaccurate and at Lavoisier's suggestion replaced it with *calorique*. This substance was in fact supposed to be the material principle of the sensation of heat, and by naming it *calorique* (caloric), Lavoisier expressed more precisely its hypothetical quality. By contrast, the name of light remained unaltered. For the other three substances, the case for a radical change of their names was supported by more compelling experimental evidence. The name *air dephlogistiqué* was probably the most controversial one, since it evoked Stahl's phlogiston theory. In 1780 Condorcet suggested a change to the more neutral *air vital*, and Lavoisier initially supported it,[161] but it was still too vague a description of the nature of this gas, and it referred only to its property of being the respirable part of the atmosphere. It was necessary to find a word that expressed the *essential* nature of this gas rather than a mere approximation of it. Lavoisier observed that the substance was present in many acids and that in combination with numerous substances it engendered the state of acidity. Extending these empirical observations into a general principle, Lavoisier created the name *oxygène*, derived from the two Greek words *oxis* (acid) and *geinomai* (to be engendered).[162] Thus the substance that was named *dephlogisticated air* by Priestley in 1774, *Feuerluft* by Scheele in the same year, *air éminemment respirable* by Lavoisier in 1777, and *air vital* by Condorcet in 1780, was now defined by the most essential of its properties, i.e., as the universal principle of acidity.

Applying the same linguistic principle, the name *gaz inflammable* was changed to one more appropriate to its inner nature. Guyton admitted that it was a property of this fluid to be inflammable, but he remarked that the property was not exclusive to it. It had recently been observed that by combining this gas with *oxygène*, water was produced, and it was therefore decided that its name should be derived from this extraordinary experiment. Accordingly the *gaz inflammable* now became *hidrogène*, from the two Greek words *idro* (water) and *geinomai* (to be engendered). As with oxygen, the name adopted by Lavoisier and Guyton to denote Priestley's inflammable air threatened a foundation stone of the tradi-

[161]Condorcet, *Histoire de l'Académie Royale des Sciences*, 1777 (1780), 20. Daumas (1955), p. 45, quoted the following passage from Lavoisier's *Registre de laboratoire*, vol. 2 (c. 1780): "air vital c'est le nom que l'historien de l'Académie donne à l'air déphlogistiqué de M. Priestley et que j'ai cru devoir adopter après lui."
[162]*Méthode* (1787), 32–33.

tional conception of matter: with his crucial experiments on the composition and decomposition of water, Henry Cavendish had demonstrated that, far from being one of the most solid axioms of chemistry, water was a combination of two elastic fluids. It can only have been a highly disorienting experience to discover that the most common element on this planet was in fact a combination of two gases whose natures were still highly controversial. We can well understand that the supporters of the phlogistonist epistemology did not find it easy to grasp and accept this extraordinary change in the view of chemical matter. Since antiquity, water had been considered a simple liquid element, and to regard it as a combination of two gases, now required a tremendous mental shift. Lavoisier's epistemology and philosophy of matter, however, could explain the discovery effectively, and embracing it provided additional evidence of the experimental validity and consistency of the new chemistry. Thus the gas that Richard Kirwan tried desperately to identify with phlogiston was now named *hidrogène* on the strength of the experiments of the phlogistonist Cavendish!

The fifth and last principle of the first class of simple, or non-decomposed, substances was Priestley's *air phlogistiqué*, a name that had already been changed by Fourcroy to *alkaligène* after its supposed property of combining with all alkalis.[163] Guyton and Lavoisier rejected both definitions: the first because it was based on phlogiston and the second because it relied on a property that was not yet sufficiently supported by experimental evidence. However, this gas possessed another property that seemed to characterize it in a more general way: in its presence, plants and minerals could not survive. Accordingly Guyton coined the term *azot* (nitrogen), from the Greek privative *alfa* and *zot* (life).[164]

The extent of the roles and the etymological precision of these five definitions were different. Beneath the semantic power of the words oxygen and hydrogen, around which Lavoisier's philosophy of matter revolved, there were less precisely determined substances, such as *azote*, whose name reflected an immediate empirical circumstance, and *calorique*, which was little more than a paraphrase of *matière de la chaleur*. The change of the name from heat to caloric was probably prompted by the different chemical function attributed to heat by Lavoisier. After 1780, when Lavoisier and Laplace published their famous *Mémoire sur la chaleur*,[165] the traditional role of caloric began to change. Lavoisier con-

---

[163] *Ibid.*, 35.

[164] *Ibid.*, 36.

[165] Lavoisier and Laplace, "Mémoire sur la chaleur," *MARS* (1780), republished in LO, vol. 2, 283–333.

sidered this substance the manifestation of a subtle elastic fluid that by combining itself with other substances brought them to the gaseous state. Gillispie remarks that this substance was the weakest in Lavoisier's system:

> . . . for caloric does not act as a chemical agent, it is excluded, absolutely excluded from the practice of chemistry, by the most elementary consideration of method. Like light, heat passes right through the walls of laboratory vessels. The bounds of the experiment are not boundaries for heat. It may not be weighed. Its effects are like some action which remains unknown to common law, not because it is nonexistent, but because of legal conventions. Just so a science which found its metric in weighed masses might not take cognizance of caloric.[166]

From this characterization of Lavoisier's caloric, some historians have drawn the incorrect inference that the introduction of caloric had functions and characteristics similar to those of phlogiston in Stahl's chemistry. In reality Lavoisier used the concept of caloric in a different way, considering it more as a mathematical hypothesis than as a real substance:

> Besides, that this expression fulfils our object in the system which we have adopted, it possesses this farther advantage, that it accords with every species of opinion, since, strictly speaking, we are not obliged to suppose this to be a *real substance*; it being sufficient, as will more clearly appear in the sequel of this work, that it be considered as the repulsive cause, whatever this may be, which separates the particles of matter from each other; so that we are still at liberty to investigate its effects in an abstract and *mathematical manner.*[167] [my italics]

This explanation speaks eloquently for itself in emphasizing the huge difference between Lavoisier's hypothesis of caloric and Stahl's phlogiston. Thus it was probably in the light of the innovative role attributed to heat that Lavoisier decided to change its name to caloric.

The second class of substances listed by Guyton consisted of 26 acidifiable bases or "*principes radicaux*" (radical principles) of the acids. In this class Guyton distinguished the acidifiable bases that were already known from those not yet isolated from their acids. In 1787 the bases that had been isolated numbered only four: *azote* (nitrogen), *charbon* (carbon), *soufre* (sulphur), and *phosphore* (phosphorus), which com-

---

[166]Gillispie (1960), 236–237.

[167]Lavoisier, *Traité élémentaire de chimie* (1789), translated into English by Robert Kerr under the title, *Elements of Chemistry* (Edinburgh, 1790), 5–6. Henceforth, Lavoisier (1790).

bined with oxygen engendered respectively nitric, carbonic, sulphuric, and phosphoric acid.[168] Interestingly the name *azote* was not retained for the acid from which it was derived. The nomenclature of the remaining acidifiable bases followed the same binomial system.

However, it was common knowledge among chemists that the same bases saturated with different quantities of oxygen produced different acids. This complication was dealt with by an ingenious method of classification. Taking the case of sulphur as an example, five different denominations were distinguished, corresponding to five different chemical states of this substance as represented in the following scheme:

Méthode (1787)	Modern Nomenclature
*acide sulfu**rique*** (or sulphur saturated with oxygen)	sulphuric acid
*acide sulfu**reux*** (or sulphur combined with a smaller quantity of oxygen)	sulphurous acid
*sul**fate*** (generic name for all salts formed from sulphuric acid)	sulphate
*sul**fite*** (generic name for all salts formed from sulphurous acid)	sulphite
*sul**fure*** (generic name for a sulphuric compound not saturated with oxygen)	sulphide[169]

The systematic introduction of suffixes, such as *-ique, -eux, -ite,* and *-ure,* was a revolutionary innovation in the syntax of eighteenth-century chemistry. The idea behind this system was in fact not only to provide different substances derived from the same base with the same denominating root, but also to give a quantitative indication of the degree of saturation of each of them. The variation of the suffix denoted a variation in the quantity of matter. Lavoisier's dream of expressing

[168]*Méthode* (1787), 37.

[169]*Ibid.*, 40–41. Useful and accurate descriptions of Guyton's nomenclature of acidifiable bases are given by Crosland (1978), 180–181, and by Albury (1972), 160–161.

chemical reactions with the rigor of mathematics came a step nearer with the effective, though still rudimentary, precision of the suffix. Another important aspect of this classification was that the constant presence of oxygen, even if it never appeared in the names of this class, was always assumed.

The third class of matter was made up of the metals. Here the authors of the *Méthode* preferred to introduce only a few changes in their nomenclature, and "following Bergman's example, they were all to be considered of the same gender in order to introduce uniformity. The words *molybdène, tungstène, manganèse* and *platine* were therefore considered as masculine to bring them into line with the others."[170]

The traditional generic name of metallic *chalk* (calx), which denoted a metal combined with oxygen, was considered too ambiguous. The name chalk denoted in fact both a calcined metal and one specific earth. Since "the principal rule of a good nomenclature is not to attribute the same appellations to the substances so essentially different," it was decided to introduce the word *oxide* (oxyd) to indicate metallic chalk. Oxide was a name that was more consistent with the principles adopted in the *Méthode,* but it was clear that the real advantage of adopting it was to extend the linguistic and chemical domain of oxygen even further. Most names of the first three classes of substances in fact indicated, directly or indirectly, the presence of oxygen.

The fourth class contained the earths, and the changes introduced were more technical than substantial. The name *terre vitrifiable* was replaced with *silice*; *argille* with *alumine*; as suggested by Bergman, the name of *baryte* was substituted for *spath pesante,* the etymology of which evoked the heaviness of the substance. The names of *chaux* (chalk) and *magnésie* were retained.

The fifth class was that of the alkalis. Despite their common use in chemical laboratories since ancient times, the names given by chemists to these substances, such as *huile de tartare* or *esprit de corne de cerf,* were obscure and equivocal. It was now decided to adopt the names given by Bergman to the alkalis (*potassinum, natrum,* and *ammoniacum*) with the exception of *natrum,* which Guyton called *soude.*[171] The naming of the alkalis posed theoretical problems to the authors of the *Méthode.* Scheele and Berthollet recognized that ammonia was not a simple substance but a compound of hydrogen and nitrogen, and that to preserve the consistency of the entire system it should have been deleted from the table of simple substances. However, the observed

[170]Crosland (1978), 182.
[171]*Méthode* (1787), 67.

fact that ammonia combined with many substances without decomposing suggested that it should nevertheless be given one name and considered a simple substance.

Guyton's memoir ended with an examination of those substances, which like ammonia were compounds but reacted chemically as simples. These were *sucre, muqueux, gluten, amidon, extractif, résine, arome, fécule, huiles, alcohol, ether*, and *savons*. Unlike ammonia these substances were ranged in a separate table and were not considered simple.

The third memoir of the *Méthode* was Fourcroy's explanation of the table of chemical nomenclature (Plate 1, following page 206). The *Tableau*[172] displayed all the 55 new and old names given to the simple substances, and it was divided into five columns corresponding to five different chemical states. If we compare it to the table of chemical nomenclature drawn up by Guyton in 1782 (see Fig. 13), we find a completely different way of framing the complexity of matter. With his table Guyton contented himself with a systematic nomenclature of inorganic chemistry, whereas the table of 1787 offered a *map* of all possible linguistic and chemical combinations. Furthermore, the table in the *Méthode* also expressed the different states of these combinations. Thus the second column indicated which names had to be assigned when the substances of the first column were combined with caloric and thus became gases. The third column indicated the names of the substances of the first column when combined with oxygen, and the fourth oxygenate substances in gaseous form. The fifth column contained the names of the substances of the first column oxygenated with bases (i.e., the neutral salts). Since the chemistry of salts was the most advanced field of investigation by 1787, the nomenclature of neutral salts was a matter of *practical* priority. Rouelle, Macquer, and Bergman provided Lavoisier and his collaborators with useful indications for the classification of salts, but the new role assumed by oxygen imposed new rules. Thus Fourcroy regarded the salts as combinations of: 1. the acidifiable bases; 2. the acidifiable principle (oxygen); 3. the bases of earths and metals. Being combinations of three substances, the denomination ought to have been expressed in three words. However, in order not to lose the advantages of the binomial nomenclature of salts outlined by Bergman and Guyton, Fourcroy proposed an ingenious solution that allowed neutral salts to be defined with just two names. The first name was derived from the acid combination and the second from the earth. In addition to this, the second name terminated with the suffix *-ate* when the salt was combined with a saturated

[172] *Ibid.*, 100.

acid and with the suffix *-ite* when it was combined with an acid not fully saturated with oxygen. Following this scheme, the name of the salt derived from the combination of *acide acétique* and *soude* became *acétate de soude* and the one derived from the combination of *acide acétaux* and *soude* was *acétite de soude.* Thus the names of neutral salts indicated both the reagents and their quantitative proportions. Fourcroy enthusiastically remarked that these rules could be extended not only to the 722 known salts but also to those that might be discovered in the future.[173]

The last column of the table of nomenclature gave the names of the simple substances combined "without being brought to the state of acidity," such as *carbonate de fer*, the *sulfures*, the *alliages*, etc. Fourcroy concluded his memoir by claiming that the new nomenclature differed from the old one only in its denominations of oxygen, hydrogen, nitrogen, and its introduction of suffixes. However, this greatly understated the innovative structure of the nomenclature, which in reality constituted an almost total overthrow of the old chemical lexicon, and it is quite surprising to note that Fourcroy overlooked the fact that oxygen represented not merely a simple substance but the entire structure of inorganic chemistry. It is difficult to decide whether Fourcroy exhibited such caution because he believed that Lavoisier was going too far in his linguistic realism or because he was sincerely convinced that the new nomenclature was compatible with the old one. Whatever the beliefs of Fourcroy on this matter, they did not prevent the *Méthode* from having a tremendous impact on the European chemical community.

The *Méthode* ended with two dictionaries: in the first all the old names, with their corresponding new names, were ranged in alphabetical order (pp. 107–143); the second, which was more complete and included far more names than the first, listed the new terms alphabetically with their old synonyms (pp. 144–237). The first dictionary contained 656 names taken from the early chemical, alchemical, pharmaceutical, and phlogistic lexicons. Significantly only the names of *acidum pingue, air putride, alkaest, mercure des métaux, causticum, phlogistique, principe inflammable,* and *principe mercuriel,* which evoked either the phlogiston theory or alchemical beliefs, found no counterpart in the new nomenclature. But all the other names were changed so radically that in most cases one could not see any semantic connection between the old one and the new. Indeed, by going through the first dictionary term by term, one perceives the enormous shift accomplished with the new nomenclature. On the left side of the dictionary, the reader was faced with names whose origins were in disparate inves-

[173]*Ibid.*, 93.

tigative traditions, sometimes indicating a theory, sometimes a therapeutic property, sometimes the name of the discoverer, sometimes astrological and metaphysical beliefs: the old names were a reflection of centuries of attempts to define chemical objects. On the right side, the new names gave no indication of the history of the old nomenclature and relied on mechanical and consistent principles, all anchored in *one* interpretation of chemical matter. In this regard the second dictionary is even more revealing: on the left side, 1055 *noms nouveaux* were listed, of which only 361 had synonyms in the old nomenclature. Two-thirds of the new names and their related substances were therefore unknown, or at least difficult to classify, in the old system.

The *Méthode* showed most tangibly that Lavoisier's concept of matter presupposed its complexity and that the increase in the number of chemical objects and the discovery of new substances were automatically provided for in the table of chemical nomenclature.

The War between the Lines

On June 13, 1787, the first report on the new nomenclature was read to the Académie des Sciences; later it was published in the *Méthode.* The authors of the *Rapport sur la nouvelle Nomenclature* were Baumé, Cadet, Darcet, and Sage, all supporters of the phlogiston theory. The first part of the *Rapport* was an objective description of the contents of the *Méthode,* in particular the table of chemical nomenclature. This was followed by the first systematic attack on Lavoisier's theory and the first explicit defence of phlogiston from within the Academy. The main criticism expressed in the *Rapport* concerned the extended and innovative meanings and general role of the new names. If the *Commissaires* agreed with Lavoisier that the phlogiston theory had many defects, they strongly disagreed with him on the necessity of introducing a completely new and different language of chemistry. The defects of the phlogiston theory could have been remedied gradually by modifications to accommodate new experimental findings without creating an entirely new language. But the major argument against the new nomenclature was that although it provided much-needed terminological reform, anyone embracing it was irresistibly forced to accept Lavoisier's ideas and hence to abandon phlogiston completely. Baumé and his collaborators saw this as a great danger and tried to defend Stahl's theory:

> The ancient theory of Stahl, which at present is attacked, we must allow to be incomplete; but are there no embarassments, no difficul-

> ties in the new one? In the ancient, several phenomena were explained one way or the other by the aid of phlogiston; and it was by the means of water, of earth, of air, and of fire, according to different abstracted orders of mixture, of composition, of supercomposition and aggregation, that were produced the acids, the alkalies, the metallic substances &c. According to the modern theory, it is oxygen united to different acidifiable bases, that produces these same acids; but can any one inform us what this oxygen is? What this radical is?[174]

This passage reveals an attitude to the new theory that may be considered typical of the reception accorded to Lavoisier's theory by phlogistonists. From Baumé's, Cadet's, Darcet's, and Sage's point of view, the fact that the formation of acids was explained more simply by Lavoisier than by the assumption of the existence of phlogiston was by no means sufficient evidence of the validity of the new hypothesis. In 1787 "Ockham's razor," or preference for simplicity in a theory, was not a part of the logic of scientific reasoning as it eventually became. The phlogistonists' perception of the complexity of matter was completely different from that of Lavoisier. The above passage makes it clear that their qualitative view of the composition of matter simply made Lavoisier's explanation to them totally untenable. It is quite obvious that in the eyes of the many supporters of phlogiston, the experiments on the decomposition of water did not represent decisive and conclusive evidence. Relying on the Stahlian concept of the mixt, and on the belief in the transmutability of chemical substances, they simply could not *see* what Lavoisier claimed to be self-evident. It is clear then that the sources of the controversy between the *Commissaires* and Lavoisier were not in the laboratory but in their different, almost diametrically opposed, ways of *perceiving* and *classifying* chemical objects.

The report on the *Méthode* ended with an appeal for caution and on a conciliatory note:

> We are therefore of the opinion that the new theory, as well as its nomenclature, should be submitted to the trial of time, to the encounter of experiments, to contending opinions: in fine, to the judgement of the public, as the only tribunal which ought, and which can determine in such an affair. It will then be no longer a theory, but a concatenation of facts or an error. In the first case; it will afford a more solid foundation to human knowledge; in the second, it must fall into oblivion as well as all former theories and systems of physic.[175]

[174]*Méthode* (1787), 245–246; *Method* (1788), 184.

[175]*Méthode* (1787), 251; *Method* (1788), 189.

When these lines were written, no one could yet clearly foresee the fate of Lavoisier's theory and nomenclature, and the atmosphere of caution and prudence restrained most European chemists from explicitly declaring themselves for or against. However, the publication of Lavoisier's first memoir and of the table of chemical nomenclature immediately provoked a critical reaction within the Academy of Sciences in Paris that led to the formation of two rival parties within the chemistry section. What was only a latent tension until 1787 exploded with the publication of the *Méthode* in fierce controversy. The phlogistonists understood that the threat was not merely to their interpretation of some restricted experiments, but to the very roots of their chemical system. Paradoxically Lavoisier had at least three reasons to be pleased with the critical reaction in the *Rapport*: first, from his point of view, the defence of the phlogiston theory confirmed his belief that it was a system largely based on a qualitative conception that he no longer found admissible; second, the passionate frontal attack of the *Rapport* certainly stimulated more curiosity about the new nomenclature than a neutral report would have done;[176] third, by recognizing the desirability of a new nomenclature, the *Commissaires* left the phlogistonists with the difficult if not impossible task of creating a phlogistic nomenclature that was equally expressive and effective.

Lavoisier had an immediate opportunity to reply to these criticisms when he was asked, together with Berthollet and Fourcroy, to write a report on the system of symbols presented by Hassenfratz and Adet in the *Méthode*. The reply, true to Lavoisier's style, was impeccably logical:

> Messrs. Hassenfratz and Adet have confined themselves in this work to the simple relation of facts, and have laid aside all hypothesis; in consequence of which they have not admitted phlogiston, whose existence did not appear to be proved, and without which all phenomena of chymistry can be explained; and they found themselves obliged by the very force of facts, to adopt what some persons please to call the new theory (*théorie nouvelle*).[177]

---

[176]In this connection Fourcroy remarked: "les critiques plus ou moins violentes et quelquefois même passionnées qu'on se permis d'en faire dans quelques ouvrages périodiques, loin de nuire à son succés, firent naître à ceux qui ne l'avoient point encore connue, le désire de la connoître, & en répandirent ainsi le goût même par la simple voie de la curiosité," Fourcroy, *Enyclopédie Méthodique—Chimie . . .*, vol. 2, article, "Chimie," Chapter 8.

[177]*Méthode* (1787), 291–292; *Method* (1788), 218–219.

Lavoisier, who probably wrote this reply himself, appended a short survey of his theory to make clear to his adversaries the major explanatory and experimental achievements that supported it. The determination of the role of caloric in the gaseous state of substances, the explanation of the gain of weight of calcined metals, and the absorption of oxygen during the combustion of phosphorus, all these experiments were backed up by sound evidence: "these relations are far from being suppositions, for every thing is proved to conviction with weight and measure."[178] Why then postulate a principle, sometimes heavy and at other times light, when the oxygen theory was supported by gravimetric evidence? Why introduce a hypothesis to explain phenomena that were better and more easily explained without it? Finally, how was it possible to reconcile the old doctrine with the extraordinary new experiments on the decomposition of water?

Replying to his own rhetorical question, Lavoisier remarked: "Nothing shows more the insufficiency of the ancient theory than the forced explanations which they are obliged to give of these experiments on water."[179] The discovery of the composition of water forced the partisans of phlogiston to make substantial modifications to the original theory outlined by Becher and Stahl and to formulate an *ad hoc* hypothesis that fitted different and unexpected experimental circumstances and preserved the consistency of the theory. The phlogiston theory was running after the experiments, instead of predicting them.

These arguments, which were very similar to those advanced by Lavoisier in his *Réflexions sur le phlogistique,* acutely emphasized the greatest weakness of phlogiston: the old doctrine was in fact no longer a coherent theory but had become a conglomeration of opinions united only by the word phlogiston. Therefore the main argument against phlogiston was coming from its own supporters rather than from the new theory. Obviously Lavoisier used this argument because it was very persuasive and did not involve any controversial experimental issue, but the strength of his belief that the phlogiston theory had to be rejected because it lost its original consistency, and that a new one had to be introduced, should not be underestimated. If his opponents did not see the utility of "Ockham's razor" in chemistry, Lavoisier clearly did, and he understood that the explanatory aspects of a scientific theory were not merely important but even essential in showing it to be true to the facts. Since modern science takes the general validity

[178] *Méthode* (1787), 297; *Method* (1788), 224.
[179] *Méthode* (1787), 299; *Method* (1788), 226.

of this logical principle for granted, Lavoisier's argument may seem somewhat banal to us, but in eighteenth-century chemistry, the call for the introduction of such a scientific methodology was far more radical and isolated than we can imagine.

To conclude his reply to the *Commissaires*, Lavoisier dismissed the *scientific* value of the criticisms raised against the *Méthode* by remarking that being rooted in a *credo* rather than in experimental evidence they had no more authority than simple individual *opinions*. The *Rapport* reflected the respected opinions of Baumé, Cadet, Darcet, and Sage, but it could not be taken as the last word on the new theory. During its history, reports read at the Academy of Sciences had simultaneously supported Newtonian and Cartesian celestial mechanics, Lemery's and Stahl's notions of fire, and Nollet's and Franklin's theories of electricity.[180] It was not surprising then to see that some members of the Academy were disappointed by Lavoisier's theory. However, Lavoisier claimed that the *real* and lasting judgment of the Academy as a whole, and consequently of the whole scientific community, usually came slowly, and certainly not immediately after the formulation of a new theory:

> However agreeable, even honorable it would have been for us to have seen our doctrine adopted by the Academy, we cannot flatter ourselves by saying it has obtained the suffrages of the assembly, and for our part we had not even the ambition to demand them. We know that the Academy is an impartial and impassible judge; that it applauds the efforts which are made under its inspection to do away with errors and prejudices, and to extend the dominion of truth; but that it is slow to pronounce. It is on account of this confidence that we have in the wisdom of its principles, that we hope it will continue to look kindly on a new doctrine nourished and formed in its bosom, which already has cost near twenty years of labor, which the force of reason and facts has obliged the most celebrated chymists to adopt, and in the favour of which a much greater number at present appear ready to decide.[181]

In other words, the individual and immediate skepticism of the *Commissaires* could not be considered a reliable indication of the Academy's ultimate verdict.

It is interesting to note that at this point in the controversy both Lavoisier and his opponents appealed to the tribunal of time and that they already did so with different degrees of confidence. As much as

---

[180] *Méthode* (1787), 310.

[181] *Méthode* (1787), 311; *Method* (1788), 235–236.

the experimental evidence that was presented, it is the contrast between the tone of the *Commissaires*, oscillating between dogmatic certainty and insecure wavering, and the consistent, resolute, and confident list of Lavoisier's arguments that attracts the reader of the *Méthode* irresistibly to the logic of the latter.

As we have seen, the *Méthode de nomenclature chimique* is not a homogeneous work, and its pages reveal all the contradictions, the genial intuitions, the controversies, and the beliefs of eighteenth-century chemistry. Paradoxically Lavoisier's initial intention of constructing a chemical lexicon based on Condillac's linguistic realism was shared only by Hassenfratz and Adet, and very tentatively by Guyton. The emphasis on creating a *method* of naming rather than a nomenclature implied a precise epistemological position. Furthermore, Lavoisier's apparently unshakable belief in the feasibility of creating a chemical language representative of reality led to an ontological realism that, as we shall see, was opposed by his collaborators. This difference of opinion is made even more important by the fact that to Lavoisier linguistic *realism* was a fundamental principle of the new nomenclature. Further evidence of this is provided by a memoir, probably written by Lavoisier around 1787,[182] in which he elaborated on some experiments on alcoholic fermentation he had begun in 1782. The primary and explicit purpose of this memoir was, however, to illuminate the reasoning that led to the experiment and to discuss the method of operation rather than the operations themselves:[183]

> I have already pointed out, in previous memoirs, that the manner of reasoning is the same for all sciences; that chemists, like geometers, can only proceed from the known to the unknown, by a true math-

[182]Lavoisier, "Mémoire sur la fermentation spiritueuse," in LO, vol. 3, 777–790. This memoir was never published during Lavoisier's lifetime, and its precise dating is particularly difficult. Albury (1972), 140, assumed that the memoir was written before 1787 because Lavoisier used the term vitriolic acid instead of sulphuric acid. Apart from this unquestionable discrepancy, however, the terminology used elsewhere in the text directly evokes that of the *Méthode*. In addition Daumas (1955), 60, remarked that the French chemist was working on alcoholic fermentation on January 20, 1787 and at the beginning of June of the same year. This evidence, though not conclusive, suggests that the memoir was probably written almost simultaneously with the *Méthode*. Daumas (1955), 62, also remarked that on November 14, 1787, Lavoisier read a memoir whose title has been not reported in the *Procès-verbaux* of the *Académie*, but which on the basis of our conjecture could be the "Mémoire sur la fermentation spiritueuse."

[183]"Ce mémoire, tel que je la présente ajourd'hui, contiendra donc moins la marche des expériences que celle des raisonnemments qui les ont dirigées, et je m'attacherai plutôt à discuter la méthode d'opérer que les opérations elles-mêmes," LO, vol. 3, 777.

> ematical analysis, and that all reasoning in scientific matters implicitly contains true equations.
>
> Let us cite an example. I suppose that I have to analyze a salt, of which I know neither the acid nor the base. I introduce a known weight of this salt into a retort; I pour vitriolic acid over it and distill; I obtain nitrous acid in the receiver and in the retort I find vitriolated tartar; I conclude that the salt which I was given to examine was nitre.
>
> But what is the mechanism of reasoning which has led me to this conclusion? An instant of reflection will soon make it known. It is first of all clear that if I wished to make an exact calculation of quantities *I had to suppose that the weight of the materials employed was the same before and after the operation and that only a change of modification took place. I therefore mentally set up an equation in which the materials existing before the operation formed the first member, and those obtained after the operation formed the second: and it was actually by the resolution of this equation that I successfully obtained the result.*[184] [my italics]

Reasoning in this manner, with X for the unknown acid and Y for the unknown base, Lavoisier formulated the following equation: X + Y + vitriolic acid = nitrous acid + vitriolated tartar = nitrous acid + fixed alkali. From this it follows that X = nitric acid; Y = fixed alkali and therefore the salt is niter.

It is true that the principle of conservation of matter had been presupposed by chemists long before Lavoisier, but the language and the method of transforming this principle into a scientific law were incontestably his. Lavoisier characteristically ended his memoir by declaring that all these "experiments" give "the chemical theory which I venture to call mine, a degree of certainty which cannot be denied."[185]

In this remarkable memoir, Lavoisier clearly declared the guidelines of his chemistry: 1. he assumed the conservation of matter as the only law for interpreting chemical combinations; 2. he raised the notion of analysis to a superior level of scientific reasoning and claimed that all chemical analysis could be mentally reduced to an equation. He followed Condillac in suggesting that to solve a chemical problem two things were required: "the statement of the data, or the state of the question; and the disengagement of the unknowns";[186] 3. the method employed in directing, analyzing, and selecting chemical experiments was more important than the experiments alone. These three assump-

---

[184]*Ibid.*, 778. English translation by Albury (1972), 140–141.

[185]"Cet ensemble de preuves, qui toutes se soutiennent et se prêtent un mutuel appui, donne à la théorie de la chimie, à celle que j'ose appeller la *mienne*, un degré de certitude auquel il est impossible de se refuser," LO, vol. 3, 785–786.

[186]Condillac (1980), 307.

tions were intimately interwoven: to ensure its successful application in chemistry, the analytic or algebraic method [3], which was nothing more than a system of identical equations [2], required that the weight of chemical reagent remained the same, before, during, and after their combination [1]. Each assumption necessarily presupposed the other. Thus Lavoisier's enunciation of the law of conservation of matter rested in a broad epistemological context and was directly associated with the project of a mathematicization of chemistry. Seen from this perspective, the historical and intellectual importance of the *Méthode* gives it a central position in the chemical revolution. As far as language and method were concerned, Lavoisier was perfectly right in claiming to have brought about a revolution in science. No one before him considered chemical problems, reactions, and nomenclature in such a comprehensive philosophical context.

Immediate Reactions, 1787–1788

Reactions to the publication of the *Méthode de nomenclature chimique* were immediate, and most of them were negative. But despite the hostile initial reception, the fierce discussion aroused by the new nomenclature contributed to its steady spread and its eventual tremendous success. Lavoisier's theory was translated for the first time into several foreign languages and his ideas became known to most European naturalists. According to Duveen and Klickstein, the *Méthode* was printed six times between 1787 and 1791; it was translated into English in 1788 and in abstract in 1789; into German in 1790 (abridged) and 1791; into Italian in 1790 and 1791; and into Spanish in 1788.[187] At the same time, many abstracts and reports of the nomenclature appeared in European scientific journals. The contrast between the wide circulation of the *Méthode* and the generally unfavorable response to it was indeed remarkable, but indirectly it confirms that the time was ripe for a scientific discussion of phlogiston. After its publication, in fact, most of Lavoisier's opponents suddenly realized the radical change in the conception of chemical objects that was implied in the new nomenclature. The instant and intense opposition to it reflected the speed of this early perception. Most of the European chemists who until 1787 had refrained from commenting on Lavoisier's theory, now realized that with the abandonment of a terminology that had parallelled the

[187]See Duveen and Klickstein (1954), 126–154.

historical and cumulative progress of chemistry the whole structure of chemistry as they knew it was threatened.

In 1787 and 1788, following the negative report by Baumé, Darcet, Cadet, and Sage, a series of critical articles appeared in Rozier's *Journal de physique.* The first to take an official position was the geologist Jean-Claude de Lamétherie, who published a faithful abstract of the new nomenclature,[188] followed by his opinion of it.[189] The beginning of de Lamétherie's essay is revealing because it contains an argument on the philosophy of language that eventually became the standard one of Lavoisier's opponents:

> Words being the object of conventions, they can represent all that one wishes; but a word adopted by an entire nation becomes appropriate to the object; it is not permissible to change it but by the destruction of all the advantages of language, which are to make all objects known by fixed signs. Thus, there is nothing else than the general consensus, which is called *use,* which has the authority to change some terms and replace them with others.[190]

Significantly, de Lamétherie began his criticism of Lavoisier's nomenclature by adopting the Lockean *credo* that language was based on human conventions and that one could only introduce changes with great caution. From this, de Lamétherie argued that the language of a science could change only gradually and that the consensus of the scientific community as a whole was necessary before a new word could be adopted generally. Chemistry, like all other sciences, perfected itself by degrees, and its names were reformed only when new discoveries were made. Accordingly de Lamétherie suggested that changes to the language of chemistry must be introduced with circumspection, that the new terms must not be too different from the ones currently in use, that new terms must not neglect the harmony and the genius of each language, and finally that "a new nomenclature cannot rely upon systematic ideas,"[191] otherwise each school of thought would adopt its own language.

Apart from the Lockean argument of the conventionalist source of scientific language, de Lamétherie's criticism revealed an epistemological attitude that had implicitly been shared by chemists for centuries.

---

[188]De Lamétherie, "Méthode de nomenclature chimique," in *Observations sur la physique,* 31 (1787), 210–219.

[189]De Lamétherie, "Essai sur la nomenclature chimique," in *Observations sur la physique,* 31 (1787), 270–285.

[190]*Ibid.,* 270.

[191]*Ibid.,* 273.

After stating his own principles of nomenclature, de Lamétherie examined in detail the 55 names proposed in the *Méthode*, and as may be expected, only the term *lumière* and a few others were found acceptable. Obviously controversial terms, such as oxygen, hydrogen, caloric, *azote*, etc., were all rejected, either because they relied on unacceptable hypotheses or because their sound was too hard and "*barbare*" and thus not in harmony with the genius of the French language. Not surprisingly, de Lamétherie found the old phlogistic nomenclature in perfect accordance with his principles.

On November 10, 1787, the French mineralogist Etienne-Clément Marivetz wrote to de Lamétherie, endorsing his criticism of the new nomenclature and appealing for extreme caution in changing scientific terminology.[192]

In an anonymous letter to de Lamétherie,[193] Locke's assertion of the conventionality of language was used to rebut the basic principles of the *Méthode*; the writer of this letter questioned the validity of the etymological criteria applied by Lavoisier in assigning new chemical names, stating that oxygen and hydrogen had in fact a literal meaning opposite to that given in the *Méthode*:

> The translation of the first word is *engendered by the acid* and not *engenderer of the acid*; that of the second, *engendered by the water* and not *engenderer of the water*.[194]

Strictly speaking, the etymology of the word oxygen did signify that the substance was produced by acid rather than being the universal principle of acidity as Lavoisier's theory assumed. For hydrogen, whether its etymology was accurate or not, its essential property of being a constituent of water seemed not to be affected.

In 1787 Baltazar Sage, a well-known mineralogist, published two letters voicing much the same criticism and coming to the conclusion that Lavoisier's nomenclature was "barbare, insignificant, et sans étymologie."[195] As a member of the chemical section of the Academy, he felt

---

[192]"Lettre de M. Le Baron Marivetz à M. De la Métherie sur la Nomenclature chimique," in *Observations sur la physique*, 31 (1787), 61–63. On the chemical theory of Marivetz, see Abbri (1984), 371–372.

[193]"Lettre aux auteurs du journal de physique sur la nouvelle Nomenclature chimique," in *Observations sur la physique*, 31 (1787), 418–424.

[194]*Ibid.*, 423.

[195]"Lettre de M. Sage a M. de la Métherie sur la Nouvelle Nomenclature," in *Observations sur la physique*, 33 (1788), 479; see also Sage's first letter (vol. 33 (1788), 268–286), which was partially translated into Italian in *Biblioteca fisica d'Europa*, 4 (1789), 47–48.

obliged to justify his dissent directly to Lavoisier, to whom he sent the following lines:

> Allow me, my dear *Confrère*, to have my religion, my doctrine, my language [. . .]. I beg you not to take it badly if I will not join your chemical confederation.[196]

More consistent opposition to the new nomenclature came from Henry Cavendish, who, after receiving a copy of the *Méthode* from Lavoisier in September 1787, sent a letter to his young colleague Balgden with the following comments:

> I have been reading La V.[oisier]'s preface. It has only served the more to convince me of the impropriety of systematic names and the great mischief which will follow from his scheme if it should come into use.[197]

Lavoisier's nomenclature rested on a general theory rather than on a collection of soundly based experiments to which the names could have been linked. The risk of basing the language of a science upon such a nomenclature was that far more new problems would arise than one could ever imagine:

> The great inconvenience is the confusion which will arise from the different hypotheses entertained by different people and the different notions which must be expected [to] arise from the improvements continually being made. If the giving of systematic names becomes the fashion it must be expected that the other chemists who differ from these in theory will give other names agreeing with their particular theories, so that we shall have as many different sets of names as there are theories and in order to understand the meaning of the names a person employs, it will be necessary first to inform yourself what theories he adopts [. . .]. I do not imagine indeed that their [Lavoisier's] nomenclature will ever come into use, but I am much afraid it will do mischief by setting people's minds afloat and increasing the present rage of name-making.[198]

---

[196]"Permettez moi, Monsieur mon cher confrere, d'avoir ma religion, ma doctrine, mon langage; jamais difference de mots, jamais difference d'opinion ne pourra influer sur les sentimens d'estime et de consideration que je vous ai voués; je vous prie donc de ne pas trouver mauvais si je n'entre pas dans votre confederation chimique," LC, vol. V, in course of publication.

[197]*The Scientific Papers of the Honorable Henry Cavendish,* Sir Edward Thorpe (ed.), vol. 2 (Cambridge, 1921), 325.

[198]*Ibid.*, 324–325.

The first part of Cavendish's letter was prophetic and identified a controversial problem that had appeared with the publication of the *Méthode*. The "rage of name-making" spread like a fever across Europe in a few years, and as we shall see, many alternative names were invented and, for a limited period, supported by different schools of chemists. What Cavendish did not foresee was that at the end of the controversy Lavoisier's nomenclature would prevail over them all. The reason for Cavendish's misunderstanding or lack of vision was that he considered all *speculation* on chemical phenomena illegitimate and, adopting an extreme Baconian position, claimed that hypothesis had no place in science. In his view Lavoisier was certainly right in rejecting those parts of phlogiston theory that became experimentally inconsistent, but that did not entitle him to create a general theory as an alternative to it. Based on the hypothesis of oxygen, Lavoisier's theory had the same defects and relied upon a speculative system rather than on indisputable facts. Like most British chemists, Cavendish was, so to speak, "allergic" to any theoretical interpretation of facts: experiments were the only authority in science and the elaboration of an entirely new and systematic nomenclature probably seemed to him a waste of time that in fact had little or no relevance to the progress of chemistry.

Cavendish's prophecy concerning the "rage of name-making" was soon to a large extent fulfilled. The Spanish physician Juan Manuel de Aréjula, who became a pupil of Fourcroy while staying in Paris in 1788,[199] published in Madrid that year a pamphlet entitled *Reflexiones sobre la nueva nomenclatura quìmica*, which was immediately translated into French.[200] It was initially meant to serve as a summarized translation of the *Méthode*, but just as it was being published, a complete translation was produced by the Spanish apothecary Pedro Gutiérriez Bueno.[201] Aréjula's work was in fact far more than a mere translation of the new nomenclature. He also suggested several terminological changes and questioned with particular insistence the validity of the linchpin of the entire *Méthode*: the name oxygen. More generally, Aréjula claimed that Lavoisier's nomenclature needed changes of two

---

[199]On Aréjula, see Ramon Gago and Juan L. Carrillo, *La introduccion de la nueva nomenclatura quìmica y el rechazo de la teoria de la acidez de Lavoisier en España* (Malaga, 1979), in which is also reprinted Aréjula's *Reflexiones sobre la nueva nomenclatura quimica* (Madrid, 1788). Henceforth Aréjula 1979. See also Ramon Gago, "The New Chemistry in Spain," in Donovan (1988), 169–192.

[200]Aréjula (1979); *idem*, "Réflexions sur la nouvelle nomenclature chimique. Pour servir d'introduction à la traduction espagnole de cette Nomenclature," in *Observations sur la physique*, 33 (1788), 262–286.

[201]Gago (1988), 179.

kinds: 1. those required for the translation of the new terminology into different languages; 2. those that concerned single substances and the theory that claimed to support them. The first kind of change did not cause any particular problems because it was only connected with the grammatical *circumstances* of each vernacular. By contrast, the second problem was clearly more controversial and interesting, involving the relationship between words and reality.

Aréjula analyzed all the 55 simple substances and the names proposed for them in the table of chemical nomenclature. The most interesting of these analyses is his examination of oxygen. "The name oxygen was derived from its constant property of forming acids,"[202] but this property, which Lavoisier regarded as essential to its definition, was not always confirmed by the facts. A major inconsistency between this assumption and the theory of acidity was significantly observed by Claude Louis Berthollet who in 1787 noticed the absence of oxygen from prussic acid (hydrocyanic acid)[203] and parenthetically declared: "nothing is doubtful about the composition of prussic acid but the proportion of its principles, which I could not yet determine. It is a combination of nitrogen, hydrogen and carbon."[204] This extraordinary empirical observation, which cast doubt on one of the basic principles of Lavoisier's theory of acidity, strengthened his opponents in their opinion that the name of oxygen was not adequate for the nature of the substance it was supposed to represent. Strangely enough, Aréjula did not develop this powerful argument, but criticized the definition of oxygen from another angle. If, he argued, each substance combined with oxygen produced an acid, then the union of oxygen and hydrogen should have produced an acid instead of water. Aréjula also argued that the oxygenated muriatic acid that was supposed to be an acid containing a great quantity of oxygen actually displayed much weaker acidic properties than other acids saturated with a smaller quantity of oxygen.

For these reasons Aréjula suggested that the name given by Lavoisier to this substance needed to be changed to one more appropriate to its nature. The Spanish physician argued that the most obvious and clearly demonstrated property of this substance was that of promoting the combustion and, more generally, the burning of substances. Accordingly Aréjula introduced the name of *Arxicayo* or *Arki-kayo*, which was rooted in the Greek *arké* (principle) and *kaio* (to burn).[205] Thus

---

[202]Aréjula (1979), 5.

[203]Berthollet, "Mémoire sur l'acide prussique," in *MARS* (1787).

[204]*Ibid.*, 30.

[205]Aréjula (1979), 23. Slight orthographic differences occur between the Spanish and the French versions.

the name of the oxygenated muriatic acid now became in Spanish *acido muriático arxicayado.* Analogously the term *cayos* (Spanish) or *kayes* (French) replaced the term "oxide" to denote metallic calces. Though not happy with the name of hydrogen, Aréjula retained it, but he changed the name *azote* to *azoe* and made some minor technical changes to the names of salts and earths. Fortunately Aréjula's nomenclature had little success, and although much of his criticism of Lavoisier was eventually repeated by other phlogistonists, none appears to have adopted his terminology. One reason for this was probably the appearance a short time earlier of the literal translation of the *Méthode* by Bueno, which became the standard work among Spanish chemists.[206]

In 1788 another objection to the contents of the *Méthode* came from the north of Europe. The Finnish chemist Johan Gadolin, one of the most promising pupils of Torbern Bergman, published a pamphlet on the new nomenclature that although it was dedicated to "de Morveau, Lavoisier, Berthollet and de Fourcroy, auctoribus Novae nomenclaturae chemicae ingeniossissimis"[207] contained penetrating criticism of it. Like most phlogistonists, Gadolin argued that the new names assumed the validity of a theory that was not yet firmly based on experimental evidence. The essential property of oxygen as the universal principle of acidity, for instance, was confirmed only by the isolation of just four radicals of the acids, and this suggested that the name oxygen was a premature generalization. Gadolin pointed out another interesting inconsistency: if, on the one hand, the *Méthode* established that the names of all acids were derived from the names of the bases, so that, for instance, vitriolic acid had to be changed to sulphuric acid, why was it decided not to change the name of nitric acid despite the fact that it was not derived from its base (*azote*).[208] To be consistent, the authors of the *Méthode* ought to have called this acid *azotic* instead of nitric. Berthollet answered this remark, which is not so pedantic as it may seem, by pointing out that *azote* was an exception since it could be the base either of nitric

[206]Gago (1988), 179–180.

[207]Johan Gadolin, *Animadversiones in Novam Nomenclaturae Chemicae Methodum* (Aboae, 1788).

[208]"Dicit Clariss. De MORVEAU adsociatos diu dubios haesisse, utrum denominationem *acidi vitriolici* usitatam retinerent, an ejus loco *acidum sulphuricum* scriberent. Rejecerunt tandem nomen antquitus receptum, ne ulla denominatio discederet a regula ipsis praescripta, qua acidum a basi, uti phosphoricum a phosophoro & sic porro, denominaretur. Attamen ipsi mox eandem non observant regulam in denominando acido nitri. Hoc enim acidum jam secundum regulam, non *nitricum,* ut ipsi proponunt, sed *azoticum* appellandum fuisset, quia azotum acidi hujus basis," Gadolin (1788), 19.

acid or of ammonia.[209] This unusual property forced them to make an exception to their linguistic rule. To Gadolin's remarks on the name of oxygen, Berthollet replied that even if only one of the substances combined with oxygen produced an acid, the name would have been legitimate and that one should not attach too much importance to etymology. This surprising position was amplified by Berthollet in the following statement:

> We have sought the etymologies which indicated known properties only as an aid to the memory, but we did not intend to indicate with the name of a substance a general and constant property of it.[210]

From this it appears that Berthollet was to some extent dissociating himself from the basic principles behind the *Méthode* as early as the spring of 1789, when these lines were written. Berthollet also admitted that the new nomenclature had many faults but said that as a whole it was an effective means of helping students to remember and made it easier for chemists to teach their science. Berthollet therefore defended the contents of the *Méthode* not because he shared Lavoisier's belief that the new chemical language followed the path of nature and indicated the *real* essence of substances, but more simply because he found it more convenient to use than the old one. This difference, as we shall see, was stated more explicitly by Berthollet some years later in his *Essai de statique chimique.*

There was more criticism of the new nomenclature. In the introduction to his English translation of the *Méthode,* James St. John expressed reservations. St. John admitted that a reform of the nomenclature was necessary:

> Chymists have for a long time complained of the perplexity and embarrassment, occasioned in the science, by the variety of names given to the same substances; many of which names were very improper and absurd. It has been universally allowed, that a reform on the chymical denominations was absolutely necessary [. . .].
>
> It appears to me, that the new language is not the result of a theory, but is founded upon a basis of plain and simple facts.[211]

However, despite expressing these positive sentiments, St. John was

[209]Letter by Berthollet to Gadolin dated April 18, 1789, published in *Johan Gadolin 1760–1852. Wissenschaftliche Abhandlungen* (Helsinki, 1910), lxxvii–lxxviii.

[210]*Ibid.*, lxxxviii.

[211]*Method of chymical Nomenclature [. . .] translated from the French, and the new chemical names adapted to the genius of English Language* (London, 1788), v–vi.

not so confident as to embrace the new theory and abandon phlogiston, and added:

> Neither shall I urge anything in opposition to the celebrated theory of Stahl. Far from it for I even think some parts of Stahl's theory reconcilable with the modern discoveries. For though the late experiments demonstrate that phlogiston does not give weight or heaviness to metals, that phlogiston does not disengage itself from the sulphur during the formation of the sulphuric acid; yet we still allow the absolute existence of a phlogiston. It is still the matter of fire, of flame, and of heat which is liberated in combustion.[212]

This kind of peculiar compromise between two theories that were in principle totally irreconcilable became more frequent among those naturalists who appreciated the convenience of the new nomenclature, on the one hand, but were unwilling to recognize Lavoisier's theory as superior to the other in all respects.

Apart from this criticism in the introduction, St. John's translation of the *Méthode* was an accurate work and, as pointed out by Crosland,[213] with the exception of a few orthographic changes, such as *sulphur* for *sulfure*, *oxygen* for *oxygène*, *potash* for *potasse*, the French names were retained. Surprisingly St. John preferred to keep the French roots of the chemical terms rather than adapt the nomenclative principles to the target language as the Germans and the Swedes did. The English translation of the *Méthode* was in fact the only one that retained the French structure. St. John's translation was quite successful in the sense that most of the British chemists who adopted Lavoisier's ideas adopted St. John's translation of the nomenclature. In 1788, however, with the exceptions of Richard Lubbock and Thomas Beddoes, there were still no English chemists who accepted either the theory or the nomenclature proposed by Lavoisier. We possess much interesting testimony to the reaction of respected British chemists to the new nomenclature. One among them who deserves mention is the Irishman Richard Kirwan, definitely one of the leading figures of eighteenth-century chemistry. Kirwan is particularly interesting because throughout his career he showed a deep interest in the reform of mineralogical terminology. A friend of Torbern Bergman, with whom he exchanged several letters on mineralogical matters,[214] Kirwan published in 1784 his *Elements of Mineralogy*, which contained a systematic description of 733 mineral speci-

---

[212] *Ibid.*, vii.

[213] Crosland (1978), 193–194.

[214] See Bergman (1965), 174–197.

mens, classified and named methodically.[215] Though largely indebted to Bergman's nomenclature, Kirwan introduced several names of his own that were soon accepted by contemporary mineralogists. Kirwan also conducted an intensive correspondence with Guyton de Morveau, whom he considered the "only philosopher in France."[216] With Guyton, Kirwan discussed a reform of chemical nomenclature and persuaded the French chemist to adopt the term *siderites* to denote a newly discovered semi-metal in his nomenclature of 1782.[217] But despite his genuine interest in a general reform of chemical nomenclature, itself very unusual in England, and regardless of the fact that his friend Guyton was involved in it, Kirwan was extremely irritated by the publication of the *Méthode.* On July 26, 1788, he wrote to Guyton, bluntly expressing his dissatisfaction as follows:

> The nomenclature introduced by the immortal Bergman which you have followed and perfected in the first vol. [of the *Encyclopédie Méthodique—Chimie,* 1786], is perfectly intelligible and does not put anyone off, the other [the *Méthode*] is untranslatable, appalling, systematic and unintelligible to most of those who cultivate the practical chemistry.[218]

In reply to this stern criticism, Guyton claimed that the new nomenclature relied on the principles stated by Bergman and himself and that most of it was identical with the nomenclature of 1782, of which Kirwan seemed to approve. The Irish chemist did not accept this argument as he had realized that between the earlier nomenclatures and that of the *Méthode,* a major theoretical shift had occurred. What Kirwan was criticizing was the method and the philosophical assumptions of the *Méthode* rather than its concrete result, and he made it clear that his main reason for finding it unacceptable was that the nomenclature was systematic rather than heuristic. Interestingly some years before, he had criticized the attitude of some prominent French naturalists, lamenting that "France has not yet recovered from

---

[215]Kirwan, *Elements of Mineralogy* (London, 1784). This work was translated into French in 1785, and a second enlarged edition in two volumes was issued between 1794 and 1796.

[216]Bergman (1965), 188.

[217]*Ibid.,* 180.

[218]"La nomenclature introduite par l'immortal Bergman et que vous avez suivi et perfectionné dans le 1er Vol. est parfaitement intelligible et ne rebutte personne, l'autre est intraduisible, revoltante, systematique et inintelligible pour la plus grande partie de ceux qui cultivent la chymie practique," Letter communicated to me by Michelle Goupil, and in course of publication.

this Cartesian *démangeaison*"[219] and warning Guyton of the dangers of combining philosophical and systematic matters with the investigation of natural phenomena.[220] With this penetrating criticism, Kirwan accurately identified one of the most typical aspects of the French scientific Enlightenment in general and of the *Méthode* in particular. Kirwan recognized that in both cases French scientists still retained their Cartesian belief in the explanatory power of scientific deduction. As an adherent of the Baconian tradition of thought, Kirwan pointed out that the investigation of nature, the testing of phenomena, and the conducting of experiments could not be combined with philosophical speculation or with systematic hypothesis. Like Priestley, Cavendish, and Black, Kirwan maintained that the only legitimacy in science was bestowed by collecting facts.

With respect to the question of nomenclature, Kirwan believed that "the language of science, not being a language apart, but a branch of that of the country [. . .], should strictly conform to the parent stock."[221] In so saying, he was clearly rejecting Lavoisier's and Condillac's belief that the language of chemistry, like the language of any other science, could be reduced to a universal and accurate language. That this linguistic realism was an aspect of French chemistry that irritated British chemists is confirmed by Marsilio Landriani, who during a visit to England in 1788 wrote to Madame Lavoisier reporting the reactions to the new theory.[222] In Birmingham, Manchester, Edinburgh, etc., Landriani reported only supporters of phlogiston and even

---

[219]Letter by Kirwan to Guyton dated April 10, 1783; in course of publication. See note 218.

[220]"Les belles lettres (à moins qu'elles ne soient cultivées par des hommes superieurs tels que Voltaire, d'Alembert ec.) et la philosophie naturelle, sont des choses trop disparates pour se concilier ensemble." Letter by Kirwan to Guyton dated March 17, 1787. See note 218.

[221]Quoted by Crosland (1978), 200.

[222]Marsilio Landriani was a well-known naturalist and statesman from Milan; he invented the word "*eudiometro*" to denote the instrument he built to detect the salubrity of air. Landriani was also secretary of the *Società Patriottica*, a society promoting the application of science in industry founded in Milan in 1776. An intimate friend of Priestley and in correspondence with the most prominent European chemists, Landriani was considered by his contemporaries one of the most authoritative pneumatic chemists. This biographical information can help to explain why Lavoisier put so much effort in trying to convert Landriani to the new chemistry. The figure of Landriani is even more interesting because he left a diary of his scientific travels in Switzerland, France, and England. Landriani's report on his travels has recently been published by M. Pessina under the title *Relazioni di Marsilio Landriani sui progressi delle manifatture in Europa alla fine del Settecento* (Milan, 1981).

"les meilleurs têtes de la Grande Bretagne,"[223] Black and Watt, remained loyal to the old theory. Landriani had the impression that the British chemists regarded the antiphlogistic theory as a systematic French construction rather than a body of knowledge borne out by facts, and that they seemed unlikely to revise this view.[224] Indeed, the British chemists would not have been able to accept Lavoisier's theory without abandoning their Baconian concentration on facts and facts alone.

This attitude is well exemplified by the Scottish chemist Joseph Black. Though he did not take up a definite position on this controversy until 1790, when he announced his conversion to Lavoisier's ideas in a sensational letter, Black's opinion of the new nomenclature was in perfect agreement with that of his countrymen. In a note to Black's posthumously published chemical lectures, one of his pupils remarked that

> Dr Black highly approved a systematic nomenclature, and thought the French one extremely ingenious, and that its many barbarisms and philosophical incongruities should be overlooked as something unavoidable, or that they should be corrected [. . .]. He disapproved however exceedingly of the entire substitution of this for all other denominations of chemical substances [. . .]. The employment of the scientific names only gives an appearance of knowledge without the reality of science.[225]

In Black's own words, the new nomenclature introduced "a great and almost total revolution," and those who decided to adopt it were irresistibly forced to adopt the system underlying it. For this reason the final judgment on the *Méthode* was negative:

> These names have evidently been contrived to suit the genius of the French language, in the first place, and then have been transferred into the Latin words; or the words which were meant to be used in the Latin language have been coined from the French ones. When changes are thus made in the names of things which are familiar to us, I believe most people find them *disgusting* at the first, on account of the shock and derangement which they give to the habits we had

[223]Letter by Landriani to Madame Lavoisier dated October 12, 1788, LC, vol. V, in course of publication.

[224]"Il paroit que l'Angleterre n'aime pas à vous ceder cette victoire, et qu'elle regarde la theorie antiphlogistique comme une theorie françoise tandis que bien peu lui manque pour etre celle de la nature," LC, vol. V, in course of publication.

[225]Black, *Lectures on the Elements of Chemistry* (Edinburgh, 1803), vol. 1, 555. Crosland (1978), 197, remarked that these lectures were probably written not later than 1790.

formed before. These latinised French words appeared to me at first very *harsh* and *disagreeable.* [my italics]

In another passage, Black continued:

There is just now a rage for system, —for complete systems. We have got such a high conceit of our knowledge, that we cannot be pleased with a system which acknowledges any imperfection: It must not leave one open link: It must not leave any thing unexplained [. . .] This kind of authority accruing to a theory from its specious and extensive application to phenomena, *is always bad*; and, with mere beginners in philosophy, it is doing an irreparable hurt. It nourishes that *itch for theory*; and it makes them unsolicitous about the first foundations of it; —thus it forms in their minds the worst of all philosophical habits.[226] [my italics]

The most remarkable thing about these two passages is that when Black wrote them he had explicitly adopted Lavoisier's theory and was teaching it in his lectures. It is clear then that Black's irritation was not provoked by the experiments that supported the oxygen theory but by the method of classifying and naming them. In this respect Black's hostility did not differ from that of those who remained faithful to the phlogiston theory.

In other parts of Europe, too, the *Méthode* met with a negative reception and aroused fierce debate. In all the states of Italy, the reaction was either lukewarm or decidedly cool. Landriani, who was invited by Lavoisier to be present at his experiment on the composition of water in 1787 and to study the new nomenclature, was initially almost "convinced of the non-existence of phlogiston."[227] While travelling in England, however, he began to change his mind and to feel less certain of the validity of the new nomenclature. In a letter to Madame Lavoisier, Landriani likened his position to that of Erasmus, who, though despising the faults of the Catholic church, was not yet prepared to embrace the doctrines of Calvin or Luther.[228] In 1788, he finally came down on the opposite side and delivered a negative report on the new nomenclature to the *Società Patriottica.*[229] After

---

[226]Black, *Lectures*, 492–493 and 548.

[227]"J'ai vu la nouvelle nomenclature chimique. Mr. Morveau ne parle que ce langage. Il m'a presque convaincu sur la non-existence du phlogistique," Letter by Landriani to Jean Senebier dated September 10, 1787, kept in BPUG Senebier's collection.

[228]Letter to Madame Lavoisier. See note 218.

[229]"Sarebbe desiderabile che questa nuova nomenclatura non fosse appoggaita ad un sistema, il quale probabilmente avrà la fortuna degli altri quando le scienze chimiche saranno più inoltrate di quello che lo sono al presente," Landriani (1981), 67.

1788 Landriani became increasingly interested in the application of chemistry to the manufactures of Lombardy and withdrew from the theoretical controversy without maintaining a firm position.

The professor of chemistry at the University of Pavia, Giovanni Antonio Scopoli, ironically referred to by Kirwan as "le Sage d'Italie,"[230] was a convinced supporter of the phlogiston theory.[231] On March 17, 1787, Scopoli wrote to Giovanni Fabbroni, defending the identity of phlogiston against all who "wish to distinguish themselves from other men by [introducing] inconsistent novelties and inventing new nomenclatures."[232]

In Florence the *Méthode* met with similar reactions. In the respected periodical *Novelle Letterarie*, Lavoisier's nomenclature was dismissed in a few lines as unlikely to achieve the success its authors hoped for.[233] This opinion seems to have been shared by all Italian scientists before 1789. In general, however, only a few naturalists committed themselves in 1787 or 1788, most of them preferring to await developments.

In Germany the situation was complicated by the fact that the phlogiston theory had been formulated by Becher and developed by Stahl, two of the luminaries of German science. The question of whether to embrace or resist Lavoisier's new theory therefore involved considerations of emerging patriotism. Initially, however, the skepticism felt by most German chemists about the new nomenclature was defended on other grounds. Immediate reactions to the *Méthode* were blunt and direct. In 1788 Sigismund Friedrich Hermbstädt, who eventually became one of Lavoisier's most loyal supporters in Germany, opposed the new system. In the heart of the learned world of Germany, the University of Göttingen, Johan Friedrich Gmelin remarked that "in Germany at least, this system is not yet regarded as proven or firmly grounded, despite the ability of those who have introduced it. Consequently the terminology runs the danger of falling in a heap with the system."[234] Other prominent members of the German chemical community, such as Leonhardi, Gren, and Wiegleb, rejected the new chemistry without hesitation. Moreover, one of the best-known

[230]Bergman (1965), 188.

[231]On Scopoli, who was among other things a friend and correspondent of Linnaeus and Bergman, see Abbri, "Tradizioni chimiche e meccanismi di difesa. G.A. Scopoli e la Chimie Nouvelle," in *Archivio di storia della cultura*, IV (1991), 75–92.

[232]Quoted by Abbri (1991), 82.

[233]*Novelle Letterarie*, 19 (1788), 190–191.

[234]Quoted by Hufbauer, *The Formation of the German Chemical Community (1725–1795)* (Berkeley-Los Angeles-London, 1982), 102.

editors of scientific journals in Europe, Lorenz Crell, began to promote and publish papers by phlogistonists criticizing Lavoisier's new theory, as early as 1787.

Even on the new continent far away, the echoes of the new nomenclature provoked some immediate reactions. Thomas Jefferson, who was far from persuaded by the new theory, wrote to James Madison in July 1788:

> . . . it is yet indeed a mere embryon. Its principles are contested. Experiments seem contradictory: their subjects are so minute as to escape our senses; and their result too fallacious to satisfy the mind. It is probably an age too soon to propose the establishment of system. The attempt therefore of Lavoisier to reform the Chemical nomenclature is premature. One single experiment may destroy the whole filiation of his terms, and his string of Sulfates, Sulfites, and Sulfures, may have served no other end than to have retarded the progress of the science by a jargon from the confusion of which time will be requisite to extricate us.
>
> Accordingly it is not likely to be admitted generally.[235]

Writing to James Curie some months later, Jefferson went further, remarking:

> You have heard of the new nomenclature endeavoured to be introduced by Lavoisier, Fourcroy, etc. Other chemists in this country, of equal note, reject it, and prove, in my opinion, that it is premature, insufficient, and false. These latter are joined by the British chemists; and upon the whole I think the new nomenclature will be rejected after doing more harm than good.[236]

This last statement is even more significant when it is borne in mind that Jefferson was politically and scientifically closer to the French than to the British cultural tradition.[237]

The "American" attitude to the new nomenclature was also indirectly confirmed by Benjamin Franklin, who was a close friend of Lavoisier. On October 23, 1788, Franklin wrote a very affectionate letter to Madame Lavoisier, but he suspended judgment on the new nomenclature, simply writing: "please present my thanks to M. Lavoi-

---

[235]Letter dated July 19, in *The Papers of Thomas Jefferson*, vol. 13 (Princeton, 1956), 381.

[236]Quoted by Duveen and Klickstein, "The Introduction of Lavoisier's Chemical Nomenclature in America," in *Isis*, 45 (1954), 280.

[237]On Jefferson and science in the United States during the second half of the eighteenth century, see John C. Greene, *American Science in the Age of Jefferson* (Iowa, 1984).

sier for the *Nomenclature Chimique* he has been so good as to send me, (it must be a very useful book)."[238]

Thus in 1787 and 1788, in both the Old World and the New, Lavoisier's nomenclature was either rejected outright or received with extreme skepticism.

On February 6, 1788, Chaptal summarized this situation perceptively when he remarked:

> I have read all that has been written against the new nomenclature [. . .]. They want the *revolution* to come about gradually.[239] [my italics]

Indeed, the publication of the *Méthode* left the chemical community in a state of shock, and the immediate reaction on all sides was to appeal for a slowing down of the process of reform. Most of the phlogistonists perceived the contents of the nomenclature as a revolutionary rupture with the past and opposed it with unusual passion. Chemists known for their serene and gentle character, such as Joseph Black, would lose their temper when they heard the new names.

To sum up: Lavoisier's contemporaries realized that something very innovative was being introduced in chemistry, and many of them already regarded it as a *revolution*.

Immediate Assents

In 1787 and 1788, the prospects for Lavoisier's theory did not look very promising. With the exception of the mathematicians and physicists of the Académie des Sciences—Vandermonde, Cousin, Laplace, Monge and, possibly, Bailly—few scientists accepted the principles of the new chemistry. At least 15 years after the first draft of the oxygen theory, Lavoisier was still struggling to find support among chemists. Fortunately the few conversions to the new chemistry were important and influential. The young naturalist Chaptal did not merely support the new system, he also had the opportunity to teach it to a young genera-

---

[238] *Memoirs of the Life and Writings of Benjamin Franklin*, 3rd edition (London, 1818), vol. 3, 241.

[239] "J'ai lu tout ce qu'on ecrit contre la nouvelle nomenclature [. . .]. On vouloit que la revolution se fit lentement: et de cette manière on y travaille depuis qu'on fait de la chimie. *On ne peut ne pas brusquer une revolution lorsqu'on a un plan arretté et que tout le monde s'y conforme, mais le plan n'existoit pas en chimie et il a fallu le presenter*" (my italics). LC, vol. V, in course of publication.

tion of naturalists at Montpellier. Thus those who studied chemistry in that city were forced to learn it through the new language. On February 6, 1788, Chaptal wrote to Lavoisier reporting the impact of the new nomenclature on this wide audience:

> This year I have obtained the most satisfying proof of the advantage which the new nomenclature affords in the study of chemistry. My course is followed with enthusiasm and I am happy to see that everybody understands me. The new words being nothing but the analysis of what they represent, each time one uses a word one shows the actual thing. . . .[240]

Chaptal also remarked that the student had no difficulty in remembering the new names; on the contrary, the memory work was eased by the systematic nature of the nomenclature. Indeed, if compared to traditional chemistry courses, which always began with a long and tedious explanation of the lexicon, the facility and the consistent logic of the new nomenclature gave both the teacher and his pupils an effective means of accelerating the understanding of chemical names. Lavoisier and Condillac seemed therefore to be right when they claimed that the logic behind an analytic language had the essential advantage of following the path of human understanding rather than being an artificial and arbitrary construction. Chaptal's letter was powerful confirmation of the pedagogic and logical advantages of the *Méthode*, and though we do not have his reply, Lavoisier's confidence in the fruitful application of Condillac's method to chemistry must surely have increased.

In 1788 Lavoisier received his first letter of support from abroad. On November 7, the Swiss naturalist Horace Bénédict de Saussure wrote to Madame Lavoisier and acknowledged that the new theory had finally overcome his lingering doubts.[241] The thoroughness of this conversion was shown by Saussure's efforts to spread the new ideas.

In addition to these two important recruits, Lavoisier received an enthusiastic letter from the British theologian Edward King, who wrote to him on January 25, 1788:

---

[240] "J'ai fait cette année l'epreuve la plus satisfaisante de l'avantage que présente la nouvelle nomenclature dans l'etude de la chimie. Mon cours est suivi avec un enthousiasme dont on a peu d'exemples et je vois avec satisfaction que tout le monde m'entendra. Les mots nouveaux n'etant que l'analyse de ce qu'ils représentent, chaque fois qu'on se sert d'un mot on presente par la même chose, et deux connoissances tres difficiles celle des mots et celle des choses, nous n'en avons plus qu'une à acquerir puisqu'elles sont confondues," LC, vol. V, in course of publication.

[241] *Ibid.*

> Sir, having received so great information from your most important chemical discoveries; and having found reason to make such singular use of them, as a means of bringing to light a very surprising part of the contents of the Holy Scriptures, and of illustrating many sublime descriptions contained therein, in a manner in which they could not be illustrated till these days in which you have lived. . . .[242]

In a country where Priestley used his investigations into the nature of gases to support his theological beliefs, it was not surprising for somebody to find confirmation of his own metaphysics in Lavoisier's theory. King was the first Englishman to speak approvingly of the new chemistry and it was probably gratitude for this unexpected success that led Lavoisier to offer to begin a regular correspondence with him.[243] Nevertheless, in 1788, no more than a dozen people supported Lavoisier's ideas, and of these only three were professional chemists.

Despite the scanty support they had obtained, the Lavoisierian *cotérie* showed a confidence in the new theory that had no counterpart in the opposite side. The attack on phlogiston begun by Lavoisier was pursued by his collaborators with the same energy and conviction. To this end, Guyton published the second part of the first volume of the *Encyclopédie Méthodique* at the beginning of 1789, and in the *Second Avertissement*, he explained why he had changed his opinion since 1786 and begun to support the new chemistry and nomenclature.[244] Guyton remarked that the old doctrine still holding sway throughout Europe was based on a hypothesis that had been refuted by the pneumatic experiments carried out in the 1770s and 1780s. "In this respect the revolution has been achieved," he said;[245] all that remained was to continue along the path clearly pointed out by the new theory. Guyton admitted that when it first appeared the concept of phlogiston provided an extremely ingenious and fruitful hypothesis that substantially contributed to the early progress of the whole science. Indeed, so successful was it that for a few decades after Stahl nobody doubted the existence and identity of phlogiston[246] and

---

[242]*Ibid.* Edward King (1735?–1807) may have enclosed with his letter his *Morsels of criticism: tending to illustrate some few passages in the Holy Scripture upon philosophical principles and an enlarged view of things* (London, 1788), in which he attempted a new interpretation of the Bible based also on scientific evidence.

[243]LC, vol. V, Letter dated April 20, 1788.

[244]*Encyclopédie Méthodique*, vol. 1, 625. Guyton also briefly presented the new nomenclature in Crell's *Chemische Annalen* (1787), II, 54–55.

[245]"A cet égard la révolution est consommée," *Encyclopédie Méthodique*, 626.

[246]On this original and important view of the phlogiston theory, see Chapter One.

prominent and accurate experimenters, such as Marggraf, Bergman, Macquer, Scheele, Black, Priestley, Kirwan, and many others, subordinated their observations to its *dogma*. To put it succinctly, "they were born to this religion."[247] Lavoisier was the first to "break this long cult," and after 20 years of experimental endeavor, he succeeded in challenging its authority and outlining a theory of the gaseous state and of acidification, which, contrary to Stahl's doctrine, "*rejette toute hypothèse.*" Significantly, in order to demonstrate the consistency of Lavoisier's theory, Guyton devoted most of his "*avertissement*" to an explanation of the new nomenclature. Here he suggested that the original idea of reforming the language of chemistry was his own and that Lavoisier, Fourcroy, and Berthollet began to collaborate with him when the project was at an advanced stage.[248] Having made this, as has already been shown, quite unfounded claim, Guyton proceeded to demolish all criticism of the new nomenclature. The objection that the new nomenclature was based on a systematic theory with inadequate foundations overlooked the essential characteristics of the new language, which was "*a method of naming rather than a nomenclature.*"[249] This meant that the name of a substance could have been assigned automatically. Obviously this argument answered only partly the criticism of the phlogistonists because the new nomenclature did not merely establish a few mechanical rules of naming but also made detailed assertions concerning the structure and definition of matter and outlined a correspondence between these interpretations and the system of names.

Guyton was more incisive in replying to the criticism levelled against specific terms. To those who claimed in the *Dictionnaire de Trévoux* that the etymology of *azote* meant the first principle of metals, Guyton answered that it was the practice to give the authority to words and not to a pedantic and rigid erudition. And the following quotation from Cicero's *De Natura Deorum* was Guyton's answer to those who objected that some words had a Greek root and some a Latin one:

---

[247]*Encyclopédie Méthodique*, vol. 1, 627.

[248]"Depuis la publication de cette première partie [1786], trois célèbres Académiciens, MM. Lavoisier, Berthollet et De Fourcroy, également convaincu de la nécessité de cette réforme, *ont bien voulu travailler avec moi à y mettre la dernière main*" (my italics), *ibid.*, 635. Although Guyton did not explicitly claim the paternity of the *Méthode* in any other works, he always implied a prominent role. In view of what has been said and of the historical evidence at our disposal, we cannot but conclude that Guyton's claim was not legitimate and that as with Bergman's nomenclature he showed a tendency to claim credit that rightly belonged to others.

[249]*Ibid.*, 635.

> The term "aether" also we may borrow, and employ it like "air" as a Latin word.[250]

The Greek word *pyro* was likewise Latinized by Pliny the Elder and became a French word in the Renaissance when the term *pyrotechnie* appeared. Thus the etymological reconstruction of a word was not legitimate if the historical changes in its meaning were not taken into account. In this connection Guyton recommended his critics to read Turgot's article *Etymologie*, written for Diderot's *Encyclopédie*, which stated:

> There is at present no language which has not been formed by a mixture or an alteration of more ancient languages.[251]

Turgot claimed that etymological research into words had to be historical rather than philological and that to understand the meaning of a word correctly one had to follow its semantic evolution. This approach implied that the etymology of a word in the strict sense could only provide a partial and relative account of its meaning. Applied to chemical nomenclature, this principle deprived words of their absolute meaning and placed them in a wider and evolutionary perspective. Guyton accordingly claimed that philological etymologies were mere suppositions if not supported by a detailed historical genealogy of the words they claimed to trace. Guyton's explanation of the principles of the *Méthode* confirms that he was not fully convinced of the strict correspondence between words and things asserted by Lavoisier. Furthermore, in order effectively to answer the critics of the *Méthode*, Guyton was forced to recognize that use and convention were the roots of scientific language. As we shall later see, this belief, which owed more to Locke than to Condillac, was eventually expressed by Guyton clearly and in detail. Guyton's claim that the *Méthode* was a mere mechanism for arranging words did little to placate Lavoisier's opponents, who saw clearly that the new language implied an entirely new vision of chemical theory. Those phlogistonists, who, like Black and Kirwan, had enthusiastically adopted the nomenclature proposed by Bergman and Guyton, reacted violently against Lavoisier's nomenclature because of the basic claim that each word corresponded to one thing and because of the semantic power assigned by Lavoisier to words such as oxygen and hydrogen. It was impossible for them to distinguish the nomenclature from its theoreti-

---

[250]The full passage is: "Hunc rursus amplectitur immensus aether, qui constat ex altissimi ignibus—mutuemur hoc quoque verbum, dicaturque tam aether latine quam dicitur aer . . .," Cicero, *De Natura Deorum*, Liber II: XXXVI.

[251]Turgot, *Etymologie*, Maurice Piron (ed.) (Bruges, 1961), 9.

cal assumptions and to separate the word from the thing to which Lavoisier had applied it. In this respect the phlogistonists understood the power and the structure of the *Méthode* better than some of its supporters.

To conclude his *avertissement,* Guyton published the table of chemical nomenclature with slight changes,[252] and thereafter the *Encyclopédie Méthodique* spoke Lavoisier's language.

THE SWANSONG OF PHLOGISTON

Simultaneously with the appearance of the *Méthode,* Richard Kirwan published an explicit and comprehensive attack on Lavoisier's theory of acidity that was one of the most significant works of theoretical chemistry to appear in the 1780s.[253] The *Essay on Phlogiston* was the first chemical treatise since Stahl's *Traité du Soufre* to consider the principles of the phlogiston theory as a whole, and it compared them with the new series of discoveries.[254] Early in his essay, Kirwan agreed with Lavoisier that Stahl's theory was based upon a hypothesis and declared:

> It must be owned [. . .] that this doctrine rested on the supposition that inflammable bodies contain some substances which uninflammable bodies do not; nor have chymists, until within these few years, been able to afford any proof that this supposition was well founded, as they have never been able to exhibit this substance singly and by itself.[255]

However, since chemists could not find any better explanation of the inflammability of bodies, they retained the principle even though they were not able to isolate it. Kirwan recognized that the recent discovery that calcined metals increased their weight and absorbed air, instead of losing the phlogiston, had inspired some chemists to seek a new explanation of the calcination of metals and "on these grounds Mr. Lavoisier reversed the ancient hypothesis."[256] Instead of supposing that inflammable bodies contained a particular substance, the French chemist assumed that at a certain degree of temperature the same bodies showed a tendency to combine with the purest part of the

---

[252]*Encyclopédie Méthodique,* vol. 1, 654–664.
[253]Richard Kirwan, *An Essay on Phlogiston and the Constitution of Acids* (London, 1787).
[254]Abbri (1984), 363–369, is one of the few historians to point out the crucial importance of this work and comment at length its contents.
[255]Kirwan (1787), 2.
[256]*Ibid.,* 4.

atmosphere (the oxygen). "Hence he at first modestly proposed his doubts, whether the supposition of such a substance, as the chymists called phlogiston, were not entirely superfluous."[257] Kirwan was thus fully aware of the consistency of Lavoisier's challenge to phlogiston and saw that a logical reply was needed. Kirwan admired the experiments performed by Priestley, which showed that by heating inflammable air (hydrogen), calces were reduced to metals and the gas was absorbed and that conversely by heating the metals inflammable air was expelled from them. These experiments suggested that the base of inflammable air had the same properties as phlogiston and that its existence was now supported by conclusive evidence. Phlogiston was finally a concrete physical substance that could be isolated by a simple chemical operation. Once the existence of phlogiston had been demonstrated, Kirwan had to reply to the arguments underlying Lavoisier's anti-phlogistic reasoning. When he did so, the Irish chemist tried to treat the whole controversy as concerning a small quantity of experimental facts rather than a choice between alternative theories of matter. By this subtle approach, Kirwan aimed to demonstrate that the facts did not conclusively support either theory. Accordingly, he declared:

> Limited as this controversy appears to a small number of bodies; it is nevertheless of great importance, if an exact arrangement of our ideas, and distinct and true view of the operations of nature be of any importance. The bodies above-mentioned [oxygen, hydrogen, etc.] are the subject of many, and the instruments of all chymical operations; without a knowledge of their composition, and a clear perception of their mode of action, it will be impossible to form even an approximation to a solid theory of this science.[258]

Like most of his British colleagues, Kirwan upheld the primacy of facts over scientific theory, but he tried to justify his general attitude with some original arguments: 1. he claimed that his phlogiston theory, which identified phlogiston with hydrogen, was supported by experimental evidence; 2. he argued that the difference between his theory and Lavoisier's lay only in restricted investigative areas; 3. he claimed that most pneumatic experiments were still not sufficiently established to give the victory to either theory; 4. he recognized the merit of Lavoisier's interpretation and the simplicity of the new theory and

[257] *Ibid.*

[258] *Ibid.*, 7.

tried to extract the experiments from the theory that engendered them and to incorporate them in the domain of phlogiston.

Kirwan acknowledged that the "antiphlogistic hypothesis" was "recommendable for its simplicity" and owed a lot to Lavoisier, who introduced "mathematical precision into experimental philosophy." However, Kirwan claimed that these qualities were far from sufficient to ensure the success and the epistemological legitimacy of oxygen:

> We must not be deluded by a false shew of simplicity; when all is well considered, the ancient doctrine will be found the more uniform of the two: in this, pure air is never said to unite to any substance, but the principle of inflammability with which it is evidently united in the deflagration of inflammable air; in the modern, without being an acid, or affording any sign of salinity, the principal prerogative of acid substances, that of uniting to almost all bodies, is assigned to it.[259]

Kirwan rejected the name oxygen because it went beyond the immediate empirical evidence. In 1787, the only thing that could be said about "dephlogisticated air," a name that was now significantly changed by Kirwan to that of "pure air," was that it combined with inflammable air. Nothing confirmed the claim that this substance took part in the combination of all acids and that it therefore had to be called oxygen. Kirwan clearly realized that behind its apparent simplicity the new theory implied the total dismantling of the old chemistry, down to the last detail, and that in accepting the name of oxygen one had to adopt all the objects, speculative hypotheses, and general conceptions that inevitably orbited around it. To Kirwan, by contrast, a *simple* theory was one that remained anchored in the evidence of the experiments and that eschewed generalizations. This difference in scientific approach is illustrated by the fact that Kirwan considered Lavoisier "plutôt grand physicien que chimiste."[260] Kirwan was actually a mineralogist by training and regarded chemistry mainly as the science of the inorganic world. When describing the development of the *modern* phlogiston theory, Kirwan traced its origins back to the "mineralogy and metallurgy" of the Germans and the Swedes and praised their approach. In contrast Lavoisier tried to introduce the "method of geometers" into chemistry and dealt with chemical phenomena by considering their physical properties rather than their chemical composition. This he

---

[259] *Ibid.*, 8.

[260] Letter from Kirwan to Guyton dated April 2, 1787 (in course of publication).

saw as the most striking characteristic of Lavoisier's approach to chemistry from an entirely new methodological perspective.

The publication of the *Essay on Phlogiston* was seen by Lavoisier as a good opportunity to answer one of the major defenders of phlogiston properly. He asked his wife to translate the *Essay* into French and invited Guyton de Morveau (as has been mentioned, a close friend of Kirwan), Laplace, Monge, Berthollet, and Fourcroy to help him answer the book point by point.[261] The idea of translating and annotating the book to refute its contents was an inspired one and led to a breakthrough in the controversy on the nature of oxygen. As an indication of how effective the publication of this book was, it is worth noting that the second English edition, which appeared with all the notes and comments of the Lavoisierian *cotérie*,[262] ultimately convinced Kirwan himself of the consistency of the new theory, and that the German translator Freidrich Wolff also came to accept the new ideas.[263]

In the preface to the French translation, Madame Lavoisier underlined the propaganda value of this work and remarked:

> The study of chymistry becomes daily more general, and its advances appear to be more particularly rapid since a philosopher [Lavoisier], well known for the scrupulous attention with which he makes his experiments, and the philosophical spirit which has directed his observations, has formed a new theory, in which nothing is admitted but established facts.[264]

Lavoisier's notes on Kirwan's objections were reprints from some of his recent papers. In reply to Kirwan's introduction, for instance, Lavoisier reprinted his report on Hassenfratz's and Adet's system of chemical symbols, written in 1787 and referred to earlier in this chapter. The resounding success of this translation convinced many that while phlogiston was under serious threat, a promising new era was dawning in chemistry.

---

[261]*Essai sur le Phlogistique, et sur la constitution des acides, traduit de l'anglois de M. Kirwan, avec des notes de MM. De Morveau, Lavoisier, de la Place, Monge, Berthollet, et de Fourcroy* (Paris, 1788).

[262]Kirwan, *An Essai of Phlogiston* . . . (London, 1789).

[263]*Antiphlogistische Anmerkungen der Herren de Morveau, Lavoisier* . . ., *zu Kirwan's Abhandlung über Phlogiston* . . . (Berlin, 1791).

[264]Kirwan, *An Essay* (1789), xii.

CHAPTER FIVE

The Chemical Revolution

"*Il n'y point de phlogistique; l'air pur est une substance simple; l'air phlogistiqué, l'air inflammable en sont d'autres; la combustion n'est qu'une combinaison de l'air pur avec les corps.* Semblable aux mots sublimes rapportés dans la Genèse, ce peu de paroles a tout éclairci, tout débrouillé; le chaos s'est arrangé, chaque fait est venu se placer, et le tout a formé le plus magnifique des tableaux."

> Georges Cuvier, *Recueil des Eloges Historiques,* vol. 1 (Paris, 1819), 207–208.

"The possibility of an ideally perfect Nomenclature is probably confined to the one case in which we are happily in possession of something approaching to it; the Nomenclature of elementary Chemistry."

> J.S. Mill, *A System of Logic,* 9th ed. (London, 1875), vol. 2, 284.

"Why this universal systematizing, systematizing, systematizing?"

> J.S. Mill, *Auguste Comte and Positivism* (London, 1865), 141.

The Permanent Revolution

The general concept of scientific revolution has given rise to so much discussion among historians and philosophers of science that it is difficult to use it without lapsing into the ambiguities of modern interpretations. Most of the secondary literature on this subject emphasizes and supports the theories of scientific change fashionable at the time when it was written, without trying to understand the historical meaning and role of the term "scientific revolution." Historians and philosophers have thus alternately endorsed and discounted modern images of science instead of examining the historical contexts to which they have referred. In his comprehensive survey of the concept of revolution in science, I. Bernard Cohen avoided this tendency and offered a different interpretative perspective which put a stronger emphasis on the *historical* role of this controversial notion.[1] In Cohen's view we can speak of a scientific revolution only in those cases when a community of scientists is aware that a rupture has occurred in the traditional way of conceiving their own science. This awareness of sudden change, which may be demonstrated by either support for the change or resistance to it, is found but rarely in the history of science. The development of science after the Renaissance seems to be a story of evolutionary progress rather than violent changes of direction. The strength of cultural traditions and the natural inclination to fit new ideas into an existing conceptual framework seem to leave little room for scientific revolutions.

However, chemistry provides one of the rare examples of the introduction of a new theory being accompanied by a general and widespread awareness of the sudden change in the theory and language of the science. Lavoisier's theory aroused a debate (see Chapter Four) in which the traditional conception of chemical substances was questioned to its foundations. What is more significant is the fact that both Lavoisier and his contemporaries themselves used the term "revolution" when speaking of this change. As Cohen pointed out:

> the Chemical Revolution has a primary place among revolutions in science in that it is the first generally recognized major one to have

[1] I. Bernard Cohen, *Revolution in Science* (Cambridge, Mass., London, 1985).

> been called revolution by its chief author, Antoine-Laurent Lavoisier. Scientists before Lavoisier had been aware that their program would lead to something entirely new and would run directly counter to established norms of accepted scientific belief; but Lavoisier, unlike the others, also had in mind the concept of revolutions in science as a particular kind of change in thinking, and he made the judgement that his own work would in fact constitute just such a revolution.[2]

This exceptional historical circumstance was appreciated by Marcelin Berthelot in 1890 in *La révolution chimique: Lavoisier*,[3] which was the first historical survey to examine the sudden theoretical shift caused by Lavoisier. Berthelot's interpretation soon became the standard one, and only a few German and Italian scholars, motivated primarily by patriotism, challenged its validity. Aldo Mieli, for instance, emphasized that Lavoisier did not contribute original experiments to the progress of pneumatic chemistry and that, gifted though he unquestionably was in organizing the experiments of others in a coherent new theory, chemistry itself was an established and autonomous science long before his theory appeared.[4]

Metzger and Meldrum developed Berthelot's picture and added new evidence of the revolutionary nature of Lavoisier's chemical work. Since then, most historians of eighteenth-century chemistry have referred to the evolution of Lavoisier's work as the "chemical revolution." Only in recent times have the claims of Mieli and Ostwald found new acceptance. Some historians have felt that "chemical revolution" is not an appropriate label for what in their view was the climax of a cumulative evolutionary process rooted in an experimental tradition that was established long before Lavoisier.[5] In their view, too much importance has been attached to the theoretical differences between Lavoisier and his adversaries and too little attention has been paid to the experimental background shared by both parties. However, the

---

[2] *Ibid.*, 229.

[3] Marcelin Berthelot, *La révolution chimique: Lavoisier* (Paris, 1890).

[4] Lavoisier "offre un esempio spiccatamente caratteristico dell'ufficio del genio organizzatore ma non inventivo. E' difficile, infatti, trovare una scoperta veramente originale di Lavoisier. Non solamente quasi tutte le sue idee, quasi tutti i suoi concetti fondamentali, erano stati emessi, applicati da altri [. . .] E se non il fondatore della chimica, cha esisteva da lungo tempo, si deve ammirare in Lavoisier l'organizzatore definitivo di una delle modificazioni nelle idee teoriche di questa scienza," A. Mieli, *Lavoisier*, 2nd ed. (Rome, 1926), 9–10.

[5] Although from different perspectives, all the following studies endorse Mieli's claims; J.B. Gough, "Lavoisier and the Fulfillment of the Stahlian Revolution," in Donovan (1988), 15–33; C.E. Perrin (1988); R. Siegfried, "Lavoisier and the Phlogistic Connection," in *Ambix*, 36 (1989), 31–40; F.L. Holmes (1989). See also Melhado (1986).

implicit aim of this interpretation has been to question the philosophical and historical legitimacy of the concept of scientific revolution rather than to investigate a specific case history. From a reading of these essays, one forms the impression that their main purpose has been to challenge the conclusions of other historians rather than to examine the sources of Berthelot's interpretation. Reversing this approach and giving absolute priority and authority to the primary sources produce surprising answers to some of the questions concerning the historical role of the chemical revolution.

The notion of chemical and scientific revolution makes its first appearance in the lexicon of the Enlightenment, and as Cohen pointed out, the meaning attributed to it during the eighteenth century was close to the one with which we are acquainted today. The first scientist to use the term was probably Fontenelle, but it was only with the publication of the *Encyclopédie* and the intellectual ferment that accompanied it that the concept of scientific revolution acquired impetus and popularity. To the *Philosophes*, the understanding of the value, the cause, and the contents of a scientific revolution meant the *historical* understanding of their own scientific activity and its relationship with that of their predecessors. As we saw in Chapter One, consciousness of the past played a significant part in their efforts to define a new role for science in society. Combining this conceptual approach with a commitment to the ideology of progress, the *philosophes* established a close connection between the notion of epoch and that of revolution. In this respect, too, the "preliminary discourse" of d'Alembert offers the most revealing example of the meaning of scientific revolution in the eighteenth century. Describing the historical role played by Cartesian physics in the history of science, d'Alembert remarked:

> . . . By that *revolt* whose fruits we are reaping today, he rendered a service to philosophy perhaps more difficult to perform than all those contributions thereafter by his illustrious successors. He can be thought of as a leader of conspirators who, before anyone else, had the courage to arise against a despotic and arbitrary power and who, in preparing a *resounding revolution*, laid the foundations of a more just and happier government, which he himself was not able to see established.[6] [my italics]

D'Alembert's passage is revealing in many respects: first, the analogy between a scientific revolution and a political one, a blow struck against despotism and in support of reason, was an innovative

[6]D'Alembert (1963), 80.

idea, and the closer connection between science and political reform after the 1750s, ultimately resulting in the French Revolution, makes d'Alembert appear something of a prophet. Second, the connection established between scientific revolutions and their historical appearance and necessity underlined the evolutionary nature of scientific knowledge and opposed the immobility of historical doctrines. Significantly Diderot shared many of these opinions, and in the article "*Encyclopédie*," he claimed that ignorance of the mechanism of human knowledge made it impossible to know its limits. Accordingly the perfectibility of human reason inevitably led to a pattern of continuous progress and revolution:

> But revolutions are necessary, they always have been and they always will be; the greatest interval between one revolution and another is fixed: this cause alone limits the extent of our works.[7]

Diderot saw the scientific revolution as inevitable, as a cyclical phenomenon, and he clearly underlined that the progress of the sciences was marked by interruption and revolution rather than by steady evolution. The cultural legacy of d'Alembert's and Diderot's ideas on the role of revolution in science undoubtedly encouraged French scientists to question the established tradition with more conviction and vigor and made them receptive to new ideas and theories. The favor enjoyed by the notion of scientific revolution is in fact a typical trait of the French Enlightenment, and it played an important role in the enormous growth in self-confidence of eighteenth-century French science.

Lavoisier subscribed to these general principles and, as we saw in Chapter Four, from his very first approach to pneumatic chemistry, defined his theory as a revolution in chemistry and physics. This constant conviction or, as Meldrum called it, "fixed idea" that he retained throughout his life, certainly played an important role in the overthrow of phlogiston and accounts for some of the antagonism shown by Lavoisier to his predecessors. The awareness of the totality of the mental shift which was involved undoubtedly inspired Lavoisier to create a new nomenclature and a new theory even when the experimental basis for such a project was clearly poor.

---

[7]"Cependant les connoissances ne deviennent et ne peuvent devenir communes, que jusq'à un certain point. On ignore, à la verité, quelle est cette limite. On ne sait jusqu'où l'espèce humaine irait, ce dont serait capable, si elle n'était point arrêtée dans ses progrès. Mais les révolutions sont nécessaires; il y en a toujours eu, et il y en aura toujours; le plus grand intervalle d'une révolution à un autre est donné: cette seule cause borne l'etendue des nos travaux," Diderot, *Oeuvres complètes*, vol. 7 (Paris, 1976), 186–187.

After 1773 Lavoisier and his contemporaries began to attach the label of revolution to their pneumatic discoveries more often.[8] On January 15, 1774, Lavoisier referred to Black's investigation of fixed air as a preparatory step towards "une revolution dans la phisique et dans la chimie."[9] Eleven years later, he announced to the geologist Henri Paul Reboul: "the revolution is on the verge of completion."[10] As early as 1788, Chaptal defended Lavoisier's theory and the scientific revolution it entailed against the phlogistonists (see Chapter Four). On August 27, 1789, having seen Lavoisier's *Traité*, Caillet applauded the "grande revolution."[11]

With the outbreak of the French Revolution, the concept of a revolution in chemistry became extremely popular. Lavoisier himself was the first to draw this analogy in a famous letter to Benjamin Franklin, dated February 2, 1790. For a full understanding of the importance and implications of this letter, it is useful to consider the whole text:

> Sir and most illustrious colleague:
>
> M. de Coullens de Caumont, who is returning to America, has been so good as to take with him two copies of a work which I published here about a year ago under the title of "Treatise on the Elements of Chemistry," one of which I beg you will accept with my best wishes, passing on the other to the "Society" in Philadelphia.
>
> In all the chemical treatises which have been published since Stahl, the authors have begun by laying down a hypothesis and afterwards have attempted to show that with this given principle all the phenomena of chemistry might be accounted for tolerably well.
>
> I believe, and a great number of scholars of today agree with me, that the hypothesis advanced by Stahl, and since modified, is erroneous, that phlogiston in the sense that Stahl understood the word does not exist, and it is principally to develop my ideas on this subject that I undertook the work which I have the honor to send you.
>
> I have sought, as you will observe in the preface, to arrive at the truth by a coordination of facts; to suppress theorizing as much as possible, which is often a false instrument and has often led us to

---

[8]On the eighteenth-century concept of chemical revolution, see H. Guerlac, "The Chemical Revolution: A Word from Monsieur Fourcroy," in *Ambix*, 23 (1976), 1–4; M.P. Crosland, "Chemistry and the Chemical Revolution," in G.S. Rousseau and R. Porter (eds.), *The Ferment of Knowledge* (Cambridge, 1980), 389–416; F. Abbri, "Teorie e strumenti: Lavoisier e la rivoluzione chimica," in S. Poggi and M. Mugnai (eds.), *Tradizioni filosofiche e mutamenti scientifici* (Bologna, 1990), 69–89.

[9]LC, vol. II, 401.

[10]*Ibid*, vol. IV, 175.

[11]*Ibid.*, vol. V (in course of publication).

stray from following the light of observation and experiment as closely as we might.

This course, which has not hitherto been followed in chemistry, has given me an occasion to classify my work in an absolutely new way, and chemistry has been brought much closer than it was to experimental physics. I sincerely hope that your leisure and your health will permit you to read the first chapter, in as much as your approval and that of those scholars of Europe who are without prejudice in matters of this sort is the object of my earnest ambition.

It seems to me that to present chemistry in this form is to render the study infinitely easier than it has been. Young men whose minds are not preoccupied with any other system seize it with avidity, but old chemists still reject it, and most of these have even more trouble in comprehending and understanding it than those who have not yet made any study of chemistry.

French scholars are divided at this moment between the old and the new doctrine. I have on my side M. de Morveau, M. Berthollet, M. de Fourcroy, M. de La Place, M. Monge, and in general the physicists of the Academy. The scholars of London and of England are also gradually dropping the doctrine of Stahl, but the German chemists still adhere to it strongly. *This then is the revolution which has occurred in an important branch of human knowledge since your departure from Europe; I look upon this revolution as well advanced and it will be complete if you will stand with us.*

Now that you have been informed as to what has transpired in chemistry, it might be well to speak of our political revolution. We look upon this as accomplished and accomplished irretrievably. There still exists, however, an aristocratic party, which makes futile efforts and whose weakness is very evident; the democratic party has on its side the greatest number and the most intelligent, able and enlightened. People of moderation, and those who have kept their heads in the general upheaval, think that events have swept us too far, that it is regrettable to have been obliged to arm the people and all the citizens, that it is impolitic to place authority in the hands of those who should be amenable to authority, and they fear that the establishment of a new constitution will meet with opposition on the part of the very ones in whose favour it has been made.

We have arrived at a most important epoch, as we are occupied at this moment, all over the realm, with the formation of the municipalities, and it appears that the elections have been conducted in a very peaceable manner. . . .[12] [my italics]

---

[12]The letter by Lavoisier was published by René Fric in the *Bulletin Historique et Scientifique de l'Auvergne,* 2nd series, IX (1924), 145–152. An English translation was made by Edgar F. Smith, *Old Chemistries* (New York, 1927), 31–32. See also Duveen and Klickstein, "Benjamin Franklin and Antoine Laurent Lavoisier," *Annals of Science,* 11 (1955), 123.

This powerful letter vividly evokes d'Alembert's analogy between a scientific and a political revolution. In both cases the connection seemed to be due to external circumstances, but it is natural to believe that both d'Alembert and Lavoisier assumed that a scientific revolution involved more than limited changes to a specific branch of knowledge and that there was a connection, though not a very clear one, between a major change in science and the progress of society in general.

In 1791 Lavoisier wrote to Chaptal, declaring: "all the young scientists adopt the new theory and I thence conclude that the revolution is accomplished in chemistry."[13] Lavoisier was not alone in considering his theory a revolutionary event in the history of chemistry; many of his adversaries argued that the new theory was unacceptable for just that reason. As was seen in the previous chapter, the publication of the *Méthode de nomenclature chimique* shook the chemical community, and it is worth noting that one of the main arguments of the phlogistonists against the new language was that its acceptance would cause a revolution in chemistry. In 1789 the British chemist James Keir declared: "the progress of Chemistry within the last twenty years has been more rapid than was ever made perhaps by any science in an equal period," and more explicitly: "Chemistry had by this time undergone or commenced a remarkable revolution."[14] A few years later, Kirwan criticized the "entire subversion of the ancient nomenclature," and his colleague Dickson defined the new chemical language as the effect of "the rage for revolution of the confederate reformists."[15] In 1796 the leading figure in the phlogistonist camp, Joseph Priestley, published a pamphlet in defence of Stahl's theory that was addressed to the principal advocates of the antiphlogistic line, "Berthollet, LaPlace, Monge, Morveau, Fourcroy, Hassenfratz and Kirwan."[16] In the introduction he declared:

> There have been few, if any, revolutions in science so great, so sudden, and so general, as the prevalence of what is now usually termed the *new system of chemistry*, or that of the *Antiphlogistians*, over the doctrine of Stahl, which was at one time thought to have been the greatest discovery that had ever been made in science.[17]

---

[13]From Grimaux (1888), 126. The translation is by Cohen (1985), 230.

[14]Keir, *A First Part of a Dictionary of Chemistry* (London, 1789), iii and i.

[15]Stephen Dickson, *An Essay on Chemical Nomenclature*, with annotations by Kirwan (London, 1796), xv and 116.

[16]Priestley, *Considerations on the Doctrine of Phlogiston, and the Decomposition of Water* (1796) (New York, 1969).

[17]*Ibid.*, 19.

In describing the overthrow of phlogiston as the greatest revolution in science, Priestley went even further than Lavoisier. In addition to this, the British chemist identified the chemical nomenclature as the main cause of the sudden rupture:

> Every year of the last twenty or thirty years has been of more importance to science, and especially to chemistry, than any ten in the preceding century. So firmly established has this new theory been considered, that a *new nomenclature*, entirely founded upon it, has been invented, and it is now almost in universal use; so that, whether [we] adopt the new system or not, we are under the necessity of learning the new language, if we would understand some of the most valuable of modern publications.[18]

The connection between the chemical revolution and the creation of an entirely new nomenclature is extremely significant, in particular because it came from an adversary of the new language. Priestley emphasized that the linguistic reforms differed from previous ones in that the new nomenclature was invented without drawing on an already established technical lexicon. The creation of such a language confronted the chemists with the new theory much more starkly. With the publication of the *Méthode*, most of the bridges with the past were cut, and an entire generation of chemists had to learn to speak a new language in order to talk about the new science. Joseph Black, who was sympathetic to Bergman's projected reform of the nomenclature, which was, however, still rooted in the lexicons of the past, declared that Bergman's nomenclature:

> . . . became very prevalent, and would, in all probability, have soon obtained the acquiescence of all the philosophical chemists, *had not the science itself sustained, at this very time, a great and almost total revolution.*[19] [my italics]

The "eclectic" German chemist Friedrich Albert Carl Gren, who made a singular attempt to reconcile Lavoisier's theory with Stahl's, finally admitted that the French chemist "engendered this great and total revolution in the concepts of chemistry, until lately based on the hypothesis of phlogiston."[20] The supporters of the new theory and

---

[18] *Ibid.*, 21.

[19] Black (1803), 488–489.

[20] "Herr Lavoisier [. . .] bewirkte diese grosse und totale Revolution in den bisherigen, auf die Annahme des Phlogistons begründeten, Vorstellungsarten in der Chemie, und erreichte ein neues System, welches, weil es das Daseyn des Phlogistons ganz leugnet, deshalb den Namen des antiphlogistischen Systems erhalten hat . . .," F.A.C. Gren, *Grundriss der Chemie*, 2nd ed. (Halle, 1800), vol. 1, 14–15.

nomenclature were even more inclined to use the term "revolution." In 1790 the Swiss naturalist M.A. Pictet wrote to Lavoisier announcing his conversion and referring to "la belle et grande revolution" in chemistry.[21] The English translator of Lavoisier's *Traité*, Robert Kerr, gave Lavoisier the credit for the new chemistry:

> The very high character of Mr Lavoisier as a chemical philosopher, and the great revolution which, in the opinion of many excellent chemists, he has effected in the theory of chemistry, has long made it much desired to have a connected account of his discoveries, and of the new theory he has founded upon the modern experiments written by himself.[22]

In 1795 the professor of chemistry at Pisa University, Giorgio Santi, declared:

> This chemical revolution is a precipitous torrent which, if it encounters obstacles, overcomes them and carries all victoriously before it.[23]

In 1798 the American naturalist Thomas P. Smith dealt with Lavoisier's work in the significantly titled *A Sketch of the Revolutions in Chemistry*,[24] while in England the physician William Babington described the introduction of the new nomenclature in chemistry as a "complete and essential change."[25] In 1796 the German chemist Girtanner announced to Fourcroy that the "revolution in chemistry" was now finally accepted by German chemists.[26] In France Fourcroy was undoubtedly the scientist who most frequently used the term "revolution" with reference to the new chemistry. On February 6, 1789, Fourcroy and Cadet de Vaux commented on the *Traité élémentaire de chimie* as follows:

> The physicists, and all who devote themselves to the study of natural philosophy, know that the revolution that chemistry has for some years been experiencing is due to the experiments by Lavoisier.[27]

---

[21]Letter by Pictet to Lavoisier kept in the Library of Clermond-Ferrand. I took this quotation from Guerlac-Perrin's manuscript notebooks n. 67.

[22]Lavoisier, *Elements of Chemistry* (London, 1790), v.

[23]"E' questa rivoluzione Chimica un torrente precipitoso, che trova ostacoli, è vero, ma che li supera, e che seco vittoriosamente tutto trasporta," quoted in Abbri (1990), 88.

[24]Thomas P. Smith, *A Sketch of the Revolutions in Chemistry* (Philadelphia, 1798), 27 and ff.

[25]William Babington, *Syllabus of a Course of Chemical Lectures Read at Guy's Hospital* (London, 1797), iv.

[26][Fourcroy], *Encyclopédie Méthodique*, vol. 3 (Paris, 1797), 617.

[27]"Les physiciens, & tous les hommes qui s'adonnent à l'étude de la philosophie naturelle, savent que c'est aux expériences de M. Lavoisier qu'est due la révolution que la Chimie a éprouvée depuis quelques années," Fourcroy and Cadet, "Extrait des Registres de la Société Royale de Médicine," in Lavoisier, *Traité* . . . (1789), vol. 2, 629.

In his *éloge* to Lavoisier, which was written in 1796, Fourcroy defined the impact of the new chemistry as a "grande revolution".[28] A year later, when tracing the history of pneumatic chemistry, he used the term "revolution in chemistry" several times and in one particularly revealing passage he declared:

> The long details contained in the preceding subchapter, and presented through a division in eight periods or epochs related to the main traits of the history of the great revolution which chemistry has enjoyed after 1755, and especially since 1766 and 1772 until 1788 [. . .] must have brought to the awareness of the attentive reader this important truth: chemistry is an absolutely new science, very different from what it was before this memorable revolution, and founded upon principles far removed from the former ones.[29]

The contrast between the experimental continuity of pneumatic chemistry during the period 1755–1788 and the revolution brought about by Lavoisier was described by Fourcroy as follows:

> A revolution was prepared in every quarter, but no one had yet undertaken to guide it, or to direct and regulate its motion. A great change in the theory was necessary, and it was in the bosom of the Academy of Sciences at Paris, that it was effected under the auspices, and by the genius, of Lavoisier.[30]

Fourcroy did not belittle the importance of the experimental sources that provided the material for the revolution, but he correctly pointed out that these experiments became relevant only when combined in a coherent theory of the gaseous state.

The list of chemists using the term chemical revolution when speaking of Lavoisier and his work could be extended, but these quotations suffice to demonstrate that the notion of chemical revolution is not, as many seem to believe today, merely an expression of Berthelot's historiographic positivism but is firmly rooted in the chemical literature of the period. What is even more significant is that this large number of chemists and naturalists, coming from different parts of the world and sometimes supporting theories in total and open opposition to Lavoisier's, all perceived Lavoisier's works as a chemical revolution, a total break with the past. From opposite points of view, Fourcroy and Priest-

---

[28]Fourcroy, *Notice sur la vie et les travaux de Lavoisier* (Paris, 1796), 36. On this, see Guerlac (1976).

[29][Fourcroy], *Encyclopédie Méthodique*, vol. 3, 715.

[30]Fourcroy, *A General System of Chemical Knowledge* (London, 1804), vol. 1, 49.

ley, Chaptal and Keir, Lavoisier and Kirwan recognized that between 1770 and 1790 a revolution had occurred: the phlogistonists, who supported a cumulative vision of science, saw in the chemical revolution, and in the sudden changes it brought about, an epistemological error; the anti-phlogistonists, who regarded the progress of chemistry as an irresistible current, applauded it as a long-awaited event.

The success of the chemical revolution spread to other disciplines and seized the imagination of other scientists, who were fascinated and inspired by its rapid and powerful impact. The astronomer Delambre, for instance, remarked:

> The revolution recently brought about in chemistry could not happen without turning many experimentalists a little out of their ordinary course, when they saw in a neighbouring science a road opened that promised more numerous discoveries.[31]

Indeed, the resounding success of Lavoisier's work attracted many scientists from other disciplines to chemistry. The physicists Laplace, Monge, Volta, and Nollet, the physicians Vicq d'Azyr, Fontana, and Moscati, the naturalists Bonnet and Spallanzani, the philosophers Condorcet and Kant, etc. perceived the innovative power of pneumatic chemistry and became involved in a discipline extraneous to their own specialization.

Another naturalist who admired the chemical revolution was Georges Cuvier, who said of Lavoisier's *Traité*:

> It is the most convenient point of departure we can choose, because it really represents one of the greatest epochs in the history of sciences [. . .] the new theory of combustion is the most important revolution enjoyed by any science during the eighteenth century.[32]

This kind of judgment, which became commonplace in the literature of the period, was epitomized in Condorcet's *Esquisse*, a work that set out to list the memorable events and epochs in the history of mankind. In the ninth epoch, Condorcet evoked, among other things, the passage from alchemy to chemistry:

> Observation of the phenomena that accompany the composition and

[31]Delambre, *Rapport historique sur les progrès des sciences mathématiques depuis 1789* . . . (1810) (Paris, 1989), 61; English translation by Crosland (1980), 389.

[32]Cuvier, *Rapport . . . sur la chimie et les sciences de la nature* . . . (1810) (Paris, 1989), 96 and 80.

the decomposition of bodies, the search for the laws governing these operations, the analysis of substances into even simpler elements acquired an increasing precision and rigour.

But we must mention in addition to this progress in chemistry some of the improvements that, affecting, as they do, a given scientific system in its entirety by extending its methods rather than by increasing its truths, foretell and prepare a successful revolution. Such was the discovery of new methods of collecting and subjecting to experiment the expansible fluids that had hitherto eluded such examination [. . .]. Another such improvement was the formation of a language in which the names designating the sciences indicate either their relations or differences between those substances having a common element, or the class to which they belong. Other such improvements were the introduction of a scientific notation in which these substances are represented by analytically combined characters that can express even the most common operations; the general laws of affinities [. . .].[33]

Condorcet said that the mere manipulation of experiment characteristic of pre-Lavoisierian chemistry differed substantially from the method introduced by Lavoisier. Like Priestley, Condorcet emphasized that the "*successful revolution*" was the analytical nomenclature of gases. Fully aware that an "increasing of truths" does not alone make a scientific revolution, Condorcet argued that its primary cause was a change in the method of investigation.

The followers of Condillac, the *idéologues*, regarded Lavoisier's revolutionary achievements as a direct result of his successful application of the analytical method to chemistry. Garat laid particular stress on the role played by Condillac's philosophy of language in Lavoisier's chemical system and wrote:

Lavoisier accomplished in chemistry a revolution so striking, so fecund in truths and happy prospects, by bringing to that science his genius, his fortune, and the principles and the language of Condillac.[34]

The physician Cabanis went so far as to speak of Lavoisier as "one of the principal authors of perhaps the most important revolution which natural science has ever experienced."[35]

---

[33]Condorcet (1955), 154–155.

[34]D.J. Garat, *Mémoires historiques sur le XVIIIe siècle et M. Suard* (Paris, 1821), vol. 2, 326; English translation by Albury (1986), 211.

[35]Quoted by Albury (1986), 213.

The numerous references to the "chemical revolution" in the literature of the period show conclusively that this notion is not a historiographic myth invented in order to celebrate the genius of Lavoisier. On the contrary, in the eyes of Lavoisier's contemporaries, the term "chemical revolution" was more often associated with specific aspects of the new pneumatic doctrine and nomenclature rather than with the cult of a single scientist. In this respect it is interesting to note that the majority of the contemporary allusions to a chemical revolution relate to the new nomenclature. It was not the experiment on the decomposition of water, nor the one on the calcination of metals, that was perceived as revolutionary, but rather the new nomenclature that defined these experiments. The chemical revolution was in fact provoked by a language that no longer reflected the history of chemistry and that forced all the chemists of Europe to become students of an entirely new chemical grammar.

Pedagogy or Science?

The first impression of Lavoisier's *Traité élémentaire de chimie* was published in March 1789, a few months before the outbreak of the French Revolution.[36] This timing was undoubtedly a fortunate coincidence, since as Daumas has clearly demonstrated, Lavoisier had been working on an elementary chemistry textbook since the beginning of the 1780s. The *Traité* was a brilliant synthesis of over 20 years of chemical investigation, theoretical reasoning, and technological achievement. After the publication of the *Méthode* and of the *Essai sur le phlogistique*, Lavoisier probably felt it urgent to produce a general work setting out the results of the new chemical system.[37] Lavoisier's collaborators Guyton and Fourcroy were already publishing chemical textbooks, and Lavoisier certainly knew that in 1788 Fourcroy was preparing a new edition of his popular *Leçons élémentaires d'histoire naturelle et de chimie*, incorporating the new theory and nomenclature, and that Chaptal intended

[36]Lavoisier, *Traité élémentaire de chimie, présenté dans un ordre nouveau et d'après les découvertes modernes* (Paris, 1789). The first printing of the first edition was published in one volume and is so rare that its existence was discovered only in the 1950s. Here I have used the second issue of the first edition printed in two volumes and the English translation by Robert Kerr that was printed in Edinburgh in 1790 under the title *Elements of Chemistry.*

[37]As Duveen and Klickstein (1954), 162, showed, the publication of the first issue of the *Traité* was printed in a hurry, and "it may well be that Lavoisier was anxious to see the finished book, not only for his own personal satisfaction, but also in order to place it before the scientific world at the earliest possible moment."

to publish an elementary textbook (*Élémens de chymie*, 1790), also based on the new principles. Feeling the pressure from his ambitious colleagues, Lavoisier wanted to demonstrate that what was beginning to be considered *la doctrine française* was in fact his own, and the *Traité* made this point very clearly.

Specialists in the subject have generally agreed that the publication of the *Traité* is one of the most important events in the history of chemistry. Paradoxically, despite this widely shared view, the contents of this seminal work have seldom been the object of systematic study. The reason for this incongruous fact is that the *Traité* has been seen rather as a synthesis of previous memoirs and the manifesto of the new theory than as an original contribution to chemistry in its own right.[38] It has been regarded as a collection of familiar experiments arranged in a rational order, and it has been assumed that the *Traité* was addressed to beginners in chemistry and that for this reason its aim was primarily pedagogic. These interpretations undoubtedly give an indication of the structure of Lavoisier's work, but they fail to explain why the publication of the *Traité* provoked such fierce discussion among experienced chemists, nor why it was used by neither teachers nor students. Is it believable that an astute tactician, such as Lavoisier, addressed his work to chemistry students but failed to get through to them? Those who have emphasized the pedagogic structure of the *Traité* have usually cited the following passage of the "Preliminary discourse":

> I have imposed upon myself, as a law, never to advance but from what is known to what is unknown; never to form any conclusion which is not an immediate consequence necessarily flowing from observation and experiment; and always to arrange the facts, and the conclusions which are drawn from them, in such an order as shall render it most easy *for beginners in the study of chemistry* thoroughly to understand them."[39] [my italics]

If read out of context, this passage certainly appears to stress the pedagogic purpose of the *Traité*, but on closer inspection, the reader finds a more complex meaning. The paraphrase of Condillac's *Logique* shows that Lavoisier certainly believed in the importance of making

---

[38]Meldrum and Guerlac concentrated on Lavoisier's writings between 1766 and 1774; Daumas studied Lavoisier's sources both experimental and technological; Partington dealt mostly with Lavoisier's memoirs; Grimaux, McKie, and Berthelot provided little more than an index to the *Traité*.

[39]Lavoisier (1790), xviii–xix.

himself understood even by beginners in chemistry, but that his primary aim was to write a chemical work that followed the path of nature and uncovered the truth. The pedagogic component could not be broken down into its ontological and naturalistic elements. The intimate relationship of these elements is well illustrated in the following passage:

> When we begin the study of any science, we are in a situation, respecting that science, similar to that of children; and the course by which we have to advanced is precisely the same Nature follows in the formation of the truth.[40]

The *Traité* was thus addressed to beginners because it followed the paths of nature and not vice versa, and it is clear that Lavoisier used the term beginner to mean a wider category of people than students alone. To make progress in science, all naturalists had to a certain extent to be beginners, in the sense that they had to follow the guidance of nature as well as that of an ordinary tutor. Lavoisier insisted throughout the *Traité* that his theory's fidelity to nature was his prime concern. This ideal, certainly inspired by Condillac and shared by most leading French scientists, was destined to have an important influence on the teaching of chemistry, but the contents of the *Traité* cannot be considered solely as a teaching manual without distorting its internal structure and historical value.

If we try to elucidate the genesis of Lavoisier's work, we obtain a more complete picture. Lavoisier began with a philosophical introduction in which he explained the sources of the *Traité*:

> When I began the following Work, my only object was to extend and explain more fully the Memoir which I read at the public meeting of the Academy of Sciences in the month of April 1787, on the necessity of reforming and completing the Nomenclature of Chemistry [. . .]. While I thought myself employed in forming a Nomenclature, and while I proposed to myself nothing more than to improve the chemical language, my work transformed itself by degrees, without my being able to prevent it, into a treatise upon the Elements of chemistry.[41]

Thus Lavoisier saw his most important work as deriving from the *Méthode* and the elaboration of a textbook in chemistry as the consequence of the reform of chemical language. There is little doubt, as Daumas has shown, that the idea of a chemical textbook existed be-

---

[40] *Ibid.*, xvi.

[41] *Ibid.*, xiii–xiv.

fore 1787, but I believe it is significant that Lavoisier produced the book only after formulating the principles of the theory to which he wanted to give clear verbal expression. This indication given by Lavoisier seems even more relevant when one considers the actual contents of the *Traité*, of which the first two parts, at least, deal with questions directly related to the new nomenclature. But we shall return to this point later.

The preliminary discourse drew attention to the controversial philosophical principles expounded in the first memoir of the *Méthode* and declared Lavoisier's support for Condillac's epistemology with even more enthusiasm. Having connected the composition of the *Traité* with that of the *Méthode*, Lavoisier confirmed his linguistic assumptions and declared:

> The impossibility of separating the nomenclature of a science from the science itself, is owing to this, that every branch of physical science must consist of three things; the series of facts which are the objects of the science, the ideas which represent these facts, and the words by which these ideas are expressed. Like three impressions of the same seal, the word ought to produce the idea, and the idea to be a picture of the fact. And, as ideas are preserved and communicated by means of words, it necessarily follows that we cannot improve the language of any science without at the same time improving the science itself; neither can we, on the other hand, improve a science, without improving a language or nomenclature which belongs to it. *However certain the facts of any science may be, and, however just the ideas we may have formed of these facts, we can only communicate false impressions to others, while we want words by which these may be properly expressed.*[42] [my italics]

A mere catalogue of observations of phenomena, together with their correct interpretation, was not enough to protect a theory from the danger of error. Without a philosophy of language suited to the nature of phenomena, science conveyed "false impressions." Lavoisier also warned his readers that this principle was not placed at the beginning of a general treatise simply as a persuasive device; on the contrary, *to the attentive reader*, it would be clear that "the first part of this treatise [would] afford frequent proofs of the truth of the above observations."[43]

Lavoisier went on briefly to describe the content of his work and his intended method of investigation. Despite his admiration for math-

[42] *Ibid.*, xiv–xv.

[43] *Ibid.*, xv.

ematics, Lavoisier regarded the efforts of Bergman, Guyton, Kirwan, and others to quantify chemical affinities as not having sufficient experimental foundation to be worth pursuing. Lavoisier was well aware that Guyton's numerical expressions giving the force of cohesion between substances were the product of arbitrary assumptions rather than the fruits of a concrete application of mathematical patterns to chemistry, and he therefore decided to cover this investigative field. Lavoisier also announced that the *Traité* would include "no chapter on the constituent and elementary parts of matter." Investigations of the primitive principles of matter, which had been characteristic of chemistry for centuries, were all based on "assertions without proofs." In this respect, too, Lavoisier was threatening the established tradition:

> I shall therefore only add upon this subject, that if, by the term *elements*, we mean to express those simple and indivisible atoms of which matter is composed, it is extremely probable we know nothing at all about them; but, if we apply the term *elements*, or *principles of bodies*, to express our idea of the last point which analysis is capable of reaching, we must admit, as elements, all the substances into which we are capable, by any means, to reduce bodies by decomposition.[44]

As was seen in the previous chapter, Lavoisier successfully replaced a conception of matter based upon a few primitive principles or elements with one centered on the notion of chemical reaction. To Lavoisier the simple substances meant little outside the general framework within which he conceived them, i.e., his theory of chemical combination.

The final part of the preliminary discourse was once again devoted to nomenclature, since the "foregoing reflections upon the progress of chemical ideas naturally [applied] to the words by which these ideas [were] to be expressed."[45] Lavoisier confirmed his faith in these linguistic principles and, while setting out his ideas on the classification of chemical substances, stated his ontological belief:

> In the *natural* order of ideas, the name of the class or genus is that which expresses a quality common to a great number of individuals: The name of the species, on the contrary, expresses a quality peculiar to certain individuals only. *These distinctions are not, as some may imagine, merely metaphysical, but are established by Nature [. . .]. This is the logic of all the sciences, and is naturally applied to chemistry.*[46] [my italics]

---

[44] *Ibid.*, xxiv.
[45] *Ibid.*, xxv.
[46] *Ibid.*, xxvi–xxvii.

These claims presented more decisively and confidently than the *Méthode* Lavoisier's naturalistic conception of scientific language. Like the experimentation presented in the *Traité*, which was true because it was faithful to nature, the new chemical nomenclature and classification were more effective because they indicated the real nature of chemical substances. The basic and explicit aim of the *Traité* was to provide a linguistic foundation for a theory based on the nature of phenomena. Lavoisier's appeal to the ultimate authority of experimentation and facts and his claim that he formed "no conclusion which [was] not fully warranted by experiment"[47] were underpinned by his belief that he had created a language that reflected reality and was structured on a syntax true to nature herself. The empiricism expressed in Lavoisier's work was therefore different in kind from that of Priestley and Black. To Lavoisier the observation of a fact alone was meaningless if not placed within a wider relational context. In this respect it is not unexpected to find Lavoisier concluding the preliminary discourse by quoting Condillac's *Logique*:

> But, after all, the sciences have made progress, because philosophers have applied themselves with more attention to observe, and have communicated to their language the precision and accuracy which they have employed in their observations: in correcting their language they reason better.[48]

Most latterday readers of the *Traité*[49] have dismissed the linguistic principles so firmly laid down by Lavoisier and claimed that there is little connection between the preliminary discourse and the rest of the *Traité*. Lavoisier has been seen as using Condillac's philosophy for rhetorical effect rather than as a fundamental principle of his chemical system. This interpretation has been so widespread that with few exceptions historians have paid little attention to the reform of chemical nomenclature. But in fact, Lavoisier claimed that his *Traité* sprang from that reform and that it was nothing more than a modification and an elaboration of the nomenclature proposed in 1787. This dis-

---

[47] *Ibid.*, xx.

[48] *Ibid.*, xxxvii.

[49] Metzger (1935); Crosland (1963), (1978), (1980); Dagognet (1969); Albury (1972) are the few historians who have recognized the importance of Condillac's philosophy to Lavoisier's chemical system. Despite the unquestionable value of these studies, the treatment of the link between Condillac and Lavoisier has been mostly descriptive. Grimaux, Berthelot, Guerlac, Partington, McKie, and the recent collection of essays edited by Donovan (1988) dismiss the theoretical and philosophical aspects of the new nomenclature in a few lines.

concerting claim has been either wholly or partially ignored, but a careful reading of the text of the *Traité* shows a striking continuity with the philosophical principles of the preliminary discourse and a consistent application of the table of chemical nomenclature presented in the *Méthode.*

The *Traité* is in three parts: in the first Lavoisier gives an account "of the formation and decomposition of Aeriform fluids—of the combustion of Simple bodies—and the Formation of Acids," in short, of his own theory; in the second part, he gives an account "of the Combination of Acids with Salifiable Bases, and the formation of Neutral Salts"; and in the last part, he provides a "description of the Instruments and Operations of Chemistry." This original arrangement will be discussed later, but it may be noted here that Lavoisier left practically no room for the traditional survey of chemical matter and that most of his *Elements of Chemistry* concentrated on pneumatic chemistry. According to Lavoisier, the elements of chemistry could only be understood in the light of the new pneumatic discoveries, as could the analysis of the animal and vegetable kingdoms. Departing from the practice followed in all chemistry textbooks printed before 1789, Lavoisier did not present the objects of the science in three separate orders (vegetable, animal, and mineral chemistry) but in a single framework. As in the table of chemical nomenclature, all the objects of chemistry were arranged in a systematics of combination, to which the main keys were oxygen and heat.

Lavoisier began the *Traité* by providing a new definition of heat. As in the *Méthode,* he argued for a change of name to the more appropriate one of *caloric.* The chemical role of caloric in Lavoisier's system was crucial, and it is not surprising to find it presented as the foundation of the entire edifice of chemistry. All interpretation of chemical relations required the identification of the role of this evanescent substance:

> It is difficult to comprehend these phenomena, without admitting them as the effects of a real and material substance, or a very subtle fluid, which, insinuating itself between the particles of bodies, separates them from each other; and, even allowing the existence of this fluid to be hypothetical, we shall see in the sequel, that it explains the phenomena of nature in a very satisfactory manner.[50]

In a system in which, in principle, only established facts were to be admitted, this description of caloric may sound like an *a priori* assumption, but this is not the case. Lavoisier claimed that if caloric were not,

---

[50]Lavoisier (1790), 4.

as he believed it to be, a real substance, its function would have been identical with those mathematical hypotheses introduced in physical systems, and its importance to his theory would have been the same. Although Lavoisier was not able to detect and isolate caloric by experiment, he needed to assume its existence to provide a consistent explanation of the change of state of substances. Lavoisier believed that "the same body becomes solid, or fluid, or aeriform, according to the quantity of caloric by which it is penetrated; or, to speak more strictly, according as the repulsive force exerted by the caloric is equal to, or stronger or weaker than, the attraction of the particles of the body it acts upon."[51] With the introduction of this concept of caloric, Lavoisier stated "as a general principle" that almost every body could exist in three different states (solid, liquid and aeriform). To be sure, the most important quality attributed to caloric was that of being the base of all kinds of gas. The particles of caloric possessed in fact the quality of "repelling" each other, and "when we are once permitted to suppose this repelling force, the *rationale* [explanations] of the formation of gases, or aeriform fluids, becomes perfectly simple: tho' we must, at the same time, allow, that it is extremely difficult to form an accurate conception of this repulsive force acting upon very minute particles at great distance from each other."[52] Lavoisier's explanation of the formation of elastic fluids in terms of the repulsive force of particles was ingenious, despite being "an explanation of elasticity, by an assumption of elasticity."[53] In assuming the existence of caloric, Lavoisier was not merely relying on a hypothesis without any experimental basis; he was providing his theory of the gaseous state with a powerful and consistent explanation that facilitated the understanding of *real* chemical reactions and, more generally, of chemical matter.

Once he had laid down the first principle of chemistry, Lavoisier went on to give an account of the chemistry of gases. He began by showing with great confidence how important this new field was to the whole of chemistry:

> These views which I have taken of the formation of elastic aëriform fluids or gases, throw great light upon the original formation of the atmospheres of the planets, and particularly that of our earth.[54]

---

[51] *Ibid.*, 7.
[52] *Ibid.*, 23–24.
[53] *Ibid.*, 22. "We must allow that this is only an explanation of elasticity, by an assumption of elasticity, and that we thus remove the difficulty one step farther, and that the nature of elasticity, and the reason for caloric being elastic, remains still unexplained."
[54] *Ibid.*, 26.

It was thanks to modern pneumatic chemistry that the atmosphere had been recognized as a compound of different kinds of airs, some of which had a tendency to take part in the combination of solid substances. Consequently Lavoisier entitled the chapter devoted to the examination of gases "Nomenclature of the Several Constituent Parts of Atmospheric Air," and as with caloric, he insisted on the necessity and importance of their accurate definition and naming. In this chapter Lavoisier also took the opportunity to clarify the criteria adopted in the *Méthode* and to answer some general criticisms, in particular that of having overthrown the traditional chemical lexicon and replaced it with abstract and hypothetical neologisms. Lavoisier made it clear that the principles behind the *Méthode* did not imply the abandonment of all the traditional names, but only of some of them:

> We have not pretended to make any alteration upon such terms as are sanctified by ancient custom; and, therefore, continue to use the words *water* and *ice* in their common acceptation: We likewise retain the word *air*, to express that collection of elastic fluids which composes our atmosphere; but we have not thought it necessary to preserve the same respect for modern terms, adopted by later philosophers, having considered ourselves as at liberty to reject such as appeared liable to occasion erroneous ideas of the substances they are meant to express.[55]

In his usual clear and concise style, Lavoisier expressed the distinction between common and scientific language. Whatever its critics may have believed, the new nomenclature was not formed to create a new language in opposition to a specific tradition or to challenge a theory of matter; most of the common words (those of metals and of many salts) were in fact retained. Only those specific terms that gave a misleading idea of the substance they were supposed to describe were replaced with new ones. Concerning the way he used Greek, which had aroused much criticism, the French chemist reminded his adversaries that the use of words taken from a dead language was practical, but that they had to be used "in such a manner as to make their etymology convey some idea of what was meant to be represented."[56] Lavoisier was forced to admit that certain names coined in the *Méthode* did not comply with the general principles he set out in his introductory memoir. The term *gas*, for instance, which was introduced by Van Helmont to denote a metaphysical entity (see Chapter Three), was retained in the

[55] *Ibid.*, 50.
[56] *Ibid.*

nomenclature of 1787. In cases like this, however, Lavoisier argued that it was better to restrict the meaning of the word than to change it. The term *gas* therefore became "a generic term, expressing the fullest degree of saturation in any body with caloric; being, in fact, a term expressive of a mode of existence."[57] Finally Lavoisier restated the principles applied to the denomination of gases:

> I should anticipate subjects more properly reserved for the subsequent chapters, were I in this place to enter upon the nomenclature of the several species of gasses: It is sufficient, in this part of the work, to establish the principles upon which their denominations are founded. The principal merit of the nomenclature we have adopted is, that, when once the simple elementary substance is distinguished by an appropriate term, the names of all its compounds derive readily, and necessarily, from this first denomination.[58]

In this passage Lavoisier spotlighted an essential feature of the new nomenclature that was entirely original and that expressed the new theory. By emphasizing the systematic and combinatory qualities of the new language, Lavoisier was in fact characterizing his philosophy of matter, in which the simple substances were considered with relation to their part in chemical reactions rather than as isolated and individual *substantives.* The "principal merit" of the new chemistry, as of the new language, was the possibility of "necessarily" deducing the entire edifice of chemistry from an "appropriate" definition of the elementary substances. The universe of language anticipated knowledge of the material universe.[59] Most of the names coined in the *Méthode* and the *Traité* expressed and defined substances that had not yet been isolated and in fact existed only on paper. As Hélène Metzger has clearly shown, Lavoisier often named substances before he could isolate them chemically.[60] In this connection Lavoisier's definition of *muriatic acid* (Hydrochloric acid—HCl) may be recalled. Muriatic acid "is capable of combining with a farther dose of oxygen, by being distilled from the oxyds of manganese, lead, or mercury, and the resulting acid, which we name *oxygenated muriatic acid,* can only, like the former, exist in the gaseous form, and is absorbed, in a much smaller

---

[57] *Ibid.*, 50–51.

[58] *Ibid.*, 53.

[59] "L'universe du langage, pour employer la pittoresque expression de M. Brunschvicg, forma momentanément écran à la connoissance de l'univers matériel," Metzger (1935), 31.

[60] *Ibid.*, 31–32

quantity of water."[61] This definition turned out to be wrong and, despite the efforts of Thenard and Gay Lussac to defend it, Davy demonstrated beyond any doubt that it is impossible to isolate any quantity of oxygen from this acid. This flaw in Lavoisier's theory of acidity called into question the very foundations of Lavoisier's philosophy of matter, in which the role and definition of oxygen were central. But oxygen meant more than "begetter of acidity," and Lavoisier assigned a wider kingdom to the name. The explanation of common operations, such as calcination (oxydation) and combustion, and of many other pneumatic phenomena, depended on the definition of oxygen. The language of oxygen was invading the organic world as well:

> We have likewise oxyds in great numbers from the vegetable and animal kingdoms; and I shall show, in the sequel, *that this new language throws great light upon all the operations of art and nature.*[62] [my italics]

Once again, it was the language that was explaining the phenomena of nature, and by developing its syntax, the chemist was able to explore even those territories that his experimental knowledge did not yet reach. The expressive power of this passage can hardly be overstated, and it is important to repeat that Lavoisier dwelt on the explanation and the function of the new nomenclature throughout the *Traité*. The nomenclature of the acids, which is given in the fifth chapter, illustrates the intrinsic character of the work. In it Lavoisier insisted on the validity of the linguistic principles, more specifically of those related to the definition of compounds. Naturally Lavoisier was happy with the introduction of the suffixes denoting different degrees of saturation of the acids. "By this may be seen," he remarked, "that the language we have adopted is both copious [fecond] and expressive."[63] Nevertheless, Lavoisier was well aware that the animal and vegetable kingdoms were still to be investigated and that the many newly discovered substances confronted chemists with a vision of matter even more complex than originally thought. The numerous vegetable acids were showing that organic chemistry was far more complex than inorganic chemistry and that the new nomenclature already needed considerable extension. Significantly this problem did not cause Lavoisier much anxiety; to solve it, he proposed to follow the principles which had inspired the *Méthode*:

---

[61] Lavoisier (1790), 234.
[62] *Ibid.* (1790), 81.
[63] *Ibid.* (1790), 80.

> It would be easy to apply the principles of our nomenclature to give names to these vegetable acids and oxyds, by using the names of the two substances which compose their bases. They would thus become hydro-carbonous acids and oxyds.[64]

The application of the suffixes produced, for example, the following names:

ELEMENTS (1790)	*TRAITÉ (1789)*
Hydro-carbonous oxyd	oxide hydro-carboneux
Hydro-carbonic oxyd	oxide hydro-carbonique
Carbono-hydrous oxyd	oxide carbone-hydreux
Carbono-hydric oxyd	oxide carbone-hydrique
Hydro-carbonous acid	acide hydro-carboneux
Hydro-carbonic acid	acide hydro-carbonique
Oxygenated hydro-carbonic acid	acide hydro-carbonique oxygéné
Carbono-hydrous acid	acide carbone-hydreux
Carbono-hydric acid	acide carbone-hydrique
Oxygenated carbono-hydric acid	acide carbone-hydrique oxygéné

Lavoisier believed that the above terms would suffice to indicate all the varieties in nature, and that, as the vegetable acids became well understood, they would naturally arrange themselves under these denominations. While admitting that the knowledge of vegetable chemistry was still "in some degree obscure," Lavoisier retained great confidence in the explanatory power of his nomenclature and predicted its application to organic chemistry. This prophecy was made when the knowledge of vegetable acids, and of vegetable chemistry in general, was still extremely poor. The fact that the nomenclature was eventually successfully applied to organic chemistry by Berzelius reiterates the importance of Lavoisier's emphasis on scientific language.

From the methodical classification and naming of vegetable acids and oxides, Lavoisier went on to discuss the chemical analysis of their compounds. In the chapter devoted to the "Decomposition of Vegetable Oxyds by the Vinous Fermentation," the French chemist finally formulated his famous *principe*:

> We may lay it down as an incontestible axiom [principe] that, in all operations of art and nature, nothing is created; an equal quantity of

[64]*Ibid.* (1790), 117–118.

> matter exists both before and after the experiment; the quality and quantity of the elements remain perfectly the same; and nothing takes place beyond changes and modifications in the combination of these elements. Upon this principle the whole art of performing chemical experiments depends: we must always suppose an exact equality [equation] between the elements of the body examined and those of the products of its analysis.[65]

This formulation of the principle of conservation of matter has given rise to a diversity of interpretations, and in various attempts to exaggerate or diminish its importance, it has often been extrapolated from the historical and logical context to which it belongs. The law of conservation of matter is in fact the logical implication of the philosophy of matter expressed by Lavoisier in his nomenclature. First of all, the principle was not inferred from any empirical observation in particular, but was conceived *a priori*. In Lavoisier's own words, one was "*obliged to suppose*" the conservation of matter regardless of the fact that empirical data might contradict it. Just as, in the table of nomenclature, the combination of names of compounds *necessarily* derived from their linguistic roots, so in a concrete chemical reaction the proportions of the constituents were regulated within the framework of the law. In Lavoisier's lexicon nothing was lost, and in each name, the primitive constituents were traceable; likewise, in the chemical laboratory, "nothing [took] place beyond changes and modifications in the combination" of the elements. In both cases *a priori* assumptions played a part, serving the same purpose. Had not Lavoisier defined the meaning of each substance so clearly, his law of conservation of matter would have been far less effective. How would it have been possible to demonstrate the conservation of matter when the properties of a substance were not apparent? Many of the names in the phlogistic nomenclature denoted ambiguous qualities that never expressed quantitative ideas. A name of a compound of two substances, for instance, was often coined *ex nihilo* without any linguistic reference to the two original names. Lavoisier constructed a nomenclature that reflected his definition of the law of conservation of matter perfectly. In the new chemical names, as in the chemical reactions, everything was detectable quantitatively, and nothing was lost. The table of chemical nomenclature of 1787 anticipated the law; here the names of simple substances and compounds were fully consistent with this law, and

[65] *Ibid.* (1790), 130–131.

every name showed that in all possible combinations "the quality and quantity of the elements [remained] perfectly the same."

It is true that the law of conservation of matter had been assumed by chemists before Lavoisier, but it was only with the creation of an entirely new nomenclature capable of demonstrating its application that it became a general and universally accepted axiom of modern chemistry. It was no coincidence, I believe, that Lavoisier was able to formulate the role of this law in chemistry with such clarity only *after* stipulating the exact structure and function of language in science.

In the second part of the *Traité*, Lavoisier presented 45 tables containing the names of the simple substances and their combinations. This methodic classification of names formed "a kind of recapitulation of the first fifteen chapters of the first Part."[66] In this part it was demonstrated clearly, from the point of view of both presentation and substance, that chemistry needed a systematization of its vocabulary. The first of the 45 tables naturally showed the simple substances (see Fig. 14). By comparison with the table presented in the *Méthode*, there were some important changes. Since Lavoisier omitted several bases and all the alkalis, the original 55 simple substances of the *Méthode* were here reduced to 33. The selection reflected Lavoisier's own definition of simple substances:

> All we dare venture to affirm of any substance is, that it must be considered as simple in the present state of our knowledge, and so far as chemical analysis has hitherto been able to show.[67]

The presence of substances like ammonia, which chemical analysis showed to be compounds, would have undermined the internal consistency of the table. On the other hand, Lavoisier admitted that there might be other changes in the future:

> As chemistry advances towards perfection, by dividing and subdividing, it is impossible to say where it is to end; and these things we at present suppose simple may soon be found quite otherwise.[68]

Thus it was clearly stated that the simple substances were no longer considered the hypostatic foundations of the chemical system as they had been in the past; chemistry was now founded on a new conception

[66] *Ibid.* (1790), 174.
[67] *Ibid.* (1790), 177.
[68] *Ibid.*

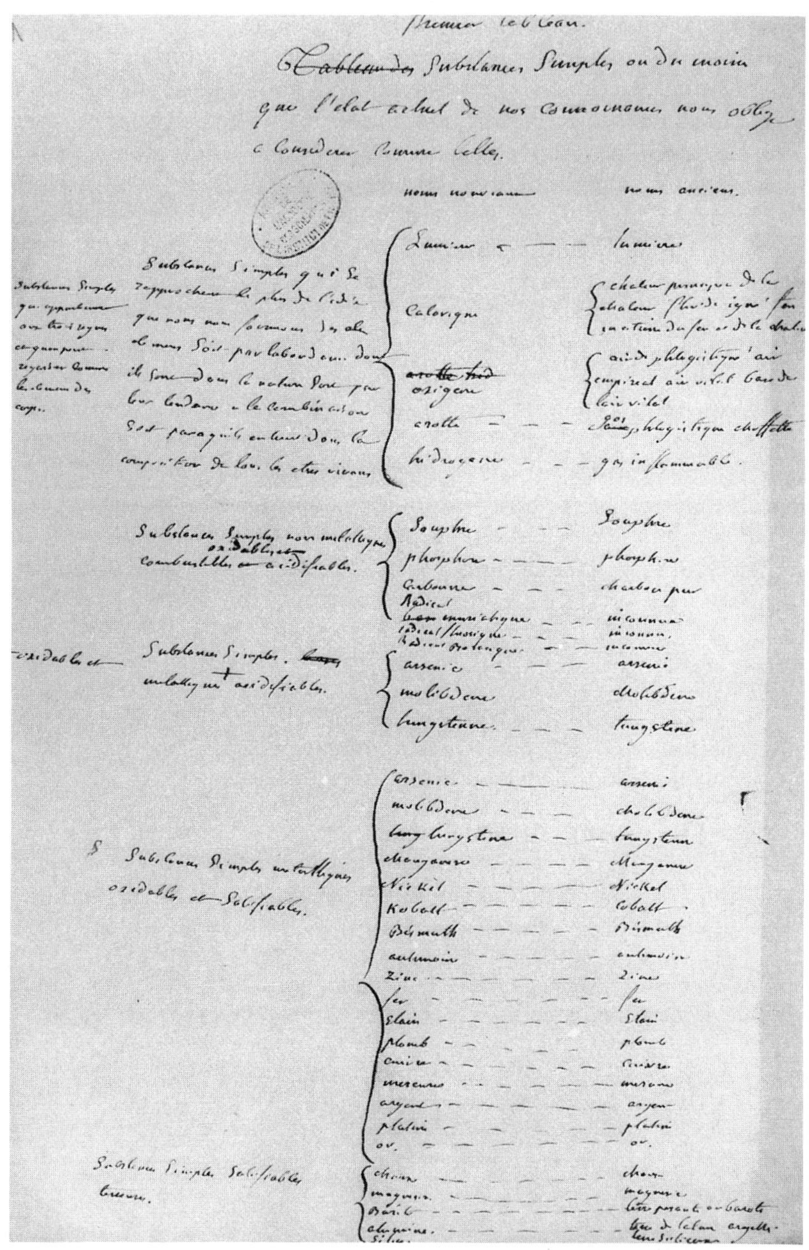

Premier tableau.

Substances simples ou du moins que l'état actuel de nos connoissances nous oblige à considérer comme telles.

noms nouveaux	noms anciens
Lumière	lumière
Calorique	
oxigene	
azotte	
hidrogene	gaz inflammable
Souphre	Souphre
phosphore	phosphore
Carbonne	charbon pur
arsenic	arsenic
molibdene	molibdene
tungstene	tungstene
arsenic	arsenic
molibdene	molibdene
tungstene	tungstene
manganese	manganese
Nickel	Nickel
Kobalt	Cobalt
Bismuth	Bismuth
antimoine	antimoine
Zinc	Zinc
fer	fer
Etain	Etain
plomb	plomb
cuivre	cuivre
mercure	mercure
argent	argent
platine	platine
or	or
chaux	chaux
magnesie	magnesie
Barite	
alumine	

FIGURE 14 A preliminary attempt by Lavoisier to write his famous table of simple substances appeared in the *Traité élémentaire de chimie* in 1789. Courtesy of Arc. Acad., Dossier Lavoisier 154.

of matter that dispensed with the traditional *principes de la matière*. Meanwhile, Lavoisier had no difficulty in envisaging a future in which, for instance, the earths ceased "to be considered as simple bodies." On the contrary, such a discovery would confirm his philosophy of matter.

The tables of nomenclature that followed listed the simple substances that entered "into the composition of the acids and the oxyds, together with the various possible combinations of these elements," and also provided a classification and nomenclature of all the "saline substances." The only difference between this nomenclature and the one presented in the *Méthode* was in the introduction of several tables in place of a single one. In the dictionaries of the *Méthode*, the names of substances and compounds were arranged in alphabetical order, whereas in the *Traité* Lavoisier had chosen to arrange them in methodic tables, each of which represented a class of binary combinations. This new arrangement underlined the importance of methodic classification and facilitated the identification of classes of names for the reader.

The third and last part of the *Traité* was devoted to the description of chemical and physical instruments. Unlike most chemical treatises of the seventeenth and eighteenth centuries, the *Traité* placed this section at the end instead of at the beginning. Since the time of Glaser and Lemery, the description of chemical instruments and operations had introduced the theoretical part of the work. This practice had certainly not come about by chance. Most seventeenth-century chemists regarded chemistry rather as an art than as a science, and the emphasis on technology was due to the belief that useful results and remedies could be obtained only through the mastery of chemical instruments and systematic experiments. The laboratory and its numerous instruments symbolized the empirical dignity acquired by chemistry. Despite this emphasis, real progress in chemistry in the seventeenth and early eighteenth centuries had been confined to metallurgy and the chemistry of salts. The portrayal of the instruments, illustrated with several engravings, was designed to lend respectability to an art that was still steeped in controversial alchemical beliefs. This pragmatic attitude had become more pronounced during the eighteenth century, when with the exception of Stahl chemists still regarded chemistry as an art, akin to mineralogy, glassmaking, etc.

In the second half of the eighteenth century, Macquer, Guyton, Bergman, and a few others began to treat chemistry as a science and to attach great importance to the solving of theoretical and methodological problems. This implied a sharper distinction between the theory and practice of the science. The description of chemical instruments began to be moved to the end of chemical textbooks, and chemistry was defined as a science in its own right. More than his contemporaries, Lavoisier considered this separation of the science of chemistry from its practical applications and instruments an important conceptual achievement. Aware of the radical change implied by the layout of the *Traité*, Lavoisier established a hierarchy in which chemical science necessarily had to come before the technological and operational art:

> Influenced by these motives, I determined to reserve, for a third part of my work, a summary description of all the instruments and manipulations relative to elementary chemistry. I considered it as a better place at the end, rather than at the beginning of the book, because I must have been obliged to suppose the reader acquainted with circumstances which a beginner cannot know, and must therefore have read the elementary part to become acquainted with.[69]

The beginner could not understand the function of new instruments until he understood the new theory and conception of chemical composition in response to which the instruments were constructed. How could a beginner understand the use of a hydrostatic balance without understanding the chemical notion of specific gravity? Several instruments, such as the *gazometer* and the *calorimeter*, were designed only when the theory of gaseous states had already been outlined. The role of instruments was thus treated by Lavoisier as demonstrative rather than scientific. This distinction is made plain by Lavoisier's claim that "in a work of appropriate reasoning, minute descriptions of processes and of plates interrupt the chain of ideas, and render the attention both difficult and tedious to the reader."[70] Since the *Traité* was just such a "work of reasoning," the role of scientific instruments was merely supplementary to the theory. This view contradicts the common opinion of historians who regard the introduction of chemical instruments of high precision as one of the prime causes of the overthrow of phlogiston and of the ultimate triumph of the chemical revolution. Indeed, much significance has been seen in Lavoisier's purchase of balances designed by Mégnié and Fortin in 1787 and 1788, of Meusnier's gazometer, etc.[71] Lavoisier was certainly always keen on accuracy and reliability in the laboratory, and he both made important improvements to existing instruments and invented new ones. This fact has, however, to be considered in the theoretical context that gave rise to Lavoisier's interest in chemical technol-

---

[69] *Ibid.* (1790), 292.

[70] *Ibid.*

[71] On this, see Daumas (1955), 131–156.

ogy. The use of the balance, which has often been claimed to be one of Lavoisier's main contributions to quantitative chemistry, was actually practiced by Agricola over 200 years earlier and was familiar in chemical laboratories by the seventeenth century. Furthermore the quantitative data provided by even the most sophisticated balances were in any case unreliable in many respects.[72] Despite his insistence on the importance of "the determination of the weights of the ingredients and products both before and after experiment,"[73] Lavoisier was well aware of this.[74] To fill the gap between experience and theory, between the approximation of the balance and the interpretation of the data, Lavoisier did not hesitate to resort to *a priori* assumptions. An example of his reasoning is found in his description of his experiment on the fermentation of wine:

> I shall finish what I have to say upon vinous fermentation, by observing, that it furnishes us with the means of analysing sugar and every vegetable matter. We may consider the substances submitted to fermentation, and the products resulting from that operation, as forming an algebraic equation; and, by successively supposing each of the elements in this equation unknown, we can calculate their value in succession, and thus verify our experiments by calculation, and our calculation by experiment reciprocally. *I have often successfully employed this method for correcting the first results of my experiments, and to direct me in the proper road for repeating them to advantage.*[75] [my italics]

Indeed, Lavoisier conceived chemical analysis as a mental experiment, and the discrepancies between the real and the ideal experiment were always resolved in favor of the latter. The correction of the empirical data to fit into a coherent conceptual framework was a typical trait of physical epistemology, and it may be useful when considering the above passage to recall the use made of instruments by Galileo. For Galileo, instruments, in particular those related to his investigation of the laws of motion, served more to convince others than to advance his theory. The inaccurate data they provided were misleading, and Galileo was forced to adjust them to the result of his mathematical calculations. As Koyré put it, it was impossible to take exact measurements during an experiment, and between the empirical data and the theoretical explanation of the phenomena, there was a gap

---

[72]Holmes (1986) reports that Lavoisier's quantitative analysis provided results which were often inaccurate.

[73]Lavoisier, *Elements* (1790), 297.

[74]Lavoisier, *Traité* (1789), vol. 2, 335 (not translated by Kerr).

[75]Lavoisier (1790), 140.

that Galileo filled with an imaginative mental effort.[76] In the specific case of the motion of bodies,

> for example, when Galileo formulates, more or less implicitly, the law of inertia, he is content to proclaim that "any velocity once imparted to a moving body will be rigidly maintained as long as the external causes of acceleration or retardation are removed, a condition which is found only on horizontal planes." This was a new affirmation [. . .]. However (contrary to what is frequently maintained), Galileo attempts no experimental proof of his proposition—indeed it would have been quite impossible at the time—but formulates it as something taken for granted. . . ."[77]

Galileo assumed that by letting two perfectly spherical balls roll on a perfectly smooth inclined plane, his theory of motion would have been demonstrated empirically. However, the obvious impossibility of reproducing such an inclined plane inevitably raised the question of the accuracy of the instruments. Since he relied largely on the consistency of his mental experiments, the experimental discrepancies inevitably accompanying the use of the real instrument did not bother Galileo, who adjusted them to the perfect regularity and accuracy of his own mathematical calculations. Likewise, the discrepancies between Lavoisier's mental "calculation" and his empirical observations were bridged by the corrective intervention of his imagination. This approach was in keeping with Lavoisier's awareness of the limitations of strict empiricism in science. While discussing the scientific legitimacy of mesmerism in 1784, for example, Lavoisier pointed out that a rigid

---

[76]"Les expériences imaginaires, que Mach avait appélés 'expériences de pensée' (Gedanken Experimente), [. . .] ont joué un rôle très important dans l'histoire de la pensée scientifique. Ce qui se comprend facilement: les expériences réelles sont, souvent, un appareillage complexe et coûteux. En outre, elles comportent, nécessairement, un certain degré d'imprécision, et donc, d'incertitude. Il est impossible, en effet, de produire une surface plane qui soit 'véritablement' plane [. . .]. Il n'y a pas, et il ne peut pas y avoir, *in rerum natura*, de corps parfaitement rigides; pas plus que de corps parfaitement élastiques; on ne peut pas exécuter une mesure parfaitement exacte [. . .]. Entre la donnée empirique et l'objet théorique, il reste, et il restera toujours, une distance impossible à franchir.

C'est là que l'imagination entre en scène. Allègrement, alle supprime l'écart [. . .] elle 'réalise' l'idéal et même l'impossible," from A. Koyré, *Études d'histoire de la pensée scientifique* (Paris, 1973), 225. See also Émile Meyerson, *Explanation in the Sciences* (1921), English translation (Dordrecht-Boston-London, 1991), 500. Meyerson's and Koyré's interpretation has recently been questioned by many historians, in particular by Stillman Drake, who claim that Galileo's use of instruments was necessary to the outline of his theory of motion.

[77]Meyerson, *Explanation in the Sciences* (1991), 500.

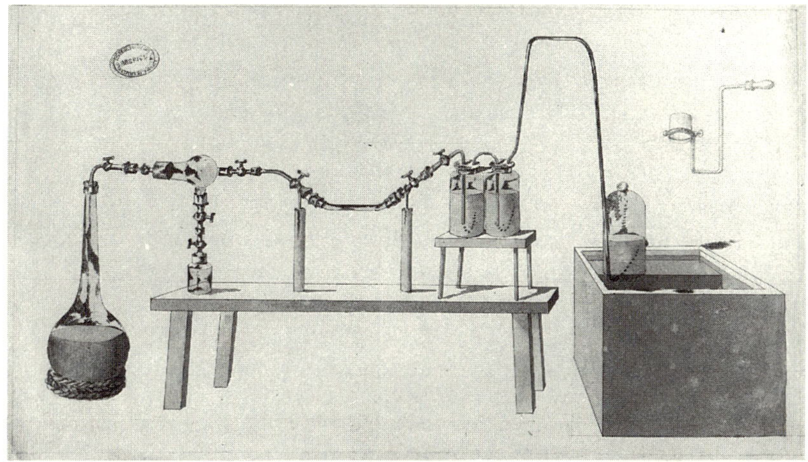

FIGURE 15 Lavoisier's apparatus for his experiment on fermentation, after an original drawing by Madame Lavoisier. Courtesy of Arc. Acad. Dossier Lavoisier 429.

adherence to immediate experience, in an effort to support their theories, was a characteristic of many charlatans. In fact there was more to the art of *scientific* experiment than simply drawing conclusions from empirical data:

> The art of inferring on the basis of experiments and observations consists in evaluating probabilities and judging whether they are strong enough or numerous enough to constitute proofs. This sort of calculation is more complicated and more difficult than one thinks; it requires great sagacity and is generally beyond the powers of the ordinary man [. . .].
>
> Their errors [the errors of ordinary men] in this type of calculation are the basis for the success of the charlatans, the sorcerers, the alchemists, as they were earlier the basis for the success of magicians, enchanters, and all those in general who delude themselves or seek to take advantage of public credulity.[78]

Lavoisier's attitude to the subordinate role of technological devices is further demonstrated by the fact that his calorimeter, pyrometer, gazometer, balances, etc. were all constructed in the 1780s, when the theory was already formulated and virtually complete. On the other hand, it is not possible to accept the opposite thesis of those who would

[78]LO, vol. 3, 508–510; English translation by Meyerson (1991), 469–470.

claim that Lavoisier's instruments served merely to convert his audience to his theory and that they fulfilled a rhetorical function rather than a scientific one. It is, of course, true that by constructing the most accurate chemical instruments of his time, Lavoisier hoped to obtain tangible confirmation of his ideas. Even if the results hitherto were not as conclusive as he had hoped, a good approximation to his mental experiment showed that he was on the right track. In this respect Lavoisier's attitude was very similar to Galileo's in the sense that both scientists made "instrumental" use of their laboratory experiments and subordinated them to their theoretical conceptions.

In 1989, analyzing the same kind of problem, the American historian Frederic L. Holmes raised the following question:

> In his published papers Lavoisier habitually claimed to have carried out numerous experiments of whatever type he was describing, giving the impression that the few he reported in detail were selected from a much larger number. My experience comparing his publications with his laboratory notebooks has, however, persuaded me that this was rhetorical exaggeration, that he actually performed relatively few experiments that he did not in one way or another incorporate into his publications. Why then, if he were the consummate experimentalist I believe he was, did he so regularly settle for one or several imperfect experiments on a given problem? Why did he go to such great lengths to salvage the data from the few he had performed rather than repeat them until he had reduced or removed the sources of error?[79]

Holmes did not answer his questions, which seem to contradict the traditional picture of Lavoisier as a great experimentalist. By contrast, within the interpretation I have proposed, Holmes' questions find a consistent answer. Unlike his colleagues Priestley and Scheele, who based their chemical opinions on their experimental abilities and performed the same experiments several times, Lavoisier conceived chemical experiments on a theoretical level before performing them in the crucible. The disposition and contents of the *Traité* make it evident that the method of conceiving, exhibiting, and naming chemical matter was far from the pure experimentalism of Priestley and Scheele.

The *Traité* was thus a theoretical work in which Lavoisier thoroughly developed the principles of nomenclature he had presented in

[79]Holmes, "Lavoisier the Experimentalist," in *Bulletin of History of Chemistry*, 5 (1989), 29.

1787. Two-thirds of the treatise was in fact devoted to the definition, naming, and classification of simple substances and their combination. The experiments described here and there in the narrative section were intended to illustrate and support the general definitions and classifications given at the beginning of each chapter. The connection between the Condillacian preliminary discourse and the rest of the work was therefore perfectly logical, and the entire composition of the *Traité* was consistent with Condillac's definition of analysis and his philosophy of language. The 45 tables of chemical nomenclature presented in the second part of the work can hardly be understood without seeing the *Logique* of Condillac as one of their primary sources. Lavoisier's receptiveness to Condillac's philosophy arose from his need for a new method of conceiving chemistry. We have further confirmation of the strength of this influence in the unpublished version of an elementary textbook in chemistry that was drafted at the end of 1792.[80] As Daumas and Bensaude-Vincent have pointed out, the contents and the disposition of the *Traité* had been changed. In this later work, Lavoisier intended to include several subjects to which he had paid little or no attention in the *Traité*. For example, there would be an examination of chemical attractions, a discussion of vegetable and animal chemistry, and an entire section devoted to the chemical "arts," namely, glassmaking, dyeing, gunpowder, etc. As in the *Traité*, the description of chemical instruments came at the end of the book, which began with observations on the teaching of chemistry, including an explanation of the changes introduced since the *Traité*. To follow the "natural order of ideas," Lavoisier suggested that an elementary textbook, as this was intended to be, ought to contain a general survey of all the sciences related to chemistry. He therefore proposed the following introductory chapters:

1. A summary of meteorology containing all experiments on air and atmosphere.
2. A survey of electricity.
3. A description of the construction of balances and of hydrostatic balances in particular.
4. A description of simple substances and of the different natural products of chemical interest in commerce.

---

[80]The genealogy of this work has been thoroughly discussed in Daumas (1955), 102–112. See also Bensaude-Vincent (1990), where an extended version of the manuscript is published.

5. A general consideration of the organization of the vegetable kingdom.
6. A general consideration of the organization of animals.[81]

Chapters 2 and 3, at least, of this schedule are reminiscent of the chemistry course planned by Lavoisier in 1764 (see Chapter Four). In the 1792 work, however, these chapters were to constitute the introduction to the science, rather than the science itself:

> These different memoirs and general discourses, will serve as an introduction to chemistry. They will form a kind of chain which ties [chemistry] to the rest of human knowledge.[82]

Lavoisier saw this introduction as an implementation of Condillac's precept that science should follow the path from the known to the unknown. A chemistry student who wanted to make progress in the science had to be well acquainted with calculation, "the elements of Euclid," experimental physics, etc. This new concern with the educational background of beginners has led Bensaude-Vincent to claim that "Lavoisier's primary concern was no longer the language of chemistry but chemistry teaching."[83] Since the 1792 version of his course is largely incomplete, it is difficult to pronounce on this question. But in the introduction to his course, Lavoisier still seemed to attach a lot of importance to the role of language in chemistry:

> By departing from the data [the prolegomena of chemistry], by proceeding from the known to the unknown, and never employing a word which has not been defined, it seems to me possible to lead a pupil through a nearly geometric path.[84]

This passage reveals that Lavoisier still regarded the principles that were laid down in the *Méthode* and the *Traité* as the guidelines of chemical science. In this respect, the logic of chemistry remained unaltered. Furthermore, the similarity with his previous works is evidenced by Lavoisier's repeated ontological commitment to the natural order of ideas.[85] In reality this outline established a closer connection between the

---

[81]Bensaude-Vincent (1990), 459.
[82]*Ibid.*
[83]*Ibid.*, 445.
[84]*Ibid.*, 459.
[85]"C'est une méthode bien contraire à l'ordre naturale des idées que de commencer un cours de chimie par une dissertation sur les élémens premier [. . .]. La marche naturelle des idées, avant de s'occuper du véritable objet de la chimie—la décomposition des corps—est de les aborder en masse et d'étudier leur propriétés générales" *ibid.*, 459.

original philosophical principles and the renewed interest in pedagogy. Bensaude-Vincent is right in stating that

> Lavoisier contrasted his disappointment with chemistry with his satisfaction with mathematics. Here, as in Descartes' works, mathematics is a model because propositions are connected in a logical order [. . .] To Lavoisier, logical order and didactic process were one and the same thing, since the chemist who was confronted with the outside world was like a young child, driven by elementary sensations.[86]

This comparison of Lavoisier's Condillacian concerns with the methods of Descartes is a productive one, and it reveals an important aspect of Lavoisier's chemical philosophy. In Condillac's case, the natural order of ideas in his linguistic philosophy was confirmed by its pedagogic merits; likewise Lavoisier's system of chemistry was endorsed by the fact that it followed the path of the teacher of the teachers, nature. Both assumptions could be traced back to Descartes' definition of method and science. This debt to Descartes was implicitly stated by Lavoisier when describing the obstacles he had to overcome in his first approach to chemistry:

> I recognized then the necessity of restarting my chemical education, of preserving of everything I have learnt only the facts, and of trying to rearrange them in my memory in a methodical order conforming more to the natural progress of nature.[87]

This passage was in harmony with a vision of science in which order and method had priority over the collection and description of facts. These general ideas, which showed the evolution of Lavoisier's thought from its very beginning, are the true source of the chemical revolution. Lavoisier's opponents understood better than anyone else that the real danger came from the epistemological background of the new theory rather than from the empirical evidence. To conclude, we may compare Lavoisier's chemistry with Descartes' physics: in both cases the real significance of their work did not rest on the interpretation of the phenomena (which in both cases turned out to be wrong),[88] but on their vision of the methodological framework within which to interpret and define the phenomena.

---

[86]*Ibid.*, 445.

[87]*Ibid.*, 458.

[88]Descartes' vortex theory was eventually superseded by Newton's celestial mechanics; likewise, Lavoisier's oxygen theory was proved to be wrong by the experiment made by Davy at the beginning of the nineteenth century.

"Voilà la France, bien déphlogistiquée"

It was not until after the publication of the *Traité* that the French scientific community began to react positively to Lavoisier's theory. An intensification of the "campaign of persuasion"[89] took place in 1789 with the foundation of a new scientific journal, *Annales de chimie*,[90] edited by Lavoisier and his *cotérie*. The primary aim of the journal was to promote the diffusion of the new chemical doctrine and nomenclature and to counterbalance the phlogistic *Journal de physique*, edited by de la Métherie. The foundation of the *Annales de chimie* enormously facilitated the circulation of the new ideas and offered a great opportunity to many young chemists eager to publish their first papers. From the first volume of the *Annales*, we gather that Lavoisier's party was joined by Coulomb, Haüy, de Dietrich, Seguin, Parmentier, Vauquelin, Van Mons, and many other chemists and naturalists who had not previously taken sides. Even members of the old guard were converted: Mathurin Jacques Brisson, professor of physics at the Collège de Navarre, began to teach the new theory in the same year.[91]

The most energetic promoters of the campaign against phlogiston were Lavoisier's closest collaborators Fourcroy, Guyton, Berthollet, and Chaptal. In 1789 Fourcroy published a revised edition of his popular *Élémens*,[92] presenting the new nomenclature and defending it from the attacks published in the *Observations sur la physique*.[93] In 1790 Chaptal published *Élémens de chimie*,[94] which immediately received a favorable review from Berthollet in the *Annales*.[95] In his introduction, Chaptal mentioned the pedagogic value of the new nomenclature, the facility with which the students grasped it, and the advantage of coining words expressing the real composition of chemical substances. Chaptal's enthusiasm for the new nomenclature was not undiluted, however: like

---

[89]See Perrin (1981), 50–57.

[90]*Annales de chimie; ou recueil de mémoires Concernant la Chimie et les Arts qui en dépendent par MM. de Morveau, Lavoisier, Monge, Berthollet, de Fourcroy, Le Baron de Dietrich, Hassenfratz & Adet* (Paris, 1789 and ff.).

[91]In 1797 Brisson published the Lavoisierian *Principes élémentaires de l'histoire naturelle et chímie des substances minérales* (Paris), but he adopted the new principles long before that date.

[92]Fourcroy, *Élémens d'histoire naturelle et de chimie*, 3rd edition, 5 vols. (Paris, 1789).

[93]*Ibid.*, vol. 1, xiv–xvii.

[94]Chaptal, *Élémens de chimie*, 3 vols. (Montpellier, 1790).

[95]*Annales de chimie*, 6 (1790), 197–203.

Gadolin before him, he complained of the linguistic discrepancy between the name of *azote* and the name of the acid derived from it (*acide nitrique*). To keep the new nomenclature true to its general principles, Chaptal proposed changing the name of *azote* to *nitrogène*.[96] In addition to making this technical criticism, Chaptal remarked that where the nomenclature of craftsmen was involved, one had to accept the old language, as it was practically impossible to bring "craftsmen to dependence on academic standards."[97] Chaptal, who was himself actively engaged in developing chemical factories and manufactures, was well aware of the difficulty of persuading craftsmen to change a technical terminology that often was the legacy of several generations. In actual fact, Chaptal did not seem convinced of the universal validity of the philosophical principles of the *Méthode*, and he suggested a more practical approach to the use of the new chemical language.

After 1791, much of the attention of French scientists turned to political events. Lavoisier, Guyton, Fourcroy, Hassenfratz, Adet, and others became increasingly engaged in the political scene, and the controversy between oxygen and phlogiston was momentarily set aside. After Lavoisier's execution in May 1794 and the end of the revolutionary terror, the scientific debate slowly revived. This long period of silence naturally favored the diffusion and the success of the new theory, which came to be known as the French doctrine. During the second half of the 1790s, Lavoisier's theory met with little opposition in France and even de la Métherie, his principal adversary, gave up the campaign he started in 1787 with the *Journal de physique*. However, the success of the new theory was not without ambiguities and contradictions. After Lavoisier's death, his ex-collaborators began to put their own interpretations on his works. It was Fourcroy and Guyton who began to refer to the oxygen theory as the French doctrine, and they claimed the principal credit for the new nomenclature for themselves. Besides making these assertions, the orphans of Lavoisier began to modify both the original meaning and the theoretical background of the new nomenclature. In an article published in 1798 in the *Annales de chimie* in reply to the attempts of Aréjula, Brugnatelli, and Dickson to devise alternatives to the new nomenclature, Guyton declared that the nomenclature had to follow the science[98] and not vice versa as

---

[96]Chaptal, *Élémens*, vol. 1, xlix.

[97]Bensaude-Vincent (1990), 447.

[98]Guyton, "Examen de quelques critiques de la nomenclature des chimistes français," in *Annales de chimie*, 25 (1798), 206.

Lavoisier seemed to believe. Guyton also stated that too much importance had been attached to etymology, whereas the first rule of language was "to avoid confusion."[99] The original linguistic realism that inspired Lavoisier in his memoirs was dismissed as if it had never existed, and in 1805 Guyton, while preparing his course in mineralogy at the *Ecole Polytechnique*, went so far as to claim that "the natural method [of nomenclature] does not exist [. . .]. The method is only useful insofar as it helps the memory."[100] Guyton was not betraying Lavoisier's philosophy of language so much as reverting to the principles of nomenclature that he had formulated in 1782, in which the relationship between words and things was conventional and determined by practical needs. Guyton's opinion was shared by all Lavoisier's pupils. In 1803 Berthollet, until then very reluctant to discuss the principles of nomenclature, expressed criticism of Lavoisier's attempt to construct a Condillacian chemical language. First of all, Berthollet pointed out that the essential property upon which Lavoisier constructed the name of oxygen had now to be considerably restricted, if not deleted from the chemical lexicon altogether. Oxygen was not, as Lavoisier believed, the universal principle of acidity, and Lavoisier's conclusion "that all acidity is caused by it, even that of the muriatic, fluoric, and boracic acids, is to push the limits of analogy too far."[101] Interestingly Berthollet recognized that the objection raised to the name oxygen by Aréjula some 20 years before had a sound experimental ground and that Lavoisier's theory of acidity was a hazardous and speculative generalization deduced from an insufficient number of facts. This anomaly in Lavoisier's theory was extremely grave because it cast doubt on its main foundation and at the same time appeared to legitimize all the criticism voiced against the nomenclature. Berthollet was fully aware of these

---

[99] *Ibid.*, 212.

[100] "Il n'y a point de méthode naturelle ce qu'on nomme l'espece ne peut etre determiné que par les produits comparés de l'analyse.

La méthode n'est utile qu'autant qu'elle soulage la mémoire," *Programme des leçons de Guyton de Morveau ecrit par lui meme*, BIP, Ms. 15, folio 8 recto.

[101] "Or, les deux propriétés qui caracterisent particulierèment l'oxigène, sont: 1°. de se combiner avec les substances qui sont inflammables [. . .]; 2°. de communiquer l'acidité aux combinations qu'il forme [. . .].

J'ai, dans d'autre occasions, résisté à cette dernière idée qui est due à Lavoisier; mais il me paraît ajourd'hui que l'on donnait trop d'extension au principe qu'on voulait établir, et que de mon côté j'y appartais trop de restriction.

En effet, vouloir conclure de ce que l'oxigène donne l'acidité à un grand nombre de substances que toute acidité en provient, même celle des acides muriatique, fluorique et boracique, c'est reculer trop loin les limites de l'analogie," Berthollet, *Essai de statique chimique* (Paris, 1803), vol. 2, 8.

dangers and to save the name oxygen he answered his own etymological criticism:

> It seems to me that the common interest of chemists requires that they should follow the same practice with denominations taken from a known property, and with those which are not so; that if they have recourse to etymology to obtain the acceptance of a new denomination which a discovery renders indispensable, they should omit it entirely as soon as it is adopted, and make no use of it in the combinations of words, but what may be called mechanical: *the chemist who establishes it must direct his attention more to Euphony, and the conveniences of nomenclature, than to the indication of a property.*[102] [my italics]

In denying any cognitive validity to etymology, claiming the superiority of convenience to truth, and suggesting that chemical substances be named euphoniously rather than according to their properties, Berthollet was dramatically reversing the basic principles laid down by Lavoisier in both the *Méthode* and the *Traité*. And as he abandoned the philosophy of language of his master, he adopted the linguistic pragmatism of the opponents of the *Méthode*. In another passage of the *Essai*, Berthollet was even less faithful to Lavoisier's principles, declaring:

> In general I attach less importance to the strict observation of the principles of nomenclature, which in reality, are only agreements, into which several considerations may be introduced: the essential point, in my opinion, is to compound words so that they leave nothing equivocal on the parts which enter into the composition of a combination, and on the relation of its characteristic properties with those of other substances.[103]

Like Guyton, Berthollet saw the principles of chemical nomenclature in a new light and now claimed that words were the mere products of agreements. The linguistic realism of the *Méthode*, the semantic power of the names given to the simple substances, and the "natural logic" to which Lavoisier ascribed such an important role, were now denied any value. The nomenclature of a science became a "mechanical" application of names established by mutual consent. Georges Cuvier, too, who admired the genius of Lavoisier's chemical works and his introduction of the quantitative method, was not so enthusiastic about his nomenclature and declared:

---

[102]*Ibid.*, vol. 1, 451; translation, *An Essay on Chemical Statics* (London, 1804), vol. 1, 450.
[103]*Ibid.*, 454; English translation (1804), 455.

> It would be ridiculous to intend it as an instrument of discovery, because it is nothing more than the expression of the discoveries already made.[104]

Cuvier agreed with Fourcroy that the new nomenclature was a useful tool in the teaching of chemistry, but he grew almost sarcastic when denying its philosophical background.

What explains this sudden and radical change in the attitude to scientific language? Had not Berthollet and Guyton been among Lavoisier's closest collaborators? And why did they borrow arguments that evoked those of the main opponents of the new theory? A possible answer to these questions is that Guyton, Berthollet, and Fourcroy had to cope with the spread of several alternative nomenclatures, mostly published after the death of Lavoisier. These alternative attempts were made by Brugnatelli, Westrumb, Dickson, Kirwan, etc., and their partial success seriously challenged the nomenclature proposed in 1787. Firmly convinced of the practical value of the results achieved by the *Méthode*, Berthollet, Guyton, and Fourcroy realized that it would be easier to save them and to get them adopted if the underlying philosophical principles were eliminated. If, while pointing out the practical advantages that accrued from the mechanical application of the term oxygen to a wide range of substances, it was recognized that the term itself was little more than a convention, the ultimate success of the nomenclature would obviously become more likely. Indeed, by marginalizing any discussion of the etymology of words and their connection with real properties, this pragmatic attitude helped the final triumph of the new nomenclature considerably.

Despite the general effectiveness and success of this new approach, some French scientists still rejected the new theory and terminology. There were many naturalists who did not share Lavoisier's quantitative vision of matter but favored Rouelle's qualitative and Stahlian approach to chemistry.

In 1796 Lamarck published a book that attempted to refute Fourcroy's *Philosophie chimique* point by point.[105] The alternative theory, which Lamarck called *pyrotique*, incorporated Rouelle's conception of

---

[104]"Il seroit ridicule de vouloir en faire un instrument de découvertes, puisqu'elle n'est que l'expression de découvertes faites; mais il est juste de voir en elle un excellente instrument d'enseignement," Cuvier (1989), 94.

[105]Lamarck, *Réfutation de la théorie pneumatique ou de la nouvelle doctrine des chimistes modernes* (Paris, 1796). For a detailed examination of Lamarck's chemical theory, see Leslie J. Burlingame, "Lamarck's Chemistry: the Chemical Revolution Rejected," in H. Woolf (ed.) (1981), 64–81.

chemical combination and a complex and qualitative definition of fire. Lamarck's work was based on a stern appeal to the ultimate authority of pure facts:

> In physics, the facts are the principal riches of the science. They are independent of any theory, and they must always be accurately distinguished.[106]

Lamarck's suspicion of any speculative reflection on the structure of matter was accompanied by the formulation of a new theory that he believed relied solely on the evidence of facts. Lamarck was naturally dissatisfied with the new nomenclature and proposed substantial changes to it. The theoretical importance of fire in his theory *pyrotique* is shown by the fact that he changed the name of oxygen to *feu acidifique.*[107] Lamarck's theory and nomenclature won no support, but they served to show that opposition to Lavoisier's system was still alive. In 1798 the apothecary Antoine Baumé, who was the most prestigious and competent scientist among the opposition, published his *Opuscules chimiques*, in which he criticized the nomenclative "*dispotisme*" and the "*reveries chimiques*" of the new theory, but failed to present any counterproposals beyond the claim that oxygen and phlogiston were identical.[108] At the same time, Monnet,[109] Sage, Opoix, de la Métherie, Demachy, and Faujas continued to oppose Lavoisier's theory in pamphlets, articles, and memoirs. Needless to say, the nomenclature was the favorite target of this anti-Lavoisierian club. As late as 1810, the mineralogist Balthazar Sage, who had been a colleague of Lavoisier in the Academy of Sciences, published a booklet protesting against the new nomenclature,[110] but this in fact marks the last gasp of phlogistic chemistry in France. Sage was unable to advance any relevant scientific arguments; instead he accused the "conventicle of the Arsenal" of having created a language based on a "superfluity of hypothesis." According to Sage, the nomenclative contagion and the revolutionary rage for introducing new ideas went even farther than the political revolution.[111] This series of insults was followed by some

---

[106]Lamarck (1796), 65.

[107]*Ibid.*, 29–30.

[108]Baumé, *Opuscules chimiques* (Paris, 1798). Baumé's criticism of the new nomenclature is on pages 196–197 and his identification of oxygen with phlogiston on page 62.

[109]Monnet, *Demonstration de la fausseté des principes de nouveau chimistes* (Paris, 1798).

[110]Sage, *Exposé des effets de la contagion nomenclative, et réfutation de paradoxes qui dénaturent la physique* (Paris, 1810).

[111]*Ibid.*, 9.

more disconcerting etymological criticism, Sage finding the name oxygen unacceptable because its literal etymological meaning would have been *fils de vinaigrier* (son of vinegar maker),[112] since the suffix *-gène* meant "son" in proper names, such as *thegène, Archigène, Diogène.* Continuing this line of reasoning, the term hydrogen meant "son of the water" (*fils de l'eau*), etc. In protest against what he dubbed linguistic *charlatanerie,* Sage proposed to continue using the nomenclature he used in his mineralogical writings, i.e., the phlogistic one. Although Sage's arguments were take up by some British chemists, his work had no relevance whatsoever, and it well reflects the fate of phlogiston at the beginning of the nineteenth century.

With these exceptions Lavoisier's nomenclature was adopted by most French chemists as early as 1794, and more importantly, chemistry was taught in Lavoisier's language in the courses given at the *Muséum d'histoire naturelle* and the *Ecole Polytechnique* in Paris and at the *Ecoles centrales* throughout France. At the beginning of the nineteenth century, the new generation of chemists—Thenard, Gay-Lussac, Chevreul, Ampère, Caventou, and Sévrin—all spoke Lavoisier's language. To do this successfully, however, they copied Berthollet's linguistic pragmatism and denied any natural or causal relations between words and things other than convention and convenience.

On the other hand, it would be a mistake to believe that Lavoisier's linguistic realism was abandoned *in toto.* As we saw in Chapter Two, the influence that the *Méthode* exerted on the reform of chemical language inspired prominent scientists, such as Vicq d'Azyr, Dumas, Duméril, Millin, Pinel, Cabanis, and others to classify the objects of their science in accordance with the analytical method and Condillacian logic. Even the concrete names of the decimal system of weights and measures followed the pattern of the chemical suffixes proposed in the *Méthode.*[113] In all their writings, the success of the *Méthode* and of Condillac's philosophy of language continued to play an important role, regardless of the increasing specialization of the individual sciences.

With respect to chemistry, the new progress made in the isolation of new acids and the rapid development of organic chemistry made it difficult to maintain Lavoisier's logical relationship between denominations and chemical substances. Nevertheless, his model was not totally forgotten, and at the beginning of the nineteenth century, Ampère was still

[112]"Le nom d'*oxide,* mot derivé d'*oxidos,* génitif d'*oxis,* qui signifie vinaigrier. *Oxide* ne signifie donc que *vinaigre; oxidé, vinaigré; oxidation* l'art de faire le vinaigre," *ibid.,* 10.
[113]C.A. Prieur de la Côte d'Or, *Note instructive sur les poids et mesures* (Paris, 1794), 11.

hoping to follow in his footsteps with a new *natural method of classification* of chemical substances (see Fig. 16).[114]

"THE PHILOSOPHERS OF LAPUTA"

The reception of Lavoisier's nomenclature in the British Isles provides a revealing illustration of the general attitude of British chemists to their science. As mentioned in Chapter Four, the immediate reactions to St. John's translation of the *Méthode* were negative but not numerous. After the publication of the *Traité* in 1789, the British scientific community reacted more strongly, and the philosophical principles underlying the new theory were thoroughly discussed and criticized. From our point of view, the reaction of the British chemists is of particular interest because of their persistent dislike of the new nomenclature.

In 1789 the English physician James Keir, a friend of Erasmus Darwin and the translator of Macquer's *Dictionnaire de chymie*, published *The First Part of a Dictionary of Chemistry*,[115] a work in which the contents of the *Traité* were discussed in England for the first time. Keir acknowledged that the antiphlogistic system was a "work of genius" and that many of its results would make an important contribution to the progress of pneumatic chemistry, but continued: "when it is announced to us as a mere exposition of facts, in which no hypothesis is admitted to fill up the chasms, we are to consider it in another point of view."[116] For example, Keir did not accept Lavoisier's claim that all gases were a combination of a specific base and the matter of heat, or caloric:

> What can be more hypothetical than the existence of this *matter of heat*, of a substance of which Mr. *Lavoisier* has candidly acknowledged, after the most scrupulous investigation, that he can discover no sensible weight.[117]

To a Baconian mind like that of Keir, Lavoisier's definition of caloric as a mathematical hypothesis could not be acceptable. Another general and methodological point criticized by Keir was Lavoisier's

---

[114]See the article "Corps" of the *Dictionnaire des sciences naturelles*, vol. 10 (Paris, 1818), 532.

[115]Keir, *The First Part of a Dictionary of Chemistry* (Birmingham, 1789).

[116]*Ibid.*, viii.

[117]*Ibid.*

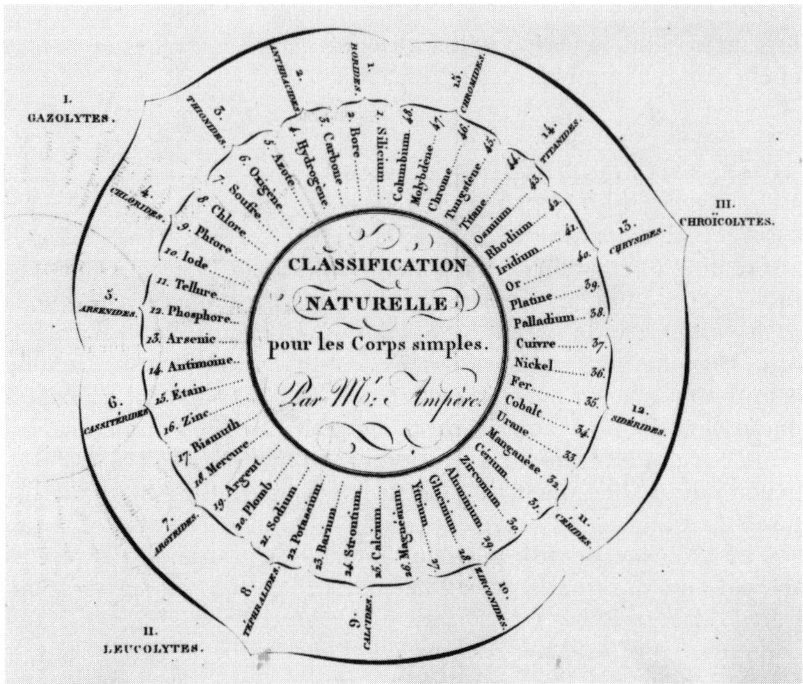

FIGURE 16 Ampère's natural classification of chemical substances. From *Dictionnaire des sciences naturelles*, vol. 10 (Paris, 1818), article "Corps."

attempt to introduce the logic of mathematical reasoning and language into chemistry:

> If it were possible to introduce into Chemistry, the precision of method employed in geometry, and the certainty of its corollaries, it would undoubtedly be an attempt worthy of the most sublime genius. But this science seems incapable of so perfect method.[118]

Supporting the argument advanced by Venel some 30 years before, Keir claimed that chemistry was a science that examined the *sensible* and visible properties of bodies, whereas mathematics was a system of "transcripts of the ideas of our own minds." Keir advised chemists to moderate their commitment to method because the only admissible theory had to be "not the *best imaginable*, but the *best applicable*." This anti-

---

[118] *Ibid.*, ix.

speculative prudence was repeated when Keir stated the general aim of his dictionary, which "endeavoured [. . .] to relate the facts which have been discovered in the respective subjects by different chemists, *without any reference to the theory*"[119] [my italics]. In this connection Keir claimed that after studying the first 100 pages of his work the reader could not possibly know whether he was a supporter of Stahl or of Lavoisier. Keir's antipathy to any form of theoretical construction in science became particularly evident when he dealt with the question of nomenclature. He admitted the necessity of changing the erroneous names, which were relics of "barbarism and alchemical mystery," but he denied the possibility of substituting for them "proper names" capable of evoking the real objects to which they referred. A language developed by gradual evolution; its legitimacy was determined by its use:

> As language is the common property of all, no authority is equal to any other alteration, however well imagined, but what is effected gradually, and as it were by universal consent.[120]

To Keir, therefore, it was hardly believable that an attempt made by just four people to "alter the whole nomenclature of a science" could have any success. Interestingly Keir compared the *Méthode* with the attempt made by the English philosopher John Wilkins in 1668 to create a universal language for use in all sciences.[121] This comparison shows that Keir clearly understood the ontological content of the *Méthode.* Like Wilkins, the French chemists intended ultimately to create a language capable of *reflecting* objects in their very essence. This aim was totally unacceptable to a consistent empiricist, such as Keir:

> A language *justly* and *completely* formed on these principles would be a truly philosophical language, and greatly conducive to the ends proposed; but it would require a *complete* and *certain* knowledge of these constituent parts and properties, and a degree of perfection of the science to which I fear it will not soon arrive. Such may perhaps be the language of the *Gods,* of which *Homer* speaks, but I think it will not be soon spoken on *Earth.*[122]

To justify his skepticism, Keir pointed out the danger implicit in the counterreforms suggested by Aréjula and Hopson. These attempts

---

[119]*Ibid.,* xii–xiii.

[120]*Ibid.,* xiv.

[121]John Wilkins, *An Essay Towards a Real Character and Philosophical Language* (London, 1668).

[122]Keir, *A Dictionary* (1789), xv.

to reform the nomenclature of 1787 clearly showed that if "every one would propose words according to his own ideas," all that would be created would be the Tower of Babel. The only possible way out of this linguistic confusion was to return to facts and experimentation, the only two authorities of science. Keir concluded his introduction with a reference to the dichotomy between words and facts:

> Whatever logical propriety these innovations may possess, who does not see that they open an endless source for confusion, that they draw our attention from science to verbal criticism; and that they are apt to make us forget the ancient saying, that WORDS *are but Daughters of Earth,* while THINGS *are the Sons of Heaven?*[123]

Keir's criticism, verging on ridicule, held out little hope of any success for the *Méthode.* Like Cavendish in 1788, Keir believed that the "rage for name-making" provoked by the new nomenclature would ultimately lead to its collapse.

After 1789 the general attitude of British chemists to Lavoisier's theory and nomenclature became increasingly negative. In 1792 the apothecary Robert White suggested the use of the Linnaean method of naming and classifying minerals but dismissed the nomenclature of the French with the following comment:

> In the present unsettled state of chemistry, there needs no apology for omitting the ingenious, yet visionary plans, and complicated principles of the antiphlogistic code. . . .[124]

Lavoisier's nomenclature was even badly received by those chemists, namely Black and Kirwan, who had been converted to the new theory. The conversions were in fact not really complete, and as we shall see, a recognition that the experiments on the decomposition of water were leading to the overthrow of phlogiston did not imply a willingness to replace Stahl's principle with Lavoisier's oxygen. This ambivalence is indicative of the difficulty of absorbing and fully understanding the new conceptual framework created by Lavoisier.

In January 1791 Richard Kirwan wrote to Berthollet, admitting the defeat of phlogiston:

> Finally I lay down my arms and abandon phlogiston. I see clearly that there is no authenticated experiment attesting to the production of fixed air by pure inflammable air; and that being so, it is impossible to

---

[123] *Ibid.*, xviii.

[124] Robert White, *An Analysis of the New London Pharmacopoeia* (New Market, 1792), v.

> support the system of phlogiston in the metals, sulphur, etc. [. . .]. I myself shall offer a refutation of my Essay on Phlogiston.[125]

In the same year, defeat was acknowledged by another chemist of high repute: Joseph Black wrote to Lavoisier telling him that he had begun to teach the new theory because he found it "simpler, more consistent, and better supported by the facts than the old one."[126] In another letter Black was even more complimentary, declaring that the new theory was "infinitely better grounded than the old one" and that his students at Edinburgh were enthusiastic about the new nomenclature.[127]

For pure and accurate experimentalists, such as Kirwan and Black, the failure to isolate phlogiston after so many experimental efforts finally overcame initial hostility to Lavoisier's theory. However, the new chemistry still had many difficulties and obstacles to surmount before being accepted throughout the British Isles. Both Black himself (whatever his students' preferences) and Kirwan rejected the new nomenclature and declared their preference for the old one. This obviously made their conversion less significant. As we saw in Chapter Four, Black and Kirwan despised the "Cartesian itch" and the "itch for theory" of the new nomenclature. This instinctive reaction persisted into the 1790s, and the English translation of the *Méthode* did little to persuade the British chemists to use the new terminology. On the contrary, Black and Kirwan opposed it systematically and proposed alternatives. The most interesting result of these initiatives was the publication of a critical examination of the French nomenclature by Kirwan and Stephen Dickson in 1796.[128] In the preface Dickson appeared to share Lavoisier's concern "that our thoughts can neither be satisfactorily adjusted, nor correctly imparted, unless we possess words aptly accommodated to them" and that "inaccuracy of expression pre-

---

[125]*Encyclopédie Méthodique,* vol. 3 (Paris, 1796), 560.

[126]*Ibid.*

[127]"Je suis convaincu que votre doctrine est infiniment mieux fondée que l'ancienne et sous ce rapport, elles ne peuvent souffrir de comparaisons, mais si le pouvoir de l'habitude empêche quelques uns des anciens Chimistes d'aprouver vos idées, les jeunes ne seront pas influencés par le même pouvoir, ils se rangerant universellement dans votre coté. Nous en avons l'experience dans cette Université où les Etudians jouissent de la plus parfaite Liberté de choix de leurs opinions scientifiques. Ils embrassent en général vôtre Système et commencent à faire usage de la nouvelle nomenclature." A copy of Black's letter to Lavoisier is kept at Arch. Ac., Dossier Lavoisier 1694.

[128]Stephen Dickson, *An Essay on Chemical Nomenclature . . . in Which Are Comprised Observations on the Same Subject, by Richard Kirwan* (London, 1796).

cludes precision of truth."[129] Unlike Lavoisier, however, Dickson believed that a language that had been found "faulty" could have been corrected by "judicious" reform, without the need for linguistic revolution. In this case the language of phlogistic chemistry needed only to be "accommodated" to the new pneumatic discoveries, and there was no necessity for it to be "new-modelled." Kirwan also recognized that the experimental circumstances that caused the chemical revolution confronted chemists with the need to reform the nomenclature, but believed that this problem could have been solved in two completely different ways. One was to reform the ancient nomenclature only when the "improprieties of denomination were prominent and notorious"; the other, which was the one chosen by the French chemists, was to bring about "the entire subversion of the ancient nomenclature."[130] The latter solution could not be accepted on either the epistemological or the experimental level. On the epistemological level, Dickson claimed that "the nomenclature must be posterior to the investigation of science"[131] and not simultaneous with its conception. The logical consequence of Dickson's first linguistic assumption was that language expressed the historical evolution of human opinions and that it was not totally "uninstructive" to retain words which had been used for hundreds of years.[132] In the specific case of chemical language, Dickson remarked:

> Let us use no word to which we do not annex a clear, distinct, and appropriate idea. If our forefathers have confounded substances, let us rectify their errour, and by precise nomenclature perpetuate just distinctions. If they knew a substance which has become for the first time the object of our examination, let us call it by that ancient name which of right belonged to it.[133]

According to Dickson, if one accepted Lavoisier's belief in a language capable of reflecting "the various production of nature," the nomenclature that resulted from it ceased "to be arbitrary" and claimed to lead to a language which was "scientifically perfect."[134] Dickson and Kirwan understood that the *Méthode* contained definite ontological commitments on the nature of scientific language. In their view, however, the project of constructing a non-arbitrary and "scientifically

[129] *Ibid.*, x and xii.
[130] *Ibid.*, xv–xvi.
[131] *Ibid.*, 1.
[132] *Ibid.*, 28 and ff.
[133] *Ibid.*, 37.
[134] *Ibid.*, 10–11.

perfect" nomenclature clashed with the basic principles of a sound empirical philosophy of science. "In fact," they concluded, "there never was a language, or could there be a language framed, capable of expressing the properties and real essences of things."[135] For British chemists it was hard to believe that Lavoisier's linguistic realism was anything more than an expression of a metaphysical and *a priori* system or that it had anything to do with real science. No method or theory could supplant the authority of empirical observations and experiments. This general difference in the perception of the role of language showed a typically Baconian and anti-speculative trait of British science. It is significant, however, that faced with the concrete problems of nomenclature British chemists nevertheless suggested reforms not substantially different from those proposed in the *Méthode*.

Dickson and Kirwan, for instance, accepted that a nomenclature should be based on the distinction of genus and species, so that when new names were needed they could be "faithfully copied from nature,"[136] that classical language should be used as a model and source for the new names, that a name should denote only one substance, that synonyms should always be avoided, etc. They also looked for a greater etymological rigor in chemical nomenclature, and claimed that "deviations in form [implied] deviations in essence."[137] This showed that the stumbling block in the *Méthode* was not the technical rules that were in fact copied here, but the philosophical assumptions on which the rules rested. Otherwise it would be difficult to understand why Richard Kirwan, who had accepted the new theory in 1791, was still so eager to criticize the new nomenclature in 1796.

The nomenclature produced by Dickson and Kirwan was a rather cumbersome mixture of new names and old. For example, fire, caloric, hydrogen, and oxygen were ranged alongside phlogiston, *mephite* (nitrogen), hepatic air, etc. Like many of Lavoisier's critics, Dickson and Kirwan complained of the inaccurate etymology of oxygen, which in fact meant "descendant" from an acid rather than "begetter," and pointed out that:

> A leading property has been mistaken for a universal one, and a name deduced from a contingent and subordinated circumstance has been attached to a principle, as if it were expressive of its fundamental power.[138]

---

[135] *Ibid.*, 42.
[136] *Ibid.*, 63.
[137] *Ibid.*, 32.
[138] *Ibid.*, 110.

Despite the inaccuracy of the etymology, Dickson and Kirwan decided to adopt the name oxygen in their system:

> Oxygen truly signifies *acid-sprung*; and in this sense the word, although of Greek descent, may be retained in chemical language; since it has already acquired considerable celebrity, and rightly interpreted has no tendency to mislead us with regard to the nature of the principle it denotes.[139]

The semantic power of the term oxygen was here reduced to little more than a mere convention, and the scientific meaning of the table of chemical nomenclature was distorted. Dickson's and Kirwan's opinion on the new nomenclature was so well known in Britain that even the 1797 edition of the *Encyclopaedia Britannica* gave a critical account of it. In the long article "chemistry," the author was still visibly inclined to favor phlogiston and the old nomenclature:

> On the subject of nomenclature it is obvious to remark that whatever may be the defects of the old one, we are ready to be involved in much greater difficulties by the introduction of a new one. Or supposing a new language to be adopted where would be the security for its permanence? That which appears most specious at one period, may still be superseded by the refinements of another; and colourable pretentions would never be wanting to successive innovators. Hence a continual fluctuation, and an endless vocabulary.[140]

Despite this conservative attitude, the author decided to reproduce the table of chemical nomenclature "for the satisfaction of [the] readers in general, as for the gratification of those in particular who may have imbibed the doctrines of its authors."[141] In reality it seems more likely that the pace of development forced the authors to give a detailed account of the new nomenclature, which by 1797 had already established itself as standard in several European countries. It was, however, difficult to accept that a terminology entirely devised by a French chemist was suited to the English language. As has been stated, British resistance to the new nomenclature was far stronger than to the new theory.

In 1795, Edward Peart published a pamphlet attacking Lavoisier,[142]

---

[139] *Ibid.*, 111.

[140] *Encyclopaedia Britannica*, 3rd ed., vol. 4 (Edinburgh, 1797), 598.

[141] *Ibid.*

[142] Edward Peart, *The Antiphlogistic Doctrine of M. Lavoisier critically examined, and demonstratively confuted* (London, 1795).

in which he defended phlogiston and criticized the antiphlogistic theory from a "new" perspective. Peart claimed that caloric was not, as Lavoisier believed, a simple substance, but a compound "of two principles, one of which combining with the earth of mercury restores it to its metallic state; the other, with oxygen forms pure air."[143] Peart also tried in vain to demonstrate that oxygen was an acid rather than the principle of acidity.[144] After stating these curious opinions, Peart rejected the new nomenclature, although he could not suggest an alternative.

In 1797 an anonymous pamphlet very critical of the new theory appeared,[145] and once again the main target was the nomenclature. Concerning the name oxygen, the author remarked:

> If the word oxygen, should at first strike every Greek scholar in a sense very opposite to that, which they assign to it, no great praise can be given to their accuracy as Nomenclators. They understand by Oxygen that which generates an acid; but the Greek scholar, who is ignorant of chemistry, will certainly render it, that which is generated by an acid.[146]

Now that the experimental basis of Lavoisier's oxygen theory was no longer challenged, the lack of etymological rigor became a classical argument against it.

Another argument against the nomenclature, often used by British chemists, concerned the practical difficulty of introducing an entirely new language in applied sciences, such as pharmacy and medicine, that constantly employed the results of chemical analysis. The fact that British pharmacopoeias and medical dispensatories printed between 1788 and 1790 reformed the language of chemistry on the lines suggested by Bergman confirmed that the systematic terminology proposed by the French was not ideal for expressing practical concepts.

Even one of the most genuine supporters of the new theory, William Nicholson, was rather cautious with respect to the language. In 1795 Nicholson, who translated the works of Fourcroy and Chaptal into English and edited an important scientific journal, published the first anti-phlogistic dictionary of chemistry to appear in Britain.[147] In his introduction Nicholson praised the simplicity of the new theory

---

[143]*Ibid.*, 34.

[144]*Ibid.*, 51.

[145]*Critical Examination of the First Part of Lavoisier's Elements of Chemistry* (London, 1797).

[146]*Ibid.*, 11.

[147]William Nicholson, *A Dictionary of Chemistry*, 2 vols. (London, 1795).

and its consistency with the latest pneumatic discoveries, but with regard to the nomenclature, he hesitated:

> I have not adopted the nomenclature of the Anti-phlogistians. We are so continually misled by words that it would, no doubt, be of great advantage if a consistent and uniform nomenclature were generally adopted. The French nomenclature, though not without its faults, appears to be more perfect than any other which has been offered; but I did not think myself at liberty to anticipate the public choice, by using it in an elementary work.[148]

Nicholson's hesitation was probably explained by the fact that the new nomenclature was still rejected with few exceptions by all British chemists. Of the exceptions George Pearson is certainly the most important. In 1794 Pearson issued a new translation of the French nomenclature[149] and became the first Englishman to defend the new language openly and unambiguously.[150] He dealt with all the objections raised by Keir in 1789,[151] and by Dickson and Kirwan in 1796,[152] and successfully adapted the table of chemical nomenclature to English requirements. With this end in view, Pearson ignored the controversial questions of the philosophical function of scientific language and its etymological accuracy and concentrated on the practical advantages to be gained from an appropriate use of the new terminology. Among them Pearson mentioned its usefulness in the teaching of chemistry:

> I must acknowledge that I have experienced greater facility in teaching, and students greater facility in learning, the new system of principles with its nomenclature than the former system of chemistry. I find the trouble of learning the meaning of new names amply compensated by a more just and extensive knowledge of things.[153]

This appeal to the practical advantages of the new language undoubtedly helped to convince the skeptical British audience, and after 1800 Lavoisier's nomenclature became the vernacular of English chemistry, pharmacy, and medicine. Once the breakthrough had been made, the new chemical language conquered the British

---

[148] *Ibid.*, vol. 1, vi–vii.

[149] George Pearson, *A Translation of the Table of Chemical Nomenclature, Proposed by Guyton formerly De Morveau, Lavoisier, Bertholet and De Fourcroy* (London, 1794).

[150] The persistence of Pearson in spreading the new nomenclature is evidenced by a second enlarged edition of the *Table* that was printed in 1799.

[151] Pearson (1794), 42 and ff.

[152] Pearson (1799), 86–96.

[153] Pearson (1794), 12.

Isles rapidly, and soon only Joseph Priestley was willing to use a phlogistic nomenclature. This is striking when we consider that British chemists disliked Lavoisier's nomenclature long after it was adopted. In 1802 Richard Kirwan wrote a memoir in which he defended the traditional nomenclature of Stahl, Bergman, Macquer, etc. against what he continued to call "the fashionable rage of coining words," and sadly added:

> Can any one be so arrogant as to pretend that these immortal authors can become unintelligible without prejudice to the Science.[154]

The success of the French nomenclature in Britain must have been painful indeed if some 20 years later Sir Humphrey Davy could still refer to it in the following terms:

> The new nomenclature was speedily adopted in France; under some modifications it was received in Germany; and after much discussion and opposition, it became the language of a new and rising generation of chemists in England. It materially assisted the diffusion of the antiphlogistic doctrine and even facilitated the general acquisition of the science [. . .] but a very slight reference to the philosophical principles of the language will evince that its foundations were imperfect [. . .]. Words should signify things, or the analogy of things, and not opinions [. . .]. A theoretical nomenclature is liable to continued alterations; *oxygenated muriatic acid* is as improper a name as *dephlogisticated marine acid.*[155]

Davy's continuing criticism exemplifies the eighteenth-century British distrust of general and speculative theories and philosophical systems. With respect to the nomenclature, English chemists wished to be the philosophers of Laputa who "instead of expressing their thought by *Words*, exhibited the *Things* themselves, of which each philosopher carried about with him a large assortment that he might never be at any loss in this truly *real* conversation."[156]

---

[154]Kirwan, "Of chymical and mineralogical nomenclature," in *The Transactions of the Royal Irish Academy*, 8 (1802), 58.

[155]H. Davy, *The Collected Works*, vol. 4 (London, 1840), 32.

[156]Keir (1789), xx. In encouraging British chemists to imitate the "Philosophers of the *Flying Island*," Keir misunderstood the satirical aim behind Swift's description. Swift in fact ridiculed the ambition of using objects to demonstrate thoughts as follows: "However, many of the most Learned and Wise adhere to the New Scheme of expressing themselves by *Things*, which hath only this Inconvenience attending it, he must be obliged in Proportion to carry a greater Bundle of *Things* upon his Back, unless he can afford one or two strong Servants to attend him," J. Swift, *Gulliver's Travels*, Part III, Chapter V.

"DIE FRANZÖSISCHE KUNSTSPRACHE IST NICHT FÜR UNS"

The reception of Lavoisier's nomenclature in Germany was substantially different from that in other European countries. Since the beginning of the eighteenth century, chemistry had been a well-established discipline in Germany, and it enjoyed great scientific autonomy. The work of Stahl had an enormous influence both on the development of chemical studies and on the creation of new chairs in chemistry at German universities.[157] Johann Friedrich Henckel, Johann Juncker, Gottfried Rothe, Caspar Neumann, and Johann Heinrich Pott, who constituted the first generation of Stahl's pupils, helped to develop this positive trend and created the base for a German school of chemistry. In consequence of this progress, most German universities boasted a chemical laboratory after 1760 and often a chair of chemistry. By comparison with England, where the first chairs of chemistry were not instituted until the beginning of the nineteenth century, Germany was extremely advanced. Quite apart from the interest in chemistry in its own right, the importance of mineralogy and the mining industry in Germany since the early Renaissance favored chemical investigations.

The existence of such a strong chemical community and the emphasis on the practical applications of the science encouraged German chemists to write in the vernacular from a very early stage. To reach a wider audience, Stahl himself wrote his main chemical works in German.

The institutional strength of German chemistry, together with the widespread use of the German language, created a *national* community of scientists aware of their tradition and confident of their scientific background. The guarded reception accorded to Lavoisier's works in Germany is thus understandable, and it is not surprising that the greatest obstacle to their success was the new nomenclature. The success of the *Méthode* in France, and of the early English, Spanish, and Italian translations that followed, threatened the universality of the phlogistic nomenclature, which after all was a language invented in Germany. Whereas in England the debate began only after St. John's translation of the *Méthode*, in Germany the new language was criti-

[157]On the institutional development of German eighteenth-century chemistry, see Hufbauer (1982). On the reception of Lavoisier in Germany, see G.W. Kahlbaum-A. Hoffmann, *Die Einführung der Lavoisier'schen Theorie im besonderen in Deutschland* (Leipzig, 1897).

cally examined long before the publication of the first translation.[158] After 1789 the discussion began to be colored by "nationalistic sentiments,"[159] and the scientific and linguistic aspects became increasingly subordinate to national and political considerations. In 1789 Crell printed an anonymous letter that well reflected the attitude of German chemists to the new theory:

> I have witnessed a most remarkable drama here [in Paris], one which to me as a German was very unexpected, and quite shocking. I saw the famous M. Lavoisier hold a ceremonial auto-da-fé of phlogiston in the Arsenal. His wife [. . .] served as the sacrificial priestess, and Stahl appeared as the *advocatus diaboli* to defend phlogiston. In the end, poor phlogiston was burned on the accusation of oxygen. Do you think I have made a droll discovery? Everything is literally true. I will not say whether the cause of phlogiston is irretrievably lost, or what I think about the issue. But I am glad that this spectacle was not presented in my fatherland.[160]

This colorful testimony had such a dramatic impact that it was quoted by Lichtenberg in his attack on the pretensions of French science in 1794, and it reflected resentment of the anti-phlogistic theory on German soil. Lavoisier's theory not only challenged a consolidated philosophy of matter, but also implied a denigration of Stahl and the German chemical tradition.

Lavoisier quickly became aware of this situation, and in his famous letter to Franklin (1790), he declared:

> The scientists of London and England are also gradually dropping Stahl's theory, but German chemists still adhere to it strongly.[161]

In 1790, however, things began to change, and the initial German sarcasm gave way to a more rational discussion of Lavoisier's theory and nomenclature. In response to pressing demands "for information about the new French nomenclature,"[162] Johann August Göttling, the editor of the successful periodical *Taschenbuch für Scheidekünstler und Apotheker*, published the first translation of the new language into German. Göttling had not actually seen the *Méthode*, and his translation was not complete, being based on the abstract given by de la Métherie

---

[158]The first German translation of the *Méthode* appeared in Vienna only in 1793.
[159]Hufbauer (1982), 97.
[160]Letter appeared in Crell's *Chemische Annalen*, I (1789); translated into English by Hufbauer (1982), 96.
[161]Quoted by Hufbauer (1982), 100.
[162]Duveen-Klickstein (1954), 145.

in 1787.[163] This partial translation was in fact a compromise between the old phlogistic nomenclature and Lavoisier's. Göttling translated the names of the five most important simple substances as follows:

GÖTTLING (1790)	*MÉTHODE*
Lichtstoff	Light
Wärmestoff	Caloric
Sauerstoff	Oxygen
Wasserstoff	Hydrogen
Tödlicherstoff	Nitrogen (azote)

With the exception of the name given to nitrogen, Göttling's translations of the names of most of the simple substances became accepted. For compounds and salts, however, Göttling preferred to use the old nomenclature. As a result, the nomenclature of acids was not uniform; there was no equivalent of the generic name oxide; and no attempt was made to render the system of suffixes in German. Göttling's nomenclature still included the name "phlogiston," not only as a simple substance but also as part of the names of compounds. Despite its lack of consistency, this translation became a significant landmark in the history of German chemical language. Unlike St. John in his English translation, Göttling decided not to retain the Greek and Latin roots of the French nomenclature but to render, for instance, the name "oxygen" with the German *sauer* (acid) and *stoff* (matter). This choice was rather unusual because all the other translations of the *Méthode* adapted the French morphology to the target language rather than translating it. Unfortunately his reasons for this choice were unstated, but it is reasonable to assume that he found the richness of the eighteenth-century German chemical lexicon naturally conducive to translation of the new French terms rather than use Germanized forms of them. Patriotism may also have played a part in this choice, but it should be noted that many of those who translated and supported the French nomenclature were Francophiles who sympathized in principle with the French Revolution.

In 1791 a young and ambitious chemist, Christopher Girtanner, produced the first systematic German translation of the table of chemical nomenclature.[164] Unlike Göttling, Girtanner explicitly tried to cre-

[163]"Alphabetisches Verzeichniss der neuen französischen Nomenklatur," in *Taschenbuch für Scheidekünstler und Apotheker* (1790), 147–155.

[164]Girtanner, *Neues Chemische Nomenklatur für die Deutsche Sprache* (Berlin, 1791).

ate a new German chemical nomenclature analogous to the French one. Accordingly he introduced several new terms and eliminated phlogiston. He retained Göttling's translation of the simple substances but decided to replace the term *Tödlicherstoff* (nitrogen) with the two alternatives *Salpeterstoff* and *Stickstoff.* In the following cases, Girtanner adapted the French terms of the *Méthode* to German:

OLD GERMAN TERM	*MÉTHODE*	*GIRTANNER (1791)*
Wasserbleikönig	Molybdène	Molybden
Braunsteinkönig	Manganèse	Magnesium (*sic*)
Weisgold	Platine	Platine
Vegetabilisches Laugensalz	Potasse	Potasche
Mineralisches Laugensalz	Soude	Soda
Flüchtiges Laugensalz	Ammoniaque	Ammoniak[165]

Girtanner introduced the term *Halbsäure* (half-acid) to denote oxide and provided a systematic nomenclature of the acids. To render the informative French suffixes in German, he proposed the following more cumbersome solution:

GIRTANNER (1791)	*MÉTHODE*
Schwefelsäure	Acide sulfurique
Schwefelsaure or Schwefelsaures	Acide sulfureux

Other difficulties arose with the nomenclature of salts. Lavoisier's *pyrolignite de chaux* was expressed in German with the considerably lengthier *Brenzlige holzsaure Kalcherde.* In translation from French into German, many of the systematic advantages were lost, and in a few cases, the new names were far less effective than the phlogistic ones. The French *Sulfure d'antimoine,* for which the German phlogistic term had been *Spiesglas,* was now given by Girtanner the accurate but longer name of *Geschwefelte Spiesglanz.*

Further attempts were made to translate the French nomenclature into German in 1792 and 1793. In 1792 Sigismund Friedrich Hermbstädt translated Lavoisier's *Traité* and instead of *Halbsäure* (oxide) suggested the term *oxideret.*[166] At about the same time, Johann Andreas

[165] *Ibid.*, 11–12.

[166] Lavoisier, *System der Antiphlogistischen Chemie* (Berlin-Stettin, 1792), vol. 1, 230.

Scherer, Georg Eimbke, and Johann Christian Remler published three partial translations of the *Méthode* that were only slight modifications of Girtanner's and that ran into the same technical difficulties in reproducing the fluency of the French.[167] All these translations bear witness to the enormous problems entailed in Göttling's original project.

In 1793 the first German translation of the complete *Méthode* was finally published by the Austrian apothecary Karl Freiherr von Meidinger.[168] Meidinger followed the general path taken by Girtanner, but he tried to devise a more effective solution to the problem of translating the French suffixes, *-ique*, *-eux*, *-ate*, *-ite*, *-ure*. For example, he translated the different states of combination of sulphur as follows:

MEIDINGER (1793)	*MÉTHODE*
Schwefelsäure	Acide sulfurique
Schwefelsaure	Acide sulfureux
Schwefelgesäurtes	Sulfate
Schwefelsaures	Sulfite
Geschwefel	Sulfure[169]

Meidinger decided as a general rule to use German adjectival forms for the French substantives to enable the French system of suffixes to be replaced with an appropriate correspondent. Meidinger retained Girtanner's term *Halbsäure* for the French oxide, introduced the term *Stickstoff* for nitrogen, and used the generic name *Radical* instead of Girtanner's *Basis* to denote a chemical base.

Meidinger's translation was not particularly successful, but it encouraged many apothecaries to use the new nomenclature, and it stimulated a constructive debate. The most prominent German chemists were, however, still unimpressed. The professor of chemistry at the University of Halle, Friedrich Albert Carl Gren, admitted that the new French language had several advantages, but:

---

[167]J.A. Scherer, *Versuch einer neuen Nomenklatur für Deutsche Chymisten* (Vienna, 1792); Georg Eimbke, *Versuch einer systematischen Nomenklatur für die phlogistische und anti-phlogistische Chemie* (Halle, 1793); J.C. Remler, *Neues chemisches Wörterbuch oder Handlexicon und allgemeine Uebersichter in neuern Zeiten entworfenen französisch=latainisch=italiänisch=deutschen chemischen Nomenklatur nach Bergmann, Berthollet, Brugnatelli, Girtanner, Hermbstädt, Jaquin, Lavoisier, Leonhardi, de Morveau, Wiegleb, Scherer* . . . (Erfurt, 1793).

[168]*Methode der chemischen Nomenklatur für das antiphlogistische System* (Vienna, 1793).

[169]*Ibid.*, 44.

> . . . on the other hand this nomenclature prevents the developing of a closer knowledge of the system because the deliberate moving from a building in which you have been living comfortably till now and which you would not leave for any reason, even if the building is not perfectly symmetrical, to another building in which you have to make an effort to find the way, does not suit the natural tendency to rest, unless you are absolutely convinced of the unshakable firmness of the building.[170]

Gren hoped to find a way of reconciling the old and the new architecture of chemistry and remarked:

> Yet whoever examines the whole, impartially and without predilection, will find that on adopting the anti-phlogistic system, there still remain chasms in the explanations of many phenomena, and especially with regard to the extrication of *light* [. . .], its fixation, its development, and its changes. It is quite in vain to deduce light from a modification of caloric. This expression, if it does not involve a modifying cause, means nothing at all. But if it be necessary to admit such a cause, modifying the caloric so as to become light; and, if it may be allowed to call this cause *Phlogiston*, it is easy to conceive, first that it is possible to reconcile the antiphlogistic system with the adoption of phlogiston, and then that the latter becomes even necessary, in order to explain satisfactorily *all* the circumstances of its various phenomena. In the sequel I shall explain the fundamental principles of this united system, which might be called *Eclectic*, as well as those of the strictly *antiphlogistic* system.[171]

In keeping with this eclectic system of chemistry, Gren favored a nomenclature that contained elements of both the phlogistic and the French terminology. Gren's main goal was to create a language that was free from any theoretical interpretation,[172] and he adopted Meidinger's nomenclature, with the main exception of the term *Phlogiston oder Brennstoff* to denote the basis of light. Gren's attempt to reconcile the French doctrine with the principles and the language of the Austrian chemist gained a few supporters, but in the long run, it did not

[170]Quoted in Kahlbaum-Hoffmann (1897), 84.

[171]Gren, *Principles of Modern Chemistry* (1796); English translation (London, 1802), vol. 1, 15.

[172]"Meine Namen sollen von aller Beziehung auf das, was hypothese ist, frey seyn. Man fängt an, der Sectirerey in der Chemie herzlich satt zu werden, und ich glaube, dass durch diese meine Bemühungen eine nähere Bereinigung der Parteyen bewirkt werden kann," Gren, *Systematisches Handbuch der gesammten Chemie*, 2nd ed., vol. 4 (Halle, 1796), 9.

satisfy either party. In fact the great majority of German chemists continued to support the phlogistic nomenclature, and Lorenz Crell, the editor of the *Chemische Annalen*, one of the most famous scientific journals in Europe, became even more determined to resist any external influence on German chemistry. In 1795 Göttling wrote a long criticism of the French nomenclature in which the main argument was as follows:

> The leading principle of most nomenclature is to concentrate all the constituents in the names of the compounds so that the names become as instructive as possible and characterize substances with the utmost exactness.
>
> This principle, if applied alone, could fail in its aim because names applied to combinations of several elements would be boring and not functional and in most cases the intended precision would be lost.[173]

Indeed, the translation of many of the new French names led to German terms that were considerably longer than the old ones.

The most vigorous attack on the *Méthode* was made by Johann Friedrich Westrumb, who saw the diffusion of Lavoisier's theory in Germany as a threat to the national linguistic tradition. To Westrumb the French nomenclature was unacceptable primarily because it was created by foreign scientists, and he appealed to patriotic values in his rejection of the French names. How could a foreign language be acceptable in a country that had a long tradition in chemistry and constantly led the way in experimental achievement? With this in mind, Westrumb devoted an entire book to the construction of a pure (*rein*) and independent (*unabhängig*) German chemical nomenclature.[174] Westrumb pointed out the inconsistencies in previous attempts to translate the French nomenclature into German and proclaimed the superiority of the German chemical language to any other. In addition he claimed that most of the French terms indicated hypothetical properties. Last but not least, a literal translation of the term oxygen would have led to the clumsy designation *Säuerbildender Stoff* rather than *Sauerstoff*.[175]

Since nothing that had been proposed by the French was acceptable, the problem Westrumb had to face was how to replace the new nomenclature with one as effective. He prefaced his work with a long historical survey of the development of chemical language and pointed out that German chemists were the first to clarify the expressions they

---

173 Quoted in Kahlbaum-Hoffmann (1897), 86.

174 Westrumb, *Kleine physikalisch-chemische Abhandlungen*, vol. 3:2 (Hanover, 1793).

175 *Ibid.*, xv–xvi.

used. According to Westrumb, the emancipation of chemistry from alchemy was due to the clarity of the terms used by German-speaking chemists, such as Stahl, Neumann, Bergman, Leonhardi, Scheele, etc. Westrumb based his "new" nomenclature on a mixture of those proposed by Bergman and Scheele. He rejected the German *Sauerstoff* (oxygen) and replaced it with the neutral *reine Luft*, introducing the name *Stickgas* for nitrogen, *Entzündbares* [*Stoff*] for hydrogen, and a new nomenclature of salts. Apart from these changes, Westrumb stuck to the phlogistic nomenclature. Westrumb's attempt to create a *pure* German nomenclature did not win favor, and his own terminology was soon forgotten, although his appeal for a defense of the German language against foreign influence fell on more receptive ears. Neither the supporters of the new theory nor those of the phlogiston theory were happy with the new language, and there is evidence that such people as Hagen, Trommsdorff, Gmelin, Crell, Wiegleb, Richter, and Lampadius opposed Lavoisier's doctrine primarily because of its nomenclature. Westrumb's patriotic arguments were so well received that as late as 1796 Wiegleb could declare that "the new Gallic nomenclature [had] been totally superseded."[176] By contrast, the experimental background of Lavoisier's work was increasingly absorbed by German chemists, and in a letter to Fourcroy in 1796, Van Mons reported:

> [Germany] counts no more among its chemical authors any partisans of the traditional phlogistic system, since I have convinced them of the presence of oxygen in oxide of mercury heated to redness by fire. They have all adopted the new doctrine without restriction or with restriction of little importance. Crell, Westrumb, Wiegleb, Trommsdorff, Gmelin, Richter, Leonhardi etc., although all trying to ally the new theory with the existence of phlogiston in combustible bodies, admit it in its entirety and its consequences.[177]

Any residual resistance was due to the difficulty of assimilating a new theory within the framework of the new nomenclature. As late as 1800, Richter, who was without doubt one of the most gifted of German chemists, referred to the question of nomenclature as follows:

> When addressing Germans, I claim the right to use German language and nomenclature whenever I can find suitable expressions, and only to seek recourse in other languages when my own is not rich enough; I hope that I shall be allowed to make this claim as a German, who likes

---

[176]Quoted by Hans-Georg Schneider, "The Fatherland of Chemistry: Early Nationalistic Currents in Late Eighteenth Century German Chemistry," in *Ambix*, 36 (1989), 17.
[177]Quoted by Partington (1962), vol. 3, 493–494.

> to be in a position to denote numerous substances by short German names, as has happened for instance, with regard to Lebenstoff—the vital element (so-called oxygen or acid-producer), Lebenstoffung—vitalisation (so-called oxydation-oxygenation), Brennstoff—the combustible element—and Brennstoffung (combustion), brennstoffen (to combust), entbrennstoffen (to devitalise), wärmestoffen (to heat) and entwärmestoffen (to decrease in heat) [. . .].
>
> Weigh, dear reader, the harshness of each of the above German expressions against the harshness of de-oxidation, nitric acid and the like, and then say which one is better suited to your German tongue and to the matter in hand.[178]

Richter, too, had little success, but as we have seen, none of the German nomenclatures presented after 1790 succeeded in prevailing over the others, and German students of chemistry were therefore confronted with a variety of terminologies. A letter received by Trommsdorff from an anonymous writer in 1796 lamented this situation in the following terms:

> Heavens how many languages will German chemists start to speak next? We already need dictionaries to understand each other. What an immense amount of work is needed for any beginner to find his way through this tangled mass of words.[179]

That this complaint was justified is shown by the immense success of Bourguet's *Chemisches Handwörterbuch*, which was published in Berlin between 1798 and 1805.[180] Bourguet tried to list almost all the terms coined by different German chemists, a task so complicated that he asked Hermbstädt, a convinced Lavoisierian, and Richter, a phlogistonist *sui generis*, to help him to collect all the "tangled mass of words." This is why his dictionary contains terms from both phlogistic and antiphlogistic nomenclatures, not to mention the eccentric neologisms of Westrumb, Göttling, and Richter.

Only after 1800 did German chemists begin to adapt their nomenclature to the uniform standard of Meidinger's translation of the *Méthode* and finally delete phlogiston from their chemical lexicon. The change was a painful one, but the adoption of a vernacular no-

---

[178]Richter, *Ueber die neuern Gegenstände der Chymie*, vol. 10 (Breslau, 1800), x–xi; English translation by Schneider (1989), 17.

[179]*Journal der Pharmacie*, 3 (1796), 312–315; quoted in Schneider, "The Threat to Authority in the Revolution of Chemistry," in *History of Universities*, 8 (1989), 143.

[180]David Ludwig Bourguet, *Chemisches Handwörterbuch nach den neuesten Entdeckungen entworfen*, 6 vols. (Berlin, 1798–1805).

menclature was considered a satisfactory compromise between the undeniable power of Lavoisier's theory and the strength of the German chemical tradition.

"SI PARLERÀ TRA NOI LINGUA BISBETICA"

Due to the political and geographic fragmentation of the Italian peninsula, the French nomenclature had a somewhat mixed reception in Italy. At the end of the eighteenth century, Italy was divided into several states, variously subject to Austrian, French, and German cultural influences. In Turin, for instance, scientists normally spoke French, and most of them, such as Lagrange, Berthollet, and Giobert, had close relations with the Parisian milieu. In Milan, Pavia, and Padua, the scientific world had stronger links with England and Germany, while Florence was relatively cosmopolitan. In Bologna, on the other hand, the influence of the Catholic church in the university still hindered an open approach to science.

Science generally, and chemistry in particular, had to contend with the obstacles posed by this political fragmentation. Whereas in Germany the authority of Stahl was sufficient to overcome all political divisions and create a national school of chemistry, in Italy there was no such unity within the scientific community, and the material conditions of chemistry were rather difficult all over the peninsula. With the exception of Bologna, where a chair of chemistry was inaugurated in 1737, there was little chemistry teaching or research at the universities.[181] Although attracting considerable notice, the important contributions of Felice Fontana,[182] Marsilio Landriani, and Alessandro Volta during the 1770s were essentially successful forays into pneumatic chemistry by scientists engaged in other investigative fields. In Italy, as in England, the professional chem-

---

[181]On Italian chemistry, see the following studies: I. Guareschi, "La chimica in Italia dal 1750 al 1800," in *Supplemento annuale alla Enciclopedia chimica*, vols. XXV and XXVI, 1909–1910; *idem*, "Lavoisier, sua vita e sue opere," in *Supplemento annuale alla Enciclopedia chimica*, XIX (1903), 307–460; G. Testi, *Storia della chimica, con particolare riguardo all'opera degli italiani* (Rome, 1940); F. Abbri, "Spallanzani e la diffusione delle teorie chimiche in Italia," in P. Rossi and G. Montalenti (eds.), *Lazzaro Spallanzani e la biologia del '700* (Florence, 1983), 121–135; and my "Gli scienziati italiani e la rivoluzione chimica," in *Nuncius*, IV (1989), 119–146.

[182]On Fontana's chemical contributions and their impact on the national and European scientific community see the study by F. Abbri, *Science de l'air. Studi su Felice Fontana* (Cosenza, 1991).

ist did not exist, and the creation of the chair of chemistry in Bologna was a nominal recognition rather than a practical one.

The disagreement between Lavoisier and Priestley on the nature of gases aroused wide interest in Italy because of the experimental problems it raised, but few scientists took up a definite position on the theoretical questions, and until 1790 most Italian naturalists supported the phlogiston theory more or less as a matter of course. The translation of the *Méthode* and of the *Traité* shattered this apparent calm, and Italian naturalists were forced to decide between the new doctrine and the old one.

The first Italian translation of the *Méthode* appeared in 1790,[183] and was the work of the Venetian apothecary Pietro Calloud. It seems to have been delayed by the initial misgivings of the *Società Veneta di Medicina,* which did not regard the new language as sufficiently soundly based.[184] Calloud's translation of the table of chemical nomenclature essentially retained the original structure and most of the Italian terminology was formed from the French. Calloud named the first five simple substances as follows:

CALLOUD (1790)	*MÉTHODE*
Luce	Lumière
Calorico	Calorique
Ossigeno	Oxygène
Idrogeno	Hydrogène
Azoto	Azote

To translate the French *radical* and *oxide,* Calloud used the generic names *Radicale* and *Ossido,* and he was able to make full use of suffixes:

CALLOUD (1790)	*MÉTHODE*
Acido solfor*ico*	Acide sulfurique
Acido solfor*oso*	Acide sulfureux
Sol*fato*	Sulfate
Sol*fito*	Sulfite
Sol*foro*	Sulfure

[183]Guyton, Lavoisier, Berthollet, Fourcroy, *Metodo di Nomenclatura chimica* (Venice, 1790); in 1791 a partial translation appeared probably by Brugnatelli, "Vocabolario della nuova nomenclatura chimica Francese, Latino, e Italiano," in *Biblioteca Fisica d'Europa,* 20 (1791).

[184]In the preface of the *Methodo,* Calloud explicitly referred to the censorship he had to face.

Calloud's translation did not have much impact on the Italian scientific community. This was probably due to the fact that Vincenzo Dandolo, who was preparing a translation of the *Traité*, claimed that Calloud's translation was full of errors and omissions and needed to be done again.[185] Actually Calloud's work was accurate enough, and although Hassenfratz's and Adet's system of chemical characters was missing, it eventually provided the essential basis of the Italian chemical language. Nevertheless, in 1791 Dandolo decided to translate the two dictionaries of the *Méthode* again.[186] They were placed after Dandolo's translation of the *Traité* and Lavoisier's memoirs on respiration,[187] and it was probably due to this fortunate editorial decision that Dandolo's translation of the French nomenclature became the most widely used reference work of Italian chemists. The changes made to Calloud's translation by Dandolo are irrelevant; Dandolo introduced *carbono* for *carbone*, *solfuro* for *solforo*, and a few other minor amendments.

Although a supporter of the new theory, Dandolo was not equally enthusiastic about the philosophical content of the *Méthode* and the *Traité*. He did not agree that language was the only and essential vehicle of human thought; he saw words only as our arbitrary symbols of reality.[188] It is interesting to note that while he criticized the assumptions underlying the new nomenclature, Dandolo nonetheless used the terms without hesitation.

Whatever its limitations, Dandolo's translation helped to spark off a lively debate on Lavoisier's theory. The first naturalist to adopt it was Giovanni Antonio Giobert, who in 1790 publicly declared his preference for the new theory and nomenclature[189] and in the same year was invited by Lavoisier to contribute to the *Annales de chimie*. Another

---

185"C'est à la célébrité de votre nom, très connu dans toute l'Italie, et a la bonté de votre caractère, qu'il faut attribuer que je prends de m'adresser directement a vous, Monsieur, pour vous faire part que votre Nomenclature chimique ayant été traduite ici en Italien, il y a un mois, pleine de fautes et sans les tables, et la description des signes chymiques," Letter by Dandolo to Lavoisier dated January 1791 that has been transcribed by Perrin during a visit to the Chabrol collection of Lavoisier's papers and that is still in the hands of the heirs.

186*Dizionarii Vecchio e Nuovo, Nuovo e Vecchio di Nomenclatura chimica* . . . (Venice, 1791).

187On the relations between Dandolo and Lavoisier, see Abbri, "Lavoisier e Dandolo: Le edizioni italiane de *Traité élémentaire de chimie*," in *Annali dell'Istituto di Filosofia dell'Università di Firenze*, VI (1984), 163–182.

188See Dandolo's commentaries on the Preliminary discourse of Lavoisier's *Trattato elementare di chimica*, 3rd edition (Venice, 1796), vol. 1, 8–32.

189"Non istupite se io comincio a servirmi della nuova nomenclatura, e se vi parlo di ossigeno e di ossigenato," Letter to Brugnatelli dated August 7, 1790 and published in *Annali di chimica* I (1790), 21.

"early" supporter of the new theory in Italy was Lazzaro Spallanzani, who enthusiastically embraced its principles in 1790. In March 1791 he wrote to Lavoisier, reporting that except by Alessandro Volta the new nomenclature was "universally adopted" in the University of Pavia.[190] This was a wild exaggeration. With the exceptions of Dandolo, Giobert, and Spallanzani, most Italian naturalists were still openly hostile to Lavoisier and far from adopting the new nomenclature. Even at Pavia, Spallanzani was the only one who accepted the new theory! Luigi Valentino Brugnatelli, the only pure chemist at the university after the death of Scopoli, was of a different opinion.[191] Brugnatelli, who founded the *Annali di chimica* and other pioneering scientific journals, quickly became one of the leading figures of Italian chemistry. The success of the *Annali di chimica*, which was read and reviewed throughout Europe, helped to create an Italian chemical community. In the editorial notes of the first volume (1790), Brugnatelli declared that he had long supported the new theory, but a few lines later, he added:

> As in many other parts of Europe, the general establishment of the nomenclature introduced by the pneumatic chemists, will be delayed.[192]

Remaining unconvinced of the merits of the new nomenclature, Brugnatelli outlined in 1795 an alternative,[193] which modified the essential achievements of the *Méthode*. Unlike Black, Dickson, and Westrumb, who criticized the new nomenclature without making their own proposals, Brugnatelli was more systematic and ambitious, and tried to create another new chemical language. In his opinion the main property of oxygen was not that it was the universal principle of acidity, as Lavoisier believed, but rather that it was the substance with the largest quantity of caloric combined with it. The presence of caloric was therefore taken to be the main characteristic of oxygen, and Brugnatelli

---

[190]L. Spallanzani, *Epistolario*, vol. 4 (Florence, 1962), 273.

[191]Brugnatelli founded the *Biblioteca Fisica d'Europa* (1788–1791), the aim of which was to translate into Italian the main scientific memoirs appearing in foreign journals and academic transactions, the *Annali di chimica* (1790–1805), the *Giornale fisico-medico* (1792–1795), and the *Giornale di fisica, chimica, e storia naturale* (1808–1827). On Brugnatelli, see my "Luigi Valentino Brugnatelli e la chimica in Italia alla fine del Settecento," in *Storia in Lombardia*, 2 (1988), 3–31.

[192]*Annali di chimica*, 1 (1790), Preface.

[193]Brugnatelli, "Prospetto di riforma alla nuova nomenclatura chimica proposta dai Sigg. Morveau, Lavoisier, Berthollet, e Fourcroy," in *Annali di chimica*, 8 (1795), 149–173, also published in the first volume of Brugnatelli's *Elementi di chimica* (Pavia, 1795–1799), 3 vols.

suggested changing its name to *termossigeno.*[194] Brugnatelli also wished to change the name of *azote* (nitrogen) and, noting the experiments by Göttling, who claimed that nitrogen was not a simple substance but a compound of oxygen and light,[195] proposed the name *fossigeno* (from the Greek "begetter of light").[196] When in 1797 Spallanzani showed Göttling's experiments to be unsound, Brugnatelli proposed another name, *septone,* derived from the Greek word for putrid, and already used by the British chemist Saltonsall.[197] Brugnatelli also recommended taking the entire chemical nomenclature from Greek and weeding out all the Latin terms. He started by changing *acid,* which came from Latin, to *ossi,* from Greek, so that, for instance, sulphuric acid became *ossi-solforico.* But Brugnatelli was also dissatisfied with some important words of Greek origin in the *Méthode.* Instead of the term *oxide,* he introduced the term *encausto,* which was also derived from Greek and meant combusted. These few, but strategic, changes radically modified the substance of the French nomenclature and resulted in an entirely new chemical language.

Brugnatelli's attempt was reasonably well received and was immediately adopted by some prominent naturalists, such as Volta, Moscati, and Mascheroni.[198] Encouraged by this initial success, Brugnatelli tried hard to spread his nomenclature elsewhere in Italy and abroad. These efforts were rewarded, and Brugnatelli's table of chemical nomenclature was even translated back into French and adopted by Van Mons[199] and his pupil Gérard.[200] The partial success of Brugnatelli's nomenclature is indirectly testified to by the critical reactions of Berthollet and Guyton, who began to fear its wider diffusion.[201]

Visiting Paris with Volta in 1801, Brugnatelli tried to persuade Berthollet, Laplace, Fourcroy, Chaptal, Haüy, Vauquelin, Guyton, and Sage to adopt his new theory of *termossigeno* and its nomenclature. But with the exception of Sage, who would have supported anything against

---

[194]Brugnatelli, *Elementi,* vol. 1, 211.

[195]J.A.F. Göttling, "Ueber das Leuchten des Phosophors in Stickluft," in *Neues Journal der Physick,* 1 (1795), 1–15.

[196]Brugnatelli, *Elementi,* vol. 1, 217.

[197]*Annali di chimica,* 15 (1798), 154.

[198]Brugnatelli also received the support of other naturalists, such as Amoretti, Notariani, and Valli. See *Annali di chimica,* 9 (1795), 1.

[199]Brugnatelli, "Synonimie des Nomenclatures chimiques modernes," in *Journal de chimie et de physique,* II (1802), 82.

[200]F. Gérard, "Des combinaisons thermoxygènes et oxigènes; de leur caractère, leur classification, et leur nomenclature," in *Journal de chimie et de physique,* II (1802), 306–320, and continued in vol. III, 86–111, 147–197, and 229–249.

[201]Guyton (1798) and Berthollet (1803), vol. 1, 447–457.

Lavoisier's theory, the response was unanimously negative. Brugnatelli's claim that oxygen was in fact constantly combined with a quantity of caloric was dismissed by Laplace as naive. While Volta was honored by Napoleon for his invention of the electric pile, Brugnatelli met with a resounding defeat. He can hardly have taken comfort from Napoleon's laconic comment to him:

> Ah! En Italie on n'est pas si fort en chimie comme on l'est en physique.[202]

Besides Brugnatelli's modifications, there was more consistent opposition from other quarters in Italy. In Milan, in addition to Landriani, the professor of mineralogy Ermenegildo Pini firmly rejected the new nomenclature and published several essays disputing its principles.[203] In Pavia Volta remained a phlogistian until he was converted to Brugnatelli's theory in 1795. In Turin all the naturalists except Giobert were still on the side of Stahl. In Padua the professor of chemistry Marco Carburi, although a fervent Jacobin, remained a convinced phlogistian. In Verona the phlogistian Antonio Lorgna managed to bring many naturalists, such as Du Pré, Viano, Merzari, Stratico, Giorgi, and Macri, together in an anti-Lavoisierian party.[204] In Tuscany Fontana, who was one of the most enlightened naturalists in Europe, supported phlogiston, along with Gioacchino Carradori and the professor of chemistry at Pisa University Giorgio Santi. In a letter to Brugnatelli, Carradori explained this hesitation in terms that typified the general attitude of Italian naturalists:

> With respect to the new *Pneumatic chemistry*, I am always embarrassed because I would like to embrace it, but it seems to me that its theory so little corresponds to the facts that I find it difficult to be persuaded.[205]

This ambivalence and pragmatism were also shown by those who decided to adopt the new theory: Giovanni Fabbroni, for instance, decided that the new theory was gaining supporters all over Europe and would ultimately prevail over phlogiston, and he therefore adopted it

---

[202]Brugnatelli, *Diario del viaggio compiuto in Svizzera e in Francia con Alessandro Volta nel 1801* (Pavia, 1953), 68.

[203]E. Pini, "Osservazioni sulla nuova teoria e nomenclatura chimica come inammissibile in mineralogia," in *Memorie di matematica e fisica della Società Italiana,* 6 (1792), 309–368; *idem, Sulla metachimica ossia sulla nuova teoria e nomenclatura chimica* (Milan, 1793).

[204]See Gmelin (1797–1799), vol. 3, 295, and Guareschi (1903), 450–458.

[205]Quoted by Guareschi (1903), 456–457.

himself. His conversion, like many others, was clearly not the result of experimental convictions.

Generally speaking, Lavoisier's nomenclature was adopted in Italy only at the very end of the eighteenth century. It is difficult to be definite about the reasons for the opposition, but it may be noted that most of the important Italian naturalists who defended phlogiston were friends of British and German chemists and had only sporadic contact with French ones. The correspondence of Fontana, Landriani, Volta, Brugnatelli, Carburi, etc. confirms the predominance of German and English scientific contacts. By contrast, Giobert, Spallanzani, Fabbroni, and Dandolo were all acquainted with the French milieu, and in the cases of Giobert and Fabbroni, there is evidence that their conversions were primarily due to their French connections.[206]

The distaste of Italian scientists for any form of speculative and theoretical controversy may well explain why their conversion or opposition to Lavoisier's theory and nomenclature depended more on external factors than on careful scientific consideration. Furthermore, in the absence of a strong chemical tradition, the interest of Italian naturalists in chemistry was more concerned with specific experimental problems than with general and theoretical questions.

The Nomenclature in Spain and Portugal

The scientific tradition of eighteenth-century Spain and Portugal was certainly not as strong as that of other European countries. Moreover, the part played by the Church in Spanish universities hindered any significant progress in science at an institutional level. It was probably because of this scientific backwardness and lack of a national scientific tradition that Lavoisier's works met with little opposition in Spain. Aréjula's attempt to reform the French nomenclature of 1788 (see Chapter Four) was overshadowed by the success of Bueno's translation of the *Méthode* into Spanish.[207] This was the first translation published anywhere, and it reveals how little resistance Spanish naturalists offered to the new system. The translator, Pedro Guitiérrez Bueno, who was professor at the Royal Laboratory of Chemistry in Madrid, decided to translate the *Méthode* to provide students with an effective and up-to-

[206]See Beretta (1989).

[207]*Méthodo de la nueva Nomenclatura química* (Madrid, 1788). On the reception of Lavoisier in Spain, see Gago (1988).

date dictionary of chemistry. This educational aim was underlined by the fact that Bueno did not support any specific chemical system, and with regard to the controversy between Lavoisier and the phlogistonists, he declared:

> We are not necessarily compelled to follow the one or the other; we explain chemical operations by whatever clarifies their comprehension.[208]

Bueno's translation of the *Méthode* was extremely faithful to the French version and, like the Italian and English ones, adapted the final syllables of the terms to the Spanish language and only rarely changed the original word. Its use in chemical courses made Bueno's translation the standard code of Spanish chemical language. Nevertheless, Lavoisier's nomenclature was not entirely unopposed. Joseph-Louis Proust, for instance, criticized it, and in 1790, commenting on a memoir by Pelletier and Donadei, he declared:

> Messrs Pelletier and Donadei, dedicated more to discovering truth than words, have used without distinction both nomenclatures; they are persuaded with justice that partisans of chalky, aerial, fixed, mephitic, etc., acid will understand matters just as well as the partisans of carbonic acid [*los carbonistas*]; for this reason they have avoided taking part in that ridiculous logomachy for which M. Bufon [*sic*] has already reviled French chemists and which Bergman was very far from foreseeing when he encouraged these chemists to make a reform in their language, one that in truth was very necessary. *Sapientia mundi tumultuosa est, et non pacifica.* Never has the intolerance and acrimony of our professors in defense of words of their own invention so justified as it does today the judgment of St. Bernard concerning human knowledge.[209]

Since he proposed no alternative other than retaining the old phlogistic language, Proust's criticism was not particularly influential, and by the end of the century, Spain, too, was totally converted to both the principles and language of Lavoisier's doctrine.

The situation in Portugal was similar to that in Spain. The lack of a sound national tradition in chemistry facilitated the diffusion of Lavoisier's theory. The first naturalist to adopt it was the Brazilian Vicente de Seabra Telles, who in 1787 wrote a dissertation on fermentation in

---

[208] Quoted by Gago (1988), 180.
[209] *Ibid.*, 175–176.

which he praised Lavoisier's theory of acidity.[210] In 1788 Seabra published the first part of a textbook in chemistry, giving the new chemistry and nomenclature his definite approval. Two years later, he provided the second part with a detailed analysis and translation of the French nomenclature. In his introduction he justified the use of the new language as follows:

> Such terminology will not fail to please any sensible person. Let the alchemists in disguise [we speak of those who disdain the system, and do not wish to adopt it] keep to themselves their insignificant and symbolic names, on which they lay their science. Wise men must express their knowledge by means of expressive words.[211]

This decisive support of the new language was followed by a translation into Portuguese. Seabra wished to reproduce the effectiveness of the *Méthode* in his own language, and with this in mind, he adapted the French nomenclature rather than translated it.

The remarkable speed with which the *Méthode* and, more generally, Lavoisier's theory as a whole were absorbed in Spain and Portugal was a sign of the weakness of the chemical tradition in those countries. As Lavoisier remarked, it was more difficult for those with a chemical background to adopt his theory than for those "who have not yet made any study of chemistry."

The New Nomenclature in Scandinavia

The introduction and translation of the new nomenclature in the Scandinavian countries faced numerous difficulties, although of a different nature from those elsewhere. In Sweden, which was undoubtedly the leading country of science in the north, chemistry had a long and consistent institutional and experimental tradition.[212] As far as nomenclature is concerned, Swedish chemists and mineralogists contributed to the establishment of a systematic terminology for inorganic

---

[210]On Seabra, see A.J.A. De Gouveia, "Vicente de Seabra and the Chemical Revolution in Portugal," in *Ambix*, 32 (1985), 97–109; Carlos L. Filgueiras, "Vicente de Seabra Telles (1764–1804), the first Brazilian Chemist," in *NTM*, 27 (1990), 27–44.

[211]Seabra, *Elementos de chimica*, Part II (Coimbra, 1790), 56; translated into English by Filgueiras (1990), 38.

[212]On the history of Swedish chemistry, see Hugo Olsson (1971); Lundgren (1988).

substances and to the reform of the alchemical lexicon very early in the eighteenth century. Explicitly addressed to European scholars, the reforms proposed by Cronstedt, Wallerius, and Bergman were all written in Latin, and there is no doubt that this contributed to their influence. Bergman declined to use the vernacular in science not only because of the universality of Latin but also, above all, because of the advantage of having a system of suffixes. Using Latin rather than Swedish names enabled Bergman to compose an extremely systematic and consistent Latin nomenclature of salts that, as we saw in Chapter Three, established the pattern for later reforms. At the same time, there existed another chemical and mineralogical tradition outside the university. Ever since the first half of the seventeenth century, chemistry in Sweden had been connected with mining, and the chemical knowledge that was required in this important industry was supplied by the *Laboratorium chemicum* of the Board of Mines, an institution founded in 1637 and embracing all ironmasters and mine inspectors. As in German mines, it was usual for these officials to have no academic background and to know only their native tongue. It is probably for this reason that most literature related to mining was published either in Swedish or in German. This tradition, vital to the economy of Sweden, led to the development of a vernacular mineralogical nomenclature. The outstanding work in Swedish was Rinman's *Bergwerks Lexicon*,[213] a two-volume dictionary that listed in alphabetical order all the technical vocabulary of mining. Being an integral part of the industry, chemical nomenclature occupied a prominent place.

Thus there were in the chemical language of Sweden two distinct traditions, the academic and the practical, respectively Latin and Swedish. Although each tradition was almost totally dominant within its own environment, they coexisted harmoniously until the end of the eighteenth century.

After the death of Torbern Bergman in 1784, there was no one to develop the reform outlined in the 1780s, his successor in the chair of chemistry at Uppsala University, Johan Afzelius, being not much concerned with the reform of chemical language. In these circumstances Scandinavian chemists were not prepared for the problems raised by the appearance of the *Méthode* in 1787. Unlike the Germans, who decided with little hesitation to adopt the vernacular when translating the new nomenclature, the Scandinavian chemists were unable to agree whether to use Swedish or to follow Bergman and continue with Latin. The Finnish chemist Johan Gadolin (see Chapter Four), who was one of Bergman's

---

[213]Sven Rinman, *Bergwerks Lexicon*, 2 vols. (Stockholm, 1788–1789).

pupils, opted for Latin and applying the principles of his master produced a nomenclature of salts.[214] However, Gadolin felt highly ambivalent about the new theory and never decided whether to accept or to oppose it. The Danish chemist N. Tychsen, who adopted Lavoisier's theory in the early 1790s, translated the French nomenclature into Danish into 1794, using Girtanner's German version as his model.[215] For the most part, Tychsen used Danish versions of the French terms, although in a few cases, he employed terms with Danish roots. For example, he named the first five simple substances:

TYCHSEN (1794)	*MÉTHODE*
Lysprincipe	Light
Warmeprincipe	Caloric
Syreprincipe	Oxygen
Brandprincipe	Hydrogen
Azotisk	Nitrogen (azote)

Tychsen retained the French *azote* to denote nitrogen and translated the rest of the names into Danish. As the general name of acid, he used the Danish *syre*, and following the example of Hermbstädt, he named oxide *oxideret* rather than coin a term *halvsyre* by analogy with Girtanner's *Halbsäure*.[216] The French nomenclature lost much of its flexibility in Danish, just as it had in German. The French *sulfite de soude* was transformed by Tychsen into the unwieldy *Ufuldkommen svovlsyret Mineralludsalt*.[217] The lack of a practical system of Danish suffixes forced Tychsen to use the adjective *ufuldkommen* to denote the French suffix *-ite*!

In 1794 Tychsen published a chemical textbook in three volumes, in which he used both the Danish and the Latin nomenclature.[218] The decision to use a Latin nomenclature may have been inspired by the difficulty of teaching the Danish one.

In 1795 two other attempts were made to translate the French nomenclature into Swedish. The first was by a well-known pupil of Linnaeus, the naturalist Anders Sparrman, who included a chemical lexicon at the end of his translation of Fourcroy's *Philosophie chimique*.[219]

---

[214]Gadolin, "De nomenclatura salium" (1807), in *Johan Gadolin—Wissenschaftliche Abhandlungen* (Helsinki, 1910), 179–191.

[215]N. Tychsen, *Fransk Chemisk Nomenklatur* (Copenhagen, 1794).

[216]*Ibid.*, 17.

[217]*Ibid.*, 25.

[218]N. Tychsen, *Chemisk Haandbog* (Copenhagen, 1794), vol. 3, 511–572.

[219]Fourcroy, *Philosophia chemica* (Stockholm, 1795), 136–169.

Influenced by his mentor, and not particularly well versed in chemistry, Sparrman favored a Latin nomenclature and used most of Bergman's lexicon with some adaptations to take account of the pneumatic discoveries. The result was a mixture of Bergman's nomenclature of salts and Lavoisier's denominations of gases. The names of the principal gases, for instance, were *gaz azoticum, hydrogenium, nitrosum, oxigenium,* etc.[220] To facilitate matters for Swedish readers, Sparrman gave the French, Swedish, Danish, German, and sometimes Italian equivalents of his terms. In the many cases where he was unable to find an appropriate Swedish equivalent for a new Latin term, Sparrman preferred to use a phlogistic term rather than invent a new one. This attitude probably reflects the feeling that the real language of science was Latin and that the use of the vernacular carried less authority.

Anders Gustaf Ekeberg, a pupil of Bergman, opted for a different solution when he translated the French nomenclature into Swedish in 1794.[221] Ekeberg's decision was motivated by the patriotic wish to give chemistry a Swedish voice, and he explained in his introduction:

> But is it necessary to give the Language of the New Chemistry a Swedish character, could one not be satisfied with the words that the French and the Germans have used? The Authors would have expected this question if it had not already been put to them. They consider it sufficiently answered with the following observations: that the translation of a nomenclature [*Konstord*] is unnecessary until it has won recognition or its use is restricted to one Science; that fortunately the time is passed when one was as careless as ignorant of the resources of language and, either forced by a need or led by a false understanding of learning, distorted the Mother Tongue with Latin or French; and that the application of scientific discoveries in daily life, which is the aim of the investigations of Men of Learning and determines the true value of their labours, is made considerably easier and is hastened by a scientific language that is understandable to the public.[222]

This is one of the few passages in which the criteria used in translating the *Méthode* into a foreign language are clearly expressed. Ekeberg justified his choice of a purely Swedish translation on two grounds: 1. Swedish had the capacity to provide an adequate scientific language—Ekeberg, who was well aware of progress in mining research and manufactures, thought it was time to make use of the rich mineralogical

---

[220] *Ibid.*, 146–147.

[221] [Ekeberg], *Försök till Svensk Nomenklatur för Chemien, lämpad efter de sednaste uptäckterne* (Stockholm and Uppsala, 1794).

[222] *Ibid.*, Introduction.

lexicon codified by Rinman in his dictionary; 2. translating the chemical language into Swedish would finally make the new theory "understandable to the public"[223] and further the primary aim of scientific investigation, which was to lead to improvement of the conditions of everyday life. This utilitarianism was a typical trait of eighteenth-century Swedish science.

The specific linguistic solutions adopted by Ekeberg in his translation were modelled on the German translation of the *Méthode*. However, the peculiar structure of Swedish facilitated the rendering of the French suffixes. The degree of saturation of the acids derived from sulphur was indicated in Swedish as follows:

EKEBERG (1794)	*MÉTHODE*
Svafvel syra	acide sulfurique
Svafvel syrlighet	acide sulfureux

Oxide was translated with the term *syrsatt*, which essentially reproduced the French etymology. Ekeberg's ability to retain the effect of the suffixes was probably the main reason for the success of his nomenclature by comparison with those of Sparrman and Gadolin. Despite its general popularity among Swedish chemists, the use of the vernacular was later opposed by Berzelius who, like Bergman, favored a return to Latin because of its universal use in the world of learning.[224]

In 1814 the Danish physicist Johan Christian Oersted published a booklet in which he presented a synthesis of the German, Swedish, Dutch, and Danish chemical nomenclature.[225] Oersted remarked that as all these languages had a common source it should be possible to devise a common nomenclature that reflected this unity; such a project would be valuable in view of the deficiencies and difficulties of the translations, i.e., the German ones, of the *Méthode*. To create a synthetic nomenclature, Oersted investigated the philological roots of the "linguae Scandinavico-Germanicae" and named the chemical substances with the original words common to all the Nordic languages (see Appendix). The results of Oersted's attempt were not totally unreasonable but the advantages of the French nomenclature were lost. Hydrogen, for instance, was translated with the term *brint*, which evokes the

[223]*Ibid.*, 17.

[224]Berzelius, *Traité de chimie*, vol. 1 (Paris, 1829), 11–32.

[225]Oersted, *Tentamen nomenclaturae chamicae omnibus linguis Scandinavico-Germanicis communis* (Copenhagen, 1814).

property of inflammability rather than its part in the decomposition of water. Oersted's attempt came too late to have any serious impact on the development of Scandinavian chemical nomenclature, but it well expresses the difficulties and questions involved in the translation of the *Méthode.*

The story of the reception and translation of the *Méthode* shows not only the attitudes of individuals to the new chemical language but also the differences in how it was perceived from the perspective of different linguistic cultural and scientific traditions.

Epilogue

Lavoisier's wish to find expressive words to represent the essential nature of chemical substances had little impact on the development of nineteenth-century chemistry. As shown in Chapter Five, his closest collaborators either abandoned or openly criticized Lavoisier's linguistic realism and opted for a conventional and pragmatic use of technical terminology. Cuvier almost derided the ingenuity of Lavoisier's beliefs and, like many historians of today, treated them as persuasive rhetoric rather than as serious scientific proposals. Indeed, the philosophical debate on the nature of scientific language that took place in the second half of the eighteenth century was almost totally forgotten by the beginning of the nineteenth. For Dalton, Berzelius, Gay-Lussac, Thenard, and Davy, the question of chemical nomenclature was no longer an issue: they all agreed that the determination of a new name was a matter of convenience and economy rather than a question of realism. Even for Michael Faraday, who had enlisted the help of the philosopher William Whewell in creating new chemical names,[1] the liaison between philosophy and science had lost the intimate and organic character that it had had during the previous century. Chemistry was becoming a specialized science, and philosophy seemed to have little to say concerning its development. Furthermore,

[1]In 1831 Faraday began to receive the assistance of Whewell in devising a new nomenclature for electro-chemistry. The results, however, were confined to the solution of specific technical problems. See *The Selected Correspondence of Michael Faraday* (Cambridge, 1971), vol. 1, 190–191, 264–272, 284, 299–302; vol. 2, 593–595.

particularly in France, any connection between philosophical doctrines and scientific endeavors was regarded with suspicion by the Napoleonic authorities. In this hostile atmosphere, the first victim of the reorganization of the French Institut National in 1803 was the class of "sciences morales et politiques," in which all the pupils of Condillac, the "*idéologues*," were keeping alive the spirit of the French Revolution. The *idéologues*, in fact, still insisted on the necessity of basing scientific endeavor on a philosophical foundation capable of putting the inductive sciences, particularly medicine, to the service of social and political reform. Napoleon fully realized the danger entailed in active collaboration between scientists and philosophers, and he efficiently dismantled the political and philosophical engagement of French science. The fundamental change that took place in French science during the period 1793–1824, which has been illuminated by Nicole and Jean Dhombres,[2] dramatically affected the ideological and political commitment of scientists. It is no coincidence that Lavoisier's linguistic realism, his debt to Condillac's analytical method, and his encyclopedic approach to science, were soon neutralized and forgotten by his former colleagues. It is significant, by contrast, that the Condillacian sources of Lavoisier's chemical system were systematically studied by philosophers during the nineteenth and early twentieth centuries and that his nomenclature was still regarded as one of the most significant manifestations of the philosophy of science. The *idéologues* were the first to assert the philosophical meaning of Lavoisier's nomenclature in reply to those scientists who began to believe that the progress of science was totally independent of philosophy. In 1796 Antoine Destutt de Tracy asked them "to open the works of the illustrious and unhappy Lavoisier, and to see if each page of the writings of this celebrated man, forever deserving our regrets, does not testify that it was the analysis of ideas which led him and his colleagues to make so much progress in the analysis of bodies."[3] In 1800 Joseph Marie Degérando

[2]Nicole and Jean Dhombres, *Naissance d'un nouveau pouvoir: science et savants en France 1793–1824* (Paris, 1989). See also Ludmilla Jordanova, "The Authoritarian Response," in Peter Hulme and Ludmilla Jordanova (eds.), *The Enlightenment and Its Shadows* (London and New York, 1990), 200–216.

[3]A. Destutt de Tracy, "Mémoire sur la faculté de penser," in *Mémoires de l'Institut National des Sciences et Arts; Sciences Morales et Politiques*, 1 (1796), 287; English translation by Albury (1986), 211. Albury convincingly underlined the political aim of this kind of assessment; in this connection, see also Acton (1959) and Robin E. Rider, "Measure of Ideas, Rule of Language: Mathematics and Language in the 18th Century," in Frängsmyr, *et al.* (1990), 113–140.

used almost the same argument,[4] and even though he was not totally persuaded of the logical validity of Condillac's linguistic realism, he claimed that Lavoisier's nomenclature was "the most perfect of which the history of science has ever furnished an example."[5] In 1815, in his history of French literature, Marie-Joseph Chénier cited Lavoisier as a model, both for his grammar and for his thinking:

> Such was the progress of Lavoisier, when he applied, as he himself said, the method of Condillac to chemistry. In renewing the nomenclature, he renewed the science.[6]

Similarly Dominique-Joseph Garat declared five years later that Lavoisier had "accomplished in chemistry a revolution so striking, so fecund in truths and happy prospects, by bringing to that science his genius, his fortune, and the principles and the language of Condillac."[7]

Lavoisier's nomenclature continued to exert a great influence throughout the nineteenth century. Comte was certainly inspired by the results achieved by the French chemist when he wrote:

> There is, in the system of the positive method, a most important, although hitherto very little appreciated, part, which chemistry was [. . .] particularly destined to raise to the highest degree of perfection. This is not the theory of classifications, which is rather poorly understood by chemists, but the general art of rational nomenclature, which is entirely independent, and of which chemistry, by the very nature of its object, must present more perfect examples than any other basic science.[8]

Like Lavoisier, Comte was convinced of the predictive power of appropriate terms and of the scientific value of systematic nomenclature in chemistry:

> The systematic nomenclature of the various compound substances

---

[4]"Ainsi, les chimistes ont une analyse mécanique, qui est comme la représentation de l'analyse qu'ils avoient d'abords fait subir à leurs idées," J.M. Degérando, *Des signes et l'art de penser considéres dans leurs rapports mutuels*, vol. 4 (Paris, 1800), 102.

[5]*Ibid.*, 177.

[6]M.J. Chénier, *Tableau historique de l'état et des progrés de la littérature française depuis 1789* (1815), new edition (Paris, 1989), 58.

[7]Dominique-Joseph Garat, *Mémoires historique sur la vie de M. Suard, sur ses écrits, et sur le XVIIIe siècle* (Paris, 1820), vol. 2, 326–327; English translation by Albury (1986), 211.

[8]Auguste Comte, *Cours de philosophie positive. La philosophie chimique* (1838), in *Oeuvres d'Auguste Comte*, vol. 3 (Paris, 1893), 43.

> will be almost enough to give, as it were spontaneously, a true initial indication of the general outcome of each chemical event.[9]

Comte went on to complain that nineteenth-century chemists were not interested in a methodical nomenclature or more generally in a philosophical systematization of their ideas.

The success of Lavoisier's nomenclature among philosophers was demonstrated by the recognition it received in Britain. British philosophers in fact unlike chemists regarded Lavoisier's nomenclature as a positive model of reasoning. The Scottish philosopher Dugald Stewart was critical of the introduction of a mathematical nomenclature in inductive sciences. However, after Lavoisier he acknowledged the part to be played by methodical languages in the progress of science:

> The technical terms, in the different sciences, render the appropriate language of philosophy a still more convenient INSTRUMENT OF THOUGHT, than those languages which have originated from popular use: and in proportion as these technical terms improve in point of precision and of comprehensiveness, they will contribute to render our intellectual progress more certain and more rapid [. . .]. The influence which these very enlightened and philosophical views [i.e., Lavoisier's nomenclature] have already had on the doctrines of chemistry cannot fail to be known to most of my readers.[10]

Even if his philosophical position was radically different from those of the French thinkers, Jeremy Bentham paid his tribute to Lavoisier as follows:

> Among the logicians, an instrument of universal empire in the regions of intelligence was supposed to have been discovered by the invention of the syllogism. Yet, in truth, what is the exploit achieved by it? The dividing an argument into three parts or members, distinguished from each other by so many names, —names, in the invention of which [. . .] not quite so much felicity has been displayed, as in those for which we are indebted to the genius of Lavoisier and Linnaeus.
>
> Characteristic names are names for the species, and for ever.[11]

William Whewell, too, compared the contribution made by Lavoisier to the language of chemistry to that of Linnaeus in botany and conceded that:

---

[9] *Ibid.*, 72.
[10] D. Stewart, *Elements of the Philosophy of the Human Mind*, 2nd ed., vol. 2 (Edinburgh, 1816), 141.
[11] *The Works of Jeremy Bentham*, vol. 6 (New York, 1962), 442.

> Though the progress of discovery, and the consequent changes of theoretical opinions, which have since gone on, appear now to require a further change of nomenclature, it is no small evidence of the skill with which this scheme was arranged, that for half a century it was universally used, and felt to be far more useful and effective than any nomenclature in any science had ever been before.[12]

According to John Stuart Mill, Lavoisier's nomenclature was an "ideally perfect"[13] example of scientific language that satisfied all requisites of the logic of inductive science, and he believed the lack of "a scientifically constructed nomenclature [was] the principal cause which [retarded] the progress of the science."[14]

Lavoisier's nomenclature did not merely inspire positivistic and empirical philosophers. The idealistic French philosopher Émile Meyerson used this historical example on several occasions to show the intrinsic ontological assumptions of any scientific endeavor.[15] Meyerson believed that science proceeded by assuming *a priori* the identity of a natural phenomenon with its interpretation, and Lavoisier's theory naturally fitted in with this conception. When asserting the formative role of language in science, the neo-Kantian philosopher Ernst Cassirer held up Lavoisier's theory of nomenclature as one of the most significant examples.[16]

Even if the evaluations of the *Méthode* by these philosophers were often inspired by motives more complex than a simple wish to provide a historical understanding of it, they constituted a challenge to the indifference shown by many historians of chemistry to the relationship between scientific changes and philosophical ideas and concepts and argued forcefully that the revolution in eighteenth-century chemistry was intimately interwoven with the change in its language.

---

[12]W. Whewell, *History of the Inductive Sciences*, vol. 3 (London, 1857), 122.

[13]"The possibility [. . .] of an ideally perfect Nomenclature, is probably confined to the one case in which we are happily in possession of something approaching to it; the Nomenclature of elementary chemistry [. . .]. The only thing left unexpressed by them was the exact proportion in which the elements were combined; and even this, since the establishment of the atomic theory, it has been found possible to express by a simple adaptation of their phraseology," J.S. Mill, *A System of Logic*, 9th ed. (London, 1875), vol. 2, 284–285.

[14]*Ibid.*, 261.

[15]See, for instance, Meyerson, *Explanation in the Sciences* (1921); English Translation (Dordrecht-Boston-London, 1991), 546–563; Lavoisier's theory is often taken as an historical example in other works by Meyerson.

[16]Cassirer correctly defined Lavoisier's nomenclature as a "chemische Begriffs-sprache." Cassirer, *Das Erkenntnissproblem in der Philosophie und Wissenschaft der neuerer Zeit*, 2nd ed., vol. 2 (Berlin, 1911), 435–441.

APPENDIX

The Role of Symbolism from Alchemy to Chemistry

"Un autre obstacle à l'avancement des sciences naquit de leur hérédité, de leur concentration dans un petit nombre de familles, que l'on peut considérer comme l'origine des castes. Toutes les vérités n'étaient pas à la portée vulgaire [. . .]. De là, les allégories, les hiéroglyphes, les langues sacrées, les emblêmes, fondement et origine de la mythologie qui, après avoir trompé et asservi les hommes, amuse maintenant l'humanité oublieuse."

Georges Cuvier, *Histoire des sciences naturelles depuis leur origine jusqu'à nos jours*, vol. 1 (Paris, 1841), 7.

"Whoever engages in chemical pursuits, and in the study of chemical authors, cannot but remark with some degree of curiosity, how extensively the use of symbols has prevailed in this science."

Martin Wall, *Dissertation on Select Subjects in Chemistry and Medicine* (Oxford, 1783), 90.

The Myth of Continuity

"Alchemy was never at any time anything other than chemistry [. . .]. Alchemy was science, it included all the industries in which chemistry was technically applied."[1] The deliverer of these two judgments, the German chemist Justus von Liebig, still clearly displayed the positivistic inclination to reduce all bodies of knowledge of the past to their *positive* and experimental principles, regardless of the historical and cultural context concerned. The established fact that the alchemists of the sixteenth and seventeenth centuries undertook their investigations in chemical laboratories and that they shared most of their mineralogical knowledge with the metallurgists and naturalists of the Renaissance seemed to provide sufficient grounds for Liebig's belief in the historical and experimental continuity between alchemy and early chemistry.

This approach to the history of early alchemy and chemistry, which dominated the history of science throughout the nineteenth and the first half of the twentieth centuries,[2] has remained standard even in more recent times. It is interesting to note that although the ideological commitment behind the reduction of early chemistry to

---

[1]"Die Alchemie ist niemals etwas anderes als die Chemie gewesen; ihre beständige Verwechslung mit der Goldmacherei des 16. und 17. Jahrhunderts ist die grösste Ungerechtigkeit. Unter den Alchemisten befand sich stets ein Kern echter Naturforscher, die sich in ihren teoretischen Ansichten häufig selbst täuschten, während die fahrenden Goldköche sich und Andere betrogen. Die Alchemie war die Wissenschaft, sie schloss alle technisch-chemischen Gewerbzweige in sich ein. Was Glauber, Böttger, Kunkel in dieser Richtung leisteten, kann kühn den grössten Entdeckungen unseres Jahrhunderts an die Seite gestellt werden," from Justus von Liebig, *Chemische Briefe* (1844), 6th edition (Leipzig and Heidelberg, 1878), 34.

[2]See, for instance, the work by Marcelin Berthelot, *Les origines de l'alchimie* (Paris, 1885), 273, in which he declared that: "L'alchimie était une philosophie, c'est-à-dire une explication rationaliste des métamorphoses de la matière." In 1922 H. Stanley Redgrove wrote, "the time is gone when it was regarded as perfectly legitimate to point to alchemy as an instance of the aberrations of the human mind [. . .]. The spontaneous change of one 'element' into another has been witnessed, and the recent work of Sir William Ramsay suggests the possibility of realising the old alchemist dream," Redgrove, *Alchemy: Ancient and Modern* (London, 1922), xi. In 1939 John Read, a historian of alchemy, still displayed the same positivistic approach when he wrote that "alchemy was the chemistry of the Middle Age," in *idem*, *Prelude to Chemistry* (1939), 2nd edition (London, 1961), 1.

alchemical thought has radically changed, the conclusions of modern historians are still similar to Liebig's. By emphasizing the fact that Renaissance alchemists adopted a wide range of *scientific* notions and *technical* implements in performing their investigations, recent historians have tried to show the decisive role that irrational, mystical, and magic beliefs played in the emergence of chemical science.[3]

This interpretative trend whose aim in principle was incompatible with positivism[4] paradoxically led to the same conclusions: alchemy was somehow related to modern science, and chemistry had its origins in alchemical thought.

Undoubtedly, both of these interpretations have improved our understanding of alchemy by promoting a rich and varied examination of its sources, but they have failed to provide a correct historical picture of the conceptual structure of alchemy as a whole. Too anxious to see their hypotheses confirmed, both the positivistic and the modern historians of alchemy have often neglected the central sphere of alchemical thought. The fundamental goal of alchemy was the discovery of the philosopher's stone, the quintessence, the potable gold, etc. through the purification and the transmutation of metals. Discrepancies between theory and practice were neglected in favor of the speculative part of the "Great Work." If we want to avoid regarding this obsessive insistence on the transmutation of metals into gold as a weakness, or worse, as a pathological state of mind, it would be better to stop regarding alchemy as a scientific or experimental endeavor, and

---

[3]In this connection, see the following works: R.J. Forbes, "Modern Chemistry and Alchemy," in *Ciba Review*, 5 (1961), 2–16; H.M.E. De Jong, *Michael Maier's Atalanta Fugiens—Sources of an Alchemical Book of Emblems* (Leiden, 1969), in which (p. 21) it is stated that "just as chemistry, alchemy is a science, in as far as their common starting-point is obtaining data from Nature, arranging them, drawing conclusions therefrom and building theories on them. *In contrast to magic practices, which tried to break through the laws of Nature by means of magic formulae, the alchemists always aimed at discovering, theoretically and practically, those laws of Nature*" (my italics). Though more prudent and refraining from such extreme generalization, other relevant examples of this interpretative trend are to be found in Allen G. Debus, *The Chemical Philosophy*, 2 vols. (New York, 1977); and the specific study by Betty J.T. Dobbs, *The Foundation of Newton's Alchemy* (Cambridge, 1975).

[4]By asserting that the only legitimate form of scientific activity was systematic experimentation and empirical induction, the positivists tried to strip alchemy of all that seemed extraneous to the scientific method and to a rational view of nature. Conversely by emphasizing the historical importance of mystical and magic factors in the laboratory experiments performed by the alchemists, more recent historians aim to underline the important role played by irrational factors in modern scientific enterprise. In both cases the supposed continuity between alchemy and chemistry offered a formidable historical example.

to trace its intellectual origins in the philosophy with which it was associated.

As Hélène Metzger, Carl G. Jung, Alexandre Koyré,[5] and more recently Robert Halleux[6] have pointed out, a historical reconstruction of alchemical thought has primarily to take into account the metaphysical background of the "Great Work" rather than its experimental results. As early as 1931, one of the most authoritative historians of magic, Grillot de Givry, pointed out that

> if the exceedingly numerous works left by the alchemists are attentively studied the reader very quickly sees that the operation of the Philosopher's Stone does not belong to the realm of pure chemistry. The method described with so remarkable a unity of doctrine excludes any idea of research or tentative procedure, and is incompatible with the abundant experimentation involved in modern chemistry.[7]

These episodic statements on the fundamental difference between alchemy and chemistry are confirmed by an analysis of the changing role of symbols in these two cultural traditions.

Symbolism in Alchemy

The origins of alchemical symbolism are as ancient as they are obscure.[8] However, it is generally believed that the first to use symbols to abbreviate the description of medical remedies were the Greeks.

---

[5]H. Metzger (1938); C.G. Jung (1944) and (1967); Alexandre Koyré (1971).

[6]Robert Halleux, "Pratique industrielle et chimie philosophique de l'Antiquité au XVIIe siècle," in *L'actualité chimique* (1987), 16–20.

[7]E.A. Grillot de Givry, *Witchcraft, Magic and Alchemy* (1931), reprinted (New York, 1971), 374. Interestingly similar was the conclusion of the German philosopher Ernst Cassirer, who in 1923 wrote: "allen alchimistischen Operationen, wie immer sie im einzelnen geartet sein mögen, liegt der Urgedanke der übertragbarkeit und der dinglichen Ablösbarkeit von Eigenschaften und Zuständen zugrunde—der gleiche Gedanke also, der sich in einem naiveren und primitiveren Stadium [. . .].

Jede besondere Beschaffenheit, die die Materie besitzt, jede Form, die sie annehmen kann, jede Wirksamkeit, die sie auszuüben vermag, wird hier zu einer besonderen Substanz, zu einem Wesen für sich hypostasiert," in Cassirer, *Philosophie der symbolischen Formen* (1923), 3rd edition (Oxford, 1958), vol. 2, 86.

[8]On alchemical symbolism, see H.C. Bolton, "History of Chemical Notations," in *Transactions of the New York Academy of Sciences*, 2 (1882–1883), 102–106; Marcelin Berthelot, *Introduction à l'étude de la chimie des anciens et du Moyen Age* (Paris, 1889), 93–126; Paul Walden, "Zur Entwicklungsgeschichte der chemischen Zeichnen," in J. Ruska (ed.), *Studien zur Geschichte der Chemie* (Berlin, 1927), 80–105; R.M. Caven and J.A. Cranston, *Symbols and Formulae in Chemistry* (London and Glasgow, 1928); Fritz Lüdy-Tenger,

Between 1887 and 1889, the French chemist Marcelin Berthelot translated and published a comprehensive collection of Greek alchemical manuscripts. Among them he included one (compiled in the eleventh century A.D.) that was richly illustrated with symbols (see Fig. 17). Although this manuscript is not particularly original from a technical point of view, it is important historical testimony to the origin of the meaning and function of alchemical symbols.

The manuscript shows the different symbols that were used to depict gold and its various compounds, and also silver, copper, iron, lead, sulphur, and their respective compounds.

At the head of the second column, we can see the first symbolic association of the seven planets with their corresponding metals:

Sun	—Gold
Moon	—Silver
Saturn	—Lead
Jupiter	—Electrum
Mars	—Iron
Venus	—Copper
Mercury	—Tin

This association had its origin in the religious meaning attached to the number seven. The colors, the metals, the planets, the notes of the musical scale, the vowels of the Greek alphabet, the sages of Greece, etc. were all seven in number. Furthermore, the association between planets and metals was strengthened by the natural resemblance in color between the two most important metals (gold and silver) and the sun and the moon, both important and ancient religious symbols.

According to the author of the manuscript reprinted by Berthelot the symbols represented were the signs of the science that were contained in the technical writings of the philosophers: above all, they

---

Alchemistische und Chemische Zeichnen (1928), reprinted (Vaduz, 1981); C.O. Zuretti, *Alchemistica Signa*, in *Catalogue des Manuscrits Alchimiques Grecs*, vol. 8 (Brussels, 1932); J.R. Partington, F. Sherwood Taylor, A.F. Titley, Gerard Heym, and D. McKie, "Report of a Discussion upon Chemical and Alchemical Symbolism," in *Ambix*, 1 (1937), 61–77; C.G. Jung, *Psychologie und Alchemie* (1944), translated into English with the title *Psychology and Alchemy* (New York, 1953); E.J. Holmyard, *Alchemy* (1957), reprinted (New York, 1990); M.P. Crosland, *Historical Studies in the Language of Chemistry* (1962), 2nd edition (New York, 1978), 227–244.

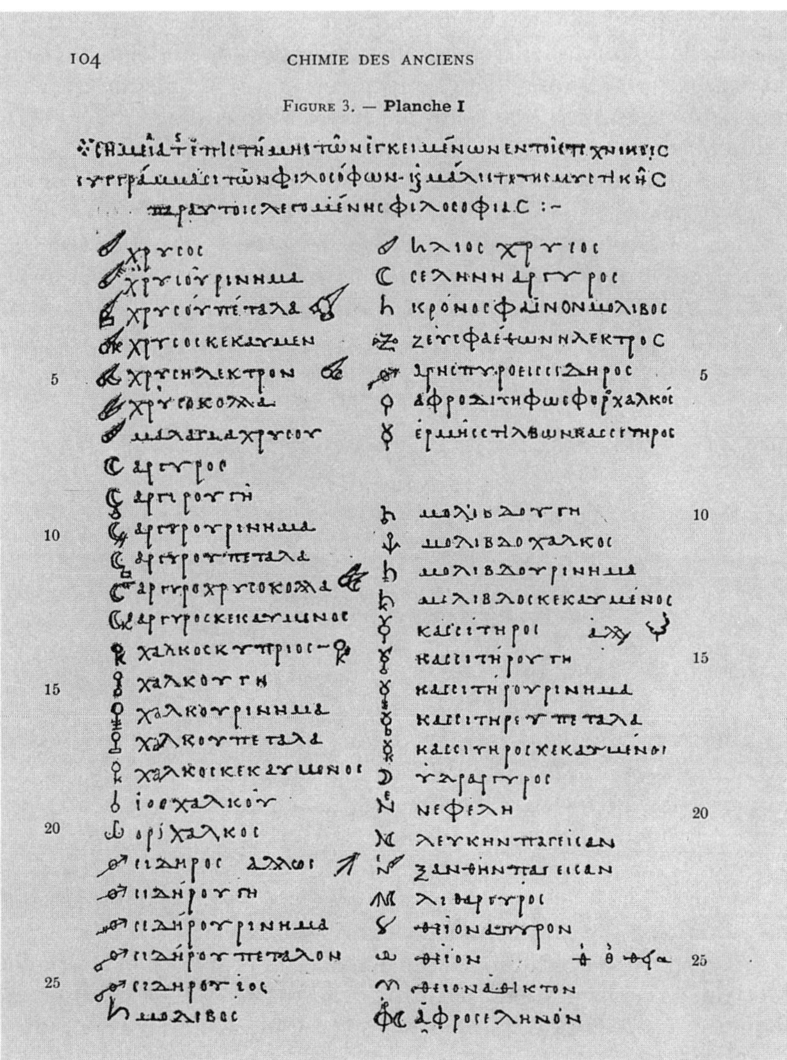

104 CHIMIE DES ANCIENS

FIGURE 3. — **Planche I**

FIGURE 17 Symbols for metals and their combinations. From Marcelin Berthelot, *Introduction à l'étude de la chimie des anciens et du Moyen Age* (Paris, 1889).

were the signs of what was called the mystical philosophy.[9] Indeed, it is no coincidence that symbols were introduced into alchemy just when alchemy was becoming far more than a mere manipulation of chemical substances.

At the beginning of the fourth century, Greek alchemists, such as Zosimos and Stephanus of Alexandria, began to employ an enigmatic nomenclature and to mix their practical knowledge with speculative notions derived from Neo-Platonism, astrology, gnosticism, and magic.[10] This mixture of different doctrines brought an entirely new philosophical perspective to the manipulation of metals and the question of the ultimate composition of matter. Most treatises on the theory and practice of early alchemy shared, though to different degrees, the belief in hylozoism, in the influence of planets and stars on the transmutation and growth of metals and in the Platonic doctrine contained in the formula "the one is the all and the all is the one" that eventually led to the conception of nature in terms of Macrocosmos and Microcosmos.

These being the main theoretical assumptions of alchemy, it is obvious that their relationship to the symbolical representation of its doctrines became extremely important.

However, it was not until the sixteenth and seventeenth centuries that alchemical symbolism flourished in the variety of forms with which we are acquainted. The rediscovery of Hermeticism and Platonism and the intellectual battle against Aristotle are among the main reasons for the philosophical and iconographical success enjoyed by alchemy during the Renaissance.[11] Being a movement of cultural opposition, alchemy had to use a language completely different from that of the Aristotelian tradition; accordingly, alchemists used pictograms, symbols, emblems, etc. instead of arguing about the etymological origins of words and their precise definition. When they were forced to use words, their meanings were not usually apparent without a knowledge of complicated unwritten rules for deciphering them. As C.G. Jung aptly puts it, "the symbolical images belong to the very essence of the alchemist's mentality."[12] Alchemical treatises shunned any direct or descriptive communication of empirical data on the composition of matter. On the contrary, by always using the same experi-

---

[9]Berthelot (1889), 105. See also *idem, Collection des anciens alchimistes Grecs*, vol. 1 (Paris, 1887), 73–85 and 92–126.

[10]Holmyard (1990), 27–28.

[11]Studying the particular case of sixteenth-century Germany, Koyré (1971) has pointed out the connection between alchemy, on one side, and mysticism and religious reformation, on the other.

[12]Jung (1953), vii.

FIGURE 18 An allegory of the alchemical "Great Work." From G.B. Nazari, *Della Trasmutatione metallica, sogni 3* (Brescia, 1599).

mental procedures, they provided a highly symbolical representation of the "Great Work." A few examples from the alchemical literature will better clarify the main features of alchemical iconography.

In 1599 the Italian alchemist Giovanni Battista Nazari published a treatise on the transmutation of metals in which several operations were described allegorically.[13] Inspired by the novel *Hypnerotomachia Poliphilo* (Venice, 1499) by Francesco Colonna, Nazari conceived his work in the form of a dream, in which pictorial allegories played a central role. The work was richly illustrated, and many of the woodcuts were evocative of Colonna's novel. In one dream Nazari encountered a *Dracone pessimo & feroce* (Fig. 18), symbolizing the complexity and difficulty of the "Great Work." By the action of fire, and the consequent purification of gold, silver, and sophic mercury, the body, soul, and spirit of the dragon would have been reunited in one whole, and the "Great Work" accomplished.[14] Nazari was content to let his allegorical language speak for itself and avoided any clear and unequivocal interpretation. As it dealt with dreams, his treatise provided the reader with a nearly infinite series of interpretative keys.

In another work published at the end of the sixteenth century, the Paracelsian physician Leonhart Thurneisser[15] associated the preparation of alchemical remedies with the reading of horoscopes. In several cases the recipes were written in exotic languages, such as Syriac, Ethiopian, Arabic, etc., and in a few cases, the code to the language was invented by Thurneisser himself (Fig. 19). Thurneisser adorned his book with several illustrations related to the "Great Work" that eloquently show its metaphysical intention.

During the seventeenth century, alchemical symbolism was at its most allegorical. In 1657 the French alchemist Annibal Barlet published an alchemical treatise[16] in which the reference to a metaphysical and cosmological conception of matter was explicit. Barlet's works contain several illustrations depicting a laboratory. Taking a positivistic viewpoint, Denis Duveen declared that these woodcuts illustrated "the diverse operations of Chemistry with a great wealth of detail, showing the various apparatus in use at the time and the general interior arrange-

---

[13]G.B. Nazari, *Della trasmutatione metallica, sogni tre* (Brescia, 1599).

[14]"Se in foco mi nutrirai, per fina che sian prima i membri miei in altra forma mutati, & poi il corpo mio purificato dal mortale veneno; et poi quando il corpo, l'anima, & lo spirito insieme vedrai congiunti: allhora sarai maggior del mondo. Chi mi ode, & non intende, consuma il viaggio, & spende il tempo senza altro fine," *ibid.*, 147–148.

[15]L. Thurneisser, *ΜΕΓΑΛΗ ΧΥΜΙΑ, vel Magna Alchymia* (Berlin, 1583). On Thurneisser, see Partington, *A History of Chemistry*, vol. 2, 152–155.

[16]A. Barlet, *Le vray et Méthodique cours de la Physique résolutive* (Paris, 1657).

Alphabetum Honorij Thebani, quo magitam scribebat, autore Petro Aponio in lib. quarto.

Alphabetum Hichi Francorum natis prisci vath.

Alphabetum secretum Iaimelis Megalopij regis sapientissimi.

Alphabetum Nortmannorum, autore D. Beda Presbytero.

Alphabetum secretum Iohannis Tritenhemij Abbatis Spanheimensis.

Alphabetum Vtopiense, teste doctiss. viro Thoma Mauro D. Erasmi Rot: pcipuo amico Anglo.

Alphabetum Caroli Magni. Teste Alcunio, Gallo, Philosoph. lib. de vita Regi:

Alphabetum exploratum Pharamundi regis Francorum per Galliam.

Alphabetum facillimum, secundum anonymum, Trittenhemio notante.

Alphabetum cuiusdam summi Alchimistæ, autore Sphanheimense.

Alphabetum Dorari q̃ Franci vsi sunt, Hunibaldo Historiographo autore.

Alphabetum Vastualdi, qui acta Regum, Ducum & principum per annos Septingentos & quinquagintaocto hisce notis primo sermone scripsit autore Hunibaldo.

Alphabetum D. Otfridi Monachi VVissenburgensis. Rabani Fuldensis discipulus.

FIGURE 19 Hermetic and alchemical codes. From Leonhart Thurneisser, *ΜΕΓΑΛΗ ΧΥΜΙΑ, Vel Magna Alchymia* (Berlin, 1583).

ments of the laboratory."[17] In reality the *fourneau cosmique* (Fig. 21) and the *fourneau astrale ou lampadaire* (Fig. 22) were improbable instruments in seventeenth-century chemistry. Furthermore, although his narrative referred to their cosmological potential, Barlet did not supply any real description of them. From even a quick glance at these illustrations, it is clear that the reference to experimental technology is, to say the least, questionable. In other works the metaphysical sources of alchemical symbolism appear even more clearly.

[17]D.I. Duveen, *Bibliotheca Alchemica et Chemica* (London, 1949), 44.

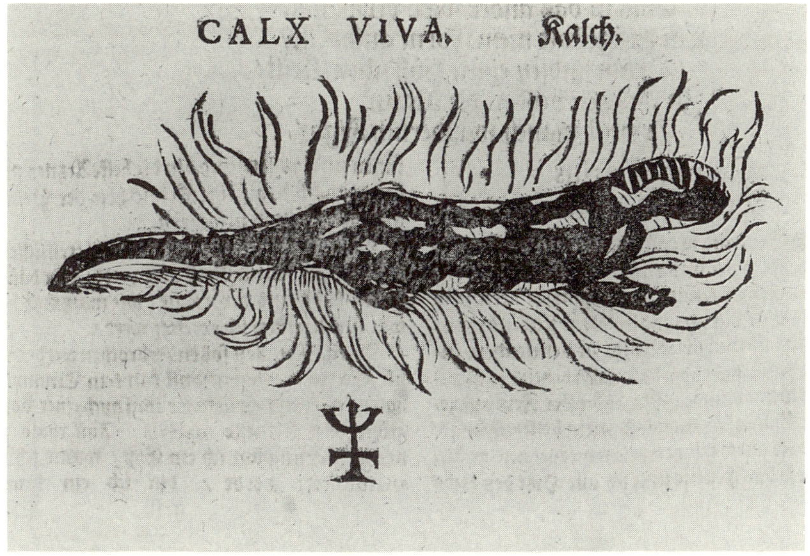

FIGURE 20 Probably because of its colors, the chalk was represented by a salamander. From Leonhart Thurneisser, *ΜΕΓΑΛΗ ΧΥΜΙΑ, Vel Magna Alchymia* (Berlin, 1584).

In 1618 the German physician Michael Maier published his famous *Atalanta Fugiens*, a work on "the new emblems of the secrets of chemistry somewhat adapted for the eye and the mind by means of copperplates and of the mottos, epigrams and annotations accompanying them."[18] Maier collected 50 emblems depicting various stages of the "Great Work," each emblem accompanied by a motto, an epigram, a commentary, and a musical *Fuga.*[19]

The emblem XLV (Fig. 23) depicts the earth at the center of the universe, the sun in orbit around it, and opposite it the moon and a dark shadow. The motto that illustrates the emblem reads: "The sun and its shadow complete the work." The epigram provides us with more details:

---

[18]Michael Maier, *Atalanta Fugiens, hoc est, EMBLEMATA NOVA DE SECRETIS NATURAE CHYMIA, Accomodata partim oculis et intellectui, figuris cupro incisis, adjectisque sententiis, Epigrammatis et notis, partim auribus et recreationi animi plus minus 50 Fugis Musicalibus trium Vocum,* . . . (Oppenheim, 1618).

[19]On Maier and his *Atalanta Fugiens*, see the detailed study by De Jong (1969), from which I have taken the English translation of Maier's text.

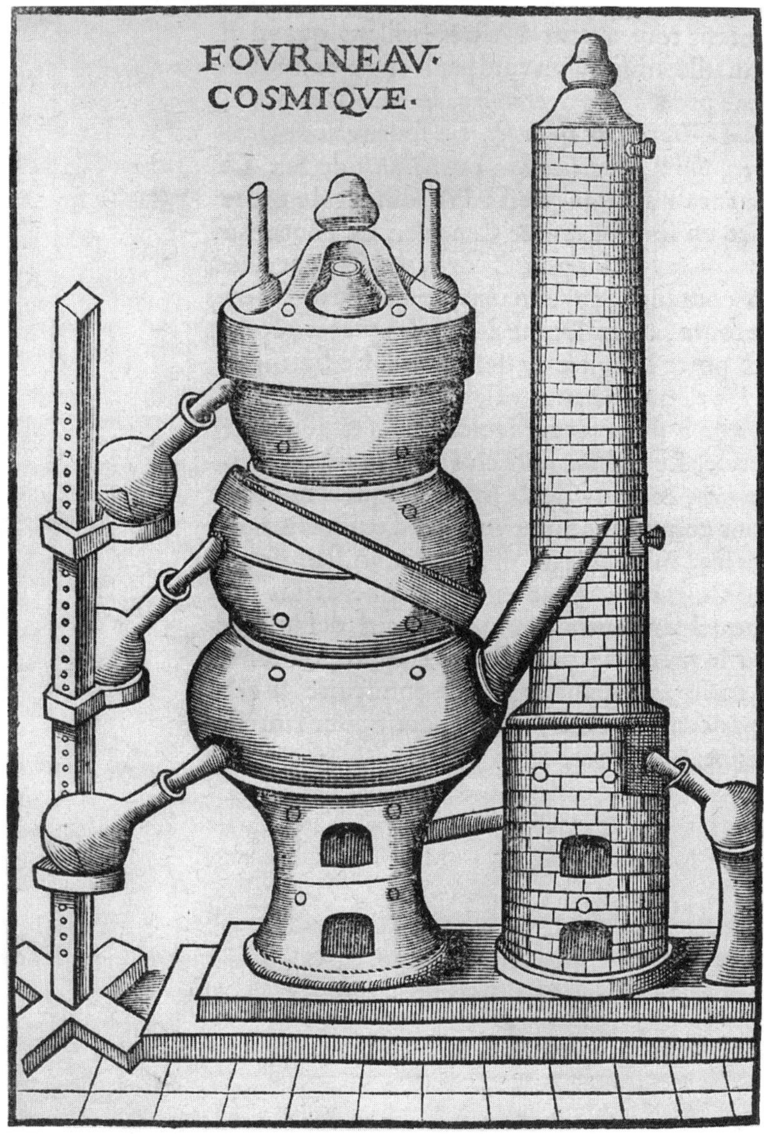

FIGURE 21 Woodcut depicting the "fourneau cosmique" (cosmic furnace). From A. Barlet, *Le vray et Méthodique cours de la Physique résolutive* (Paris, 1657).

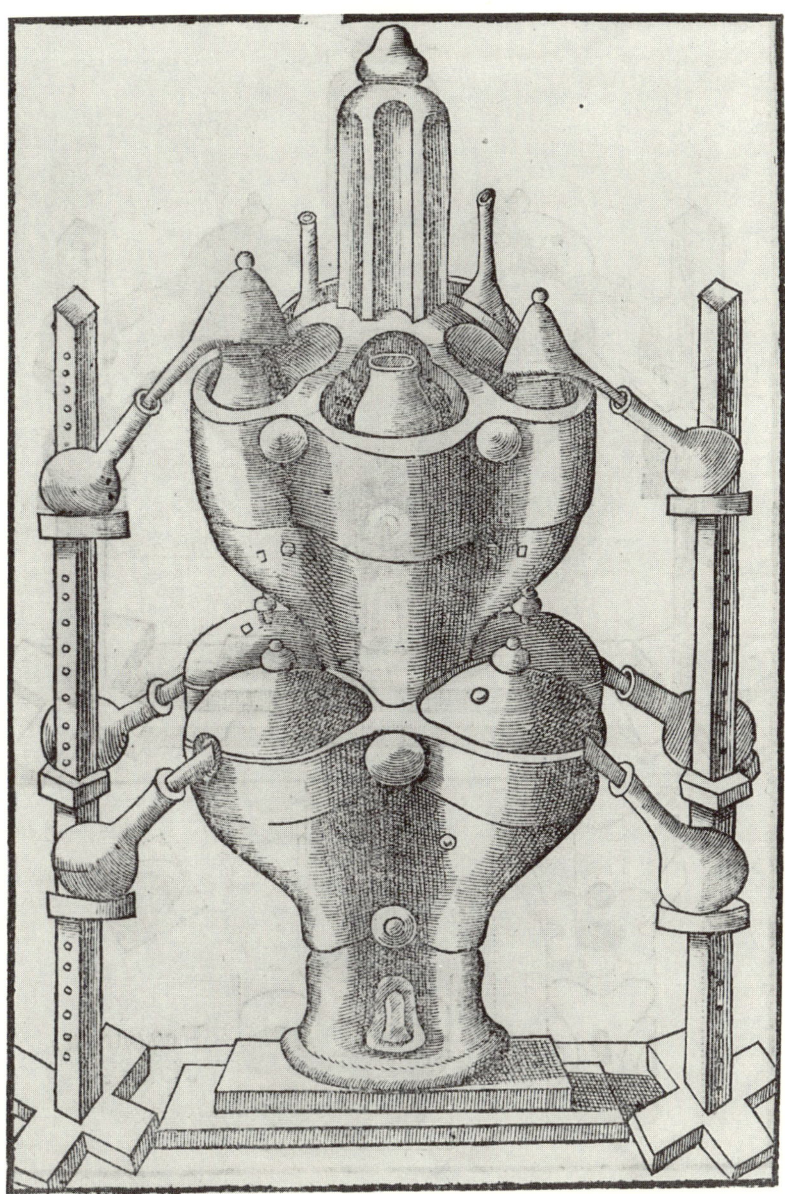

FIGURE 22 Woodcut representing the "fourneau astral ou lampadaire" (astral furnace). From A. Barlet, *Le vray et Méthodique cours de la Physique résolutive* (Paris, 1657).

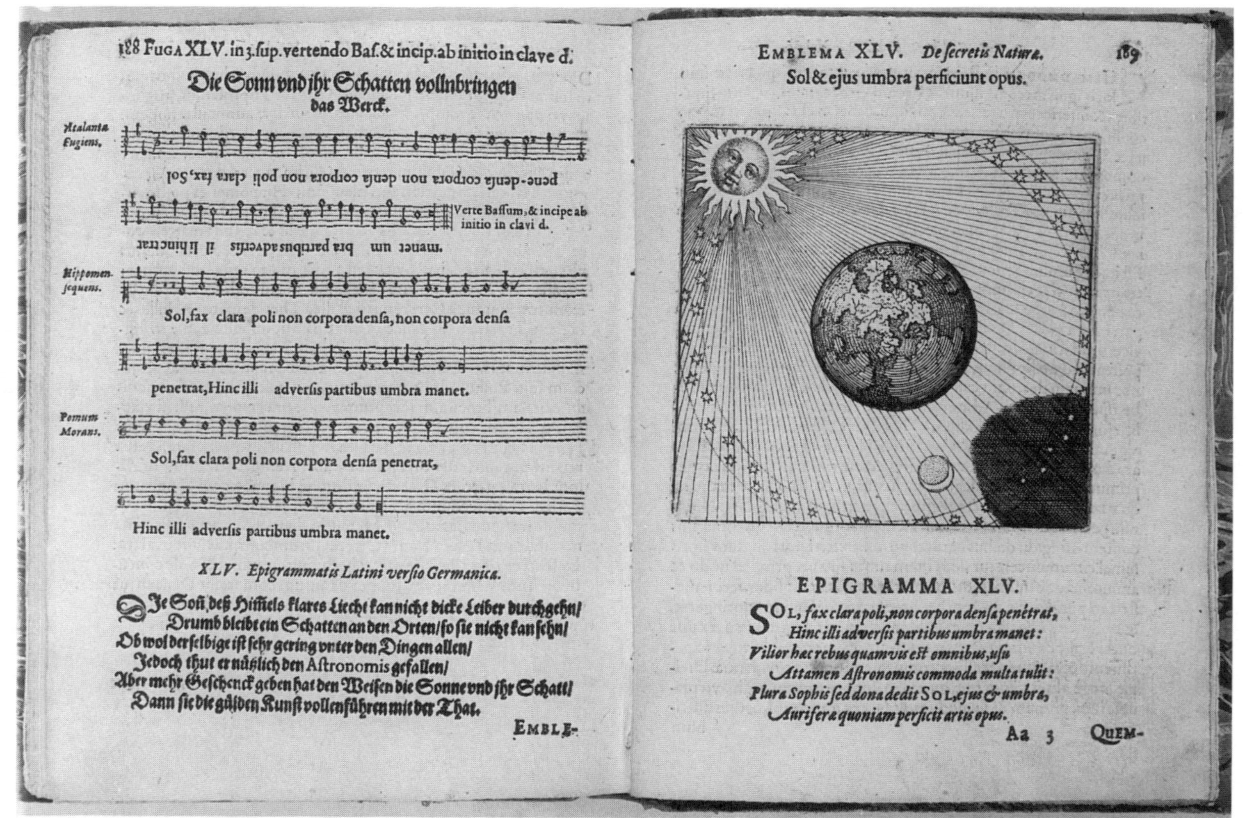

FIGURE 23 Maier's emblem depicting the marriage between the sun and the shadow. From Michael Maier, *Atalanta fugiens* (Oppenheim, 1618).

> The Sun, the bright torch of Heaven does not penetrate dense bodies, That is why there remains shadow on the parts turned away from it; Although the shadow is the most insignificant of all things,
> It has been of much use to the Astronomers: But Sol and its shadow give more gifts to the Philosophers
> Because it means the completion of the art of making gold.[20]

The immediate meaning that may be inferred from Maier's epigram on the emblem is that there is only limited power in the sun, which is the symbol of open and rational investigation of natural things but which on its own can do little to assist in the accomplishment of the "Great Work." Maier's emphasis on the importance of the shadow underlines the common belief among seventeenth-century alchemists that the philosopher's stone could only be reached by a more devious and obscure path. The difficulty of the art, its obscurity, and its darkness were accordingly major signs of the deeper meaning of alchemical endeavor.

In the chemical laboratory, Maier's emblem had another meaning: it depicted heat, symbolized by the sun, the *philosophical shadow* or the stone at the opposite corner, and the four elements in the middle, represented by the earth. The sun and the shadow also represented the ever-returning marriage of gold and silver (where silver is represented both by the moon and the shadow).

Since these emblems were intentionally ambiguous and deliberately chosen by Maier, it would be impossible to give an unequivocal interpretation of them.

Although they certainly provided a powerful expression of alchemical thought, emblems were not the only form of symbolism used by seventeenth-century alchemists. Another common and powerful way of depicting the process of the "Great Work" was through allegory.

The next picture (Fig. 24) is taken from the *Musaeum Hermeticum*,[21] a collection of alchemical tracts first printed in 1625. This beautiful engraving symbolizes the accomplishment of the "Great Work" and its place in the universe. It depicts the encounter between the Microcosmos of the alchemist and the Macrocosmos of God. The movement through the picture is from "what is above" to "what is below" through a hierarchical gradation of symbols.

In the upper part, we can see the name of God inscribed in the

---

[20] *Ibid.*, 278 and ff.

[21] *Musaeum Hermeticum reformatum et amplificatum* (1625), 2nd edition (Frankfurt, 1678). Partial interpretations of this picture are to be found in Read (1961), 83–84 and De Givry (1971), 350–351.

FIGURE 24 A famous allegory of the alchemical "Great Work." From *Musaeum Hermeticum* (Frankfurt, 1678).

Tetragrammaton and surrounded by the planets, the stars, and the angels. The lower and most complex part of the engraving illustrates the terrestrial world and its elements.

On the left is Adam, holding the sun, the symbol of gold; on the right stands Eve on the Hermetic stream that indicates *sophic mercury* holding the moon, the symbol of silver, in her left hand and a bunch of grapes, the emblem of fertility, in her right. Both Adam and Eve represent the reagents of the alchemical operation as a whole, and they are in fact chained to Macrocosmos.[22] At the center of the scene, we see the garden of Hermes, or the garden of alchemy, planted with trees bearing the symbols of the seven metals and others representing the chemical principles, tartar, niter, and salt ammoniac.

At the bottom of the picture are the elements: fire and air are

[22]This interpretation is taken from De Givry (1971), 351.

symbolized by a phoenix on the left, water and earth by an eagle on the right. In the middle the double-bodied lion, the symbol of the fusion between the two sulphurs or the quintessence, is surmounted by the alchemist who is depicted as the fusion of day and night. Just above him we notice the point of contact between the Microcosmos and the Macrocosmos, which is represented by the tree bearing the symbol of gold. The tree seems to be growing out of the alchemist's head, thus making him the connecting link between the two cosmos.

Although the whole scene is of a more complex nature, what I have described may hopefully suffice to illustrate the philosophical intentions behind the alchemical endeavor. Such allegorical representations were common during the seventeenth century and effectively reveal the tendency of the alchemist to convey his message through enigmatic and symbolic imagery. This characteristic is equally evident in those texts where alchemists used words rather than pictures. In these cases they never used words as a conventional means of communication; on the contrary, they intended their words to evoke something more than their immediate meaning. We may even say that in some cases alchemists used words purely allegorically. Philalethes, for instance, defined mercury in his tract on the metamorphosis of metals as follows:

> Mercury is our doorkeeper, our balm, our honey, oil, urine, maydew, mother, egg, secret furnace, oven, true fire, venomous Dragon, Theriac, ardent wine, Green Lion, Bird of Hermes, Goose of Hermogenes, two-edged sword in the hand of the Cherub that guards the Tree of Life etc. etc.; it is our true, secret vessel, and the Garden of the Sages, in which our Sun rises and sets.[23]

Behind this flood of words, allegories, and enigmas, there is little chemistry, indeed. On the other hand, the text is an extremely rich source of religious, mystical, and philosophical references. Here it is important to keep in mind that alchemy in the sixteenth and seventeenth centuries was in total opposition to both the Aristotelian doctrine and the conventional interpretation of Holy Scripture. The language of alchemy derived mainly from authorities other than the logic of Aristotle or scholastic philosophy, and it often incorporated unknown or neglected languages, such as Egyptian hieroglyphics.

As Ben Jonson put it in his comedy *The Alchemist* (1610):

---

[23]*Musaeum Hermeticum*, translated into English by Arthur Edward Waite with the title *The Hermetic Museum* (1893), reprinted (Yorkbeach, 1991), vol. 2, 243.

Was not all the knowledge
Of the Egyptians writ in mystic symbols?
Speak not the Scriptures oft in parables?
Are not the choicest fables of the poets
That were the fountains and first spring of wisdom
Wrapped in perplexed allegories?[24]

The mysticism and magic of the Egyptians, the reinterpretation of the Bible, and the allegories of the poets and the Platonists were indeed an integral and significant part of alchemical thought.

To reiterate the metaphysical nature of alchemical symbolism, we will look at another example, this time from a practical treatise on chemistry written by the German physician Johann Joachim Becher, who is usually regarded as the father of phlogiston theory.

In addition to his chemical and mineralogical investigations, Becher made serious efforts to develop alchemical research. For this purpose he wrote several treatises setting out the main features of alchemy. In 1689 he published *Tripus Hermeticus*, in which the "Great Work" was depicted with numerous symbols, illustrating different stages of the alchemical art.[25] Although an experienced naturalist and metallurgist, Becher did not introduce any substantial changes in alchemical symbolism. What he did was to incorporate the new knowledge acquired in chemistry and mineralogy into the traditional iconography of alchemy.

The first copperplate (Fig. 25) shows ten classes of alchemical symbols, which correspond to different genera of chemical substances and compounds. It is interesting that to keep the alchemical association between metals and planets intact, Becher classified (see the third row) metals, such as bismuth and zinc, under the genus *Mineralia*, despite the fact that more than a century earlier Georg Agricola decided bismuth was a metal. It is also interesting to note the combination of symbols and words that is so typical of the alchemical tradition.

The second plate (Fig. 26) depicts 64 instruments that were used in Becher's laboratory. Apart from typical vessels, furnaces, crucibles, alembics, assaying and ordinary balances, several implements for accomplishing the "Great Work" are shown. Number 51, for instance, is the philosophical egg (*ovum philosophicum*), a furnace that was sup-

---

[24]Ben Jonson, *The Alchemist*, in *The Oxford Authors—Ben Jonson*, Ian Donaldson (ed.) (Oxford, 1985), 145.

[25]J.J. Becher, *Tripus Hermeticus, pandens Oracula chymica: seu laboratorium portatile* (1689), in, *idem, Opuscula chymica rariora* (Nuremberg and Altorfii, 1719).

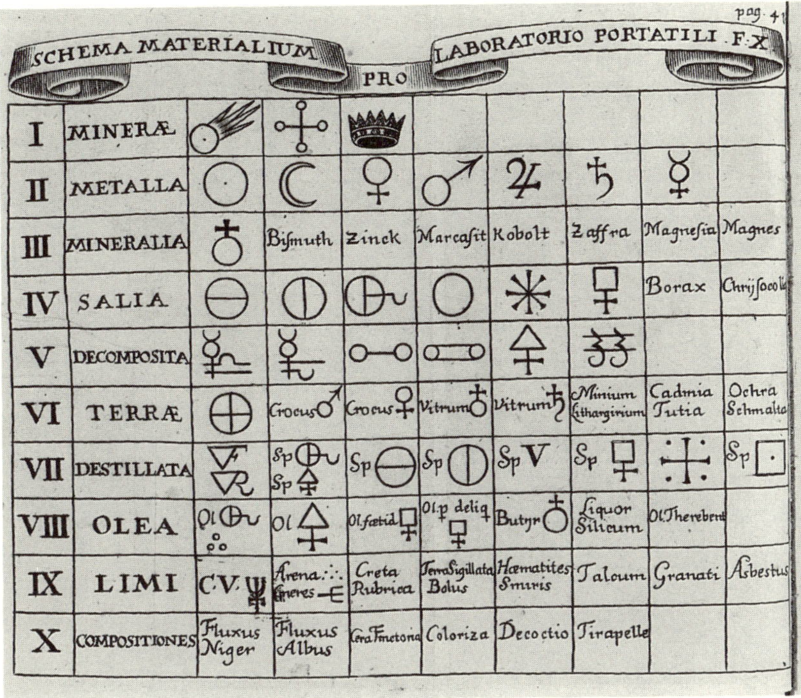

FIGURE 25 Chemical and alchemical symbols used by Becher. From J.J. Becher, *Tripus Hermeticus* (Nuremberg and Altorfii, 1719).

posed to engender the transmutation of metals. Number 58 is a set of the needles that were thought to be capable of detecting the presence of silver and gold.

In addition to these introductory plates, Becher added some interesting engravings that illustrated the different stages of the alchemical endeavor allegorically. Fig. 27 shows a chemical image of the quintessence formed by the combination of Salt and Sulphur.

EARLY CHEMICAL SYMBOLISM

During the second half of the seventeenth century, interest in alchemy was on the wane, mainly because the results obtained from the manipulation of chemical substances were obviously and repeatedly disappointing. In consequence alchemists became more and more speculative and

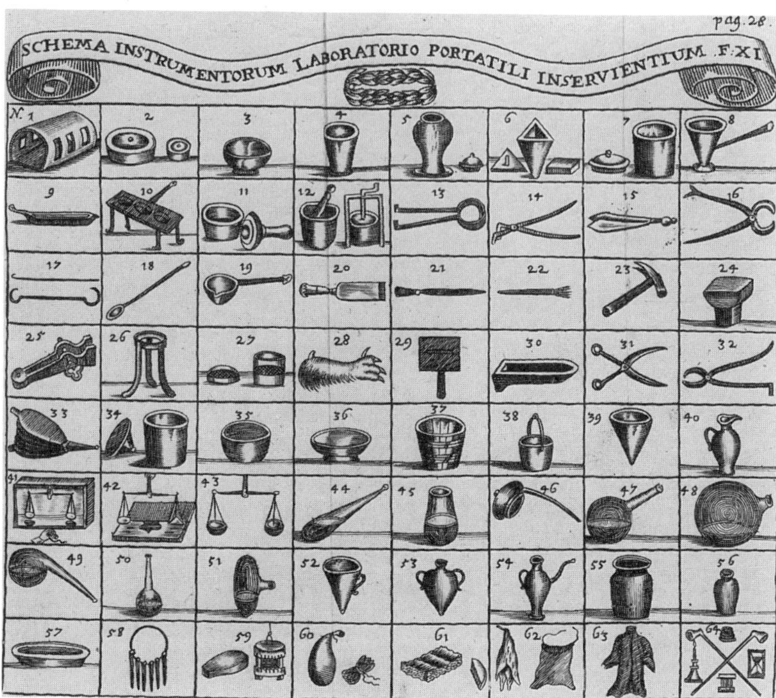

FIGURE 26 Symbols and descriptions of chemical and alchemical instruments. From J.J. Becher, *Tripus Hermeticus* (Nuremberg and Altorfii, 1719).

gradually abandoned their concrete claims regarding the manipulation of matter.

The use of alchemical symbolism lasted, however, far longer; it was not until the end of the eighteenth century that serious efforts were made to eradicate it from chemistry. Although the persistence of alchemical symbolism might be regarded as a paradox, the radical change in attitude to the function and meaning of the symbols that took place at the beginning of the eighteenth century indicates the marked discontinuity between alchemy and early chemistry. In this connection mention should be made of the well-known *Cours de Chymie* by the French apothecary Nicolas Lemery. This textbook, which first appeared in 1675, became one of the most successful manuals in the history of

FIGURE 27 A symbolic representation of the quintessence. From J.J. Becher, *Tripus Hermeticus* (Nuremberg and Altorfii, 1719).

early chemistry. With its 11 editions, the final one in 1757,[26] Lemery's work exerted a considerable influence on the emergence of chemistry as an autonomous discipline and provided the standard technical lexicon and nomenclature for more than half a century.

The chemical language of Lemery's *Cours de Chymie* represents an interesting critical reworking of the alchemical heritage. In the preface Lemery severely criticizes the obscurity of the language of alchemy:

> Most of the authors who have spoken of chemistry have written of it with such obscurity that they seem to have done their best not to be understood; and in this we might say that they have been too successful. . . .[27]

The French chemist held strong views on alchemy, which he regarded as "an art without art, whose principle is to lie, whose means is to manipulate, and whose aim is to beg,"[28] and he firmly rejected the astrological association between planets and metals that was claimed by most of the alchemists of his time.

But despite this uncompromising position, alchemical symbolism appeared in Lemery's *Cours*, and a table depicting all chemical characters was presented in the middle of the work (see Fig. 28).

As we can see, the association of planets and metals that Lemery rejected throughout the text was still an essential feature of the table. Furthermore, most of the symbols for substances derived directly or indirectly from alchemical iconography. Although this table might be regarded as a striking contradiction of Lemery's professed rejection of alchemy, an inspection of its contents and its role in Lemery's system reveals that much had changed in the pictorial representation of chemical substances.

Unlike many of his predecessors, Lemery did not use symbolism in the written text and confined its function primarily to the representation of chemical substances, such as metals, salts, and solvents that were in common use in chemical laboratories. More importantly, Lemery also introduced several new symbols to denote chemical operations and technical instruments. Significantly the symbols for operations, such as distillation, precipitation, and calcination, and for instruments, such as the alembic, were now presented alongside the symbols for chemical

---

[26] N. Lemery, *Cours de Chymie contenant la maniere de faire les operations qui sont en usage dans la Medicine, par une Méthode facile* (1675), 11th edition (Paris, 1757).
[27] *Ibid.*, xi.
[28] "Ars sine arte, cujus principium mentiri, medium laborare et finis mendicare," *ibid.*, 46.

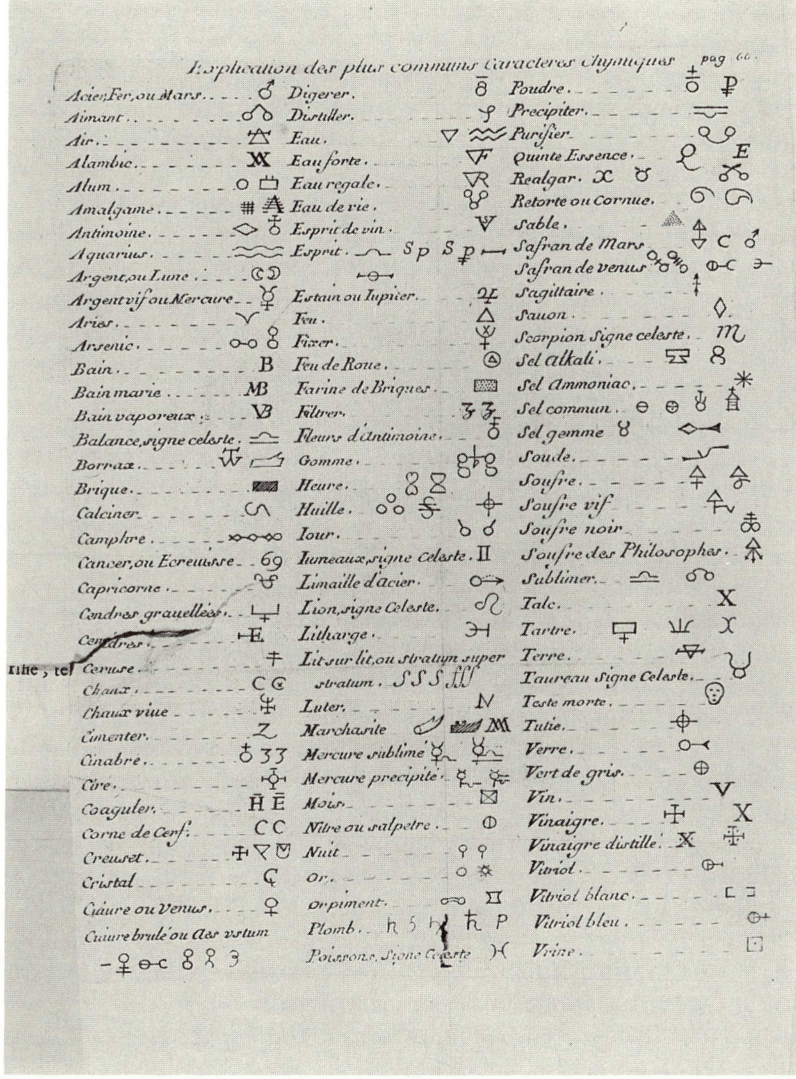

Explication des plus communs Caracteres Chymiques pag. 66.

Acier, Fer, ou Mars
Aimant
Air
Alambic
Alum
Amalgame
Antimoine
Aquarius
Argent, ou Lune
Argent vif ou Mercure
Aries
Arsenic
Bain B
Bain marie MB
Bain vaporeux VB
Balance, signe celeste
Borrax
Brique
Calciner
Camphre
Cancer, ou Ecrevisse 69
Capricorne
Cendres gravellées
Cendres E
Ceruse
Chaux C
Chaux vive
Cimenter
Cinabre
Cire
Coaguler H E
Corne de Cerf C C
Creuset
Cristal
Cuivre ou Venus
Cuivre brulé ou Aes vstum

Digerer
Distiller
Eau
Eau forte
Eau regale
Eau de vie
Esprit de vin
Esprit Sp Sp
Estain ou Iupiter
Feu
Fixer
Feu de Roue
Farine de Briques
Filtrer
Fleurs d'antimoine
Gomme
Heure
Huille
Iour
Iumeaux, signe celeste II
Limaille d'acier
Lion, signe Celeste
Litharge
Lit sur lit, ou stratum super stratum S S S
Luter
Marchasite
Mercure sublimé
Mercure precipité
Mois
Nitre ou salpetre
Nuit
Or
Orpiment
Plomb P
Poissons, Signe Celeste

Poudre
Precipiter
Purifier
Quinte Essence E
Realgar
Retorte ou Cornue
Sable
Safran de Mars
Safran de venus
Sagittaire
Savon
Scorpion Signe celeste
Sel alkali
Sel Ammoniac
Sel commun
Sel gemme
Soude
Soufre
Soufre vif
Soufre noir
Soufre des Philosophes
Sublimer
Talc X
Tartre
Terre
Taureau Signe Celeste
Teste morte
Tutie
Verre
Vert de gris
Vin V
Vinaigre X
Vinaigre distillé
Vitriol
Vitriol blanc
Vitriol bleu
Vrine

FIGURE 28 Lemery's table of chemical symbols. From N. Lemery, *Cours de Chymie*, 11th edition (Paris, 1757).

substances. This emphasis on the practice of chemistry is the main characteristic of Lemery's *Cours de Chymie*, which was basically a rich collection of pharmaceutical remedies. Accordingly the symbols were confined to one table and were used purely for practical convenience, rather than to suggest any deeper or hidden meaning. Yet even acknowledging this radical change, which indeed represents a definitive shift from alchemical thought, it is difficult to explain the persistence of alchemical symbolism in early chemical textbooks. In my opinion the most plausible answer to this controversial question is that by the end of the seventeenth and the first half of the eighteenth centuries chemistry had not yet become an autonomous discipline comparable to other sciences, such as astronomy, medicine, mathematics, and physics.

Nevertheless, Lemery succeeded in divesting this symbolism of all its allegorical and pictorial significance by restricting it to a purely practical function in his pharmaceutical lexicon.

During the eighteenth century, chemical textbooks, such as those by Boerhaave,[29] Malouin,[30] Wallerius,[31] and Erxleben,[32] used explanatory tables of chemical symbols similar to Lemery's.

There were also other attempts to standardize the chemical meanings of alchemical symbols. The German physician Johann Christoph Sommerhoff, for example, published a detailed chemical and pharmaceutical lexicon in 1701[33] in which he tried to provide an inventory of all the alchemical and chemical symbols and their meanings[34] (see Fig. 29). Unlike alchemical lexicons, such as the Paracelsian Ruland's *Lexicon Alchemiae* (1612), Sommerhoff's collection of symbols had a primarily descriptive and communicative aim. These more practical and straightforward aims were shared by Diderot, whose *Encyclopédie* included a collection of plates presenting a table of chemical symbols and their meanings in alphabetical order[35] (see Fig. 30).

Since most textbooks and dictionaries of eighteenth-century chemistry included symbols only in this type of descriptive table, we may say that the role of symbolism was generally rather small, and certainly not comparable to its part in alchemical works. There was, however, at least

[29]Hermann Boerhaave, *Elementa Chemiae*, 2 vols. (Lugduni Batavorum, 1732).

[30]Paul-Jacques Malouin, *Traité de Chimie* (Paris, 1734).

[31]Johan Gottschalk Wallerius, *Chemia Physica*, 2 vols. (Stockholm, 1759 and 1765).

[32]Johann Christian Polykarp Erxleben, *Anfangsgründe der Chemie* (Göttingen, 1775).

[33]Johann Christoph Sommerhoff, *Lexicon Pharmaceutico-chymicum, Latino-Germanicum et Germanico-Latinum, Continens Terminorum Pharmaceuticorum et Chymicorum* . . . (Nuremberg, 1701).

[34]"Characteres Rerum Chymicorum," in Sommerhoff, Part 2, 100 and ff.

[35]*Recueil de Planches sur les Sciences et les Arts*, vol. 3 (Paris, 1768), "Characteres de chymie."

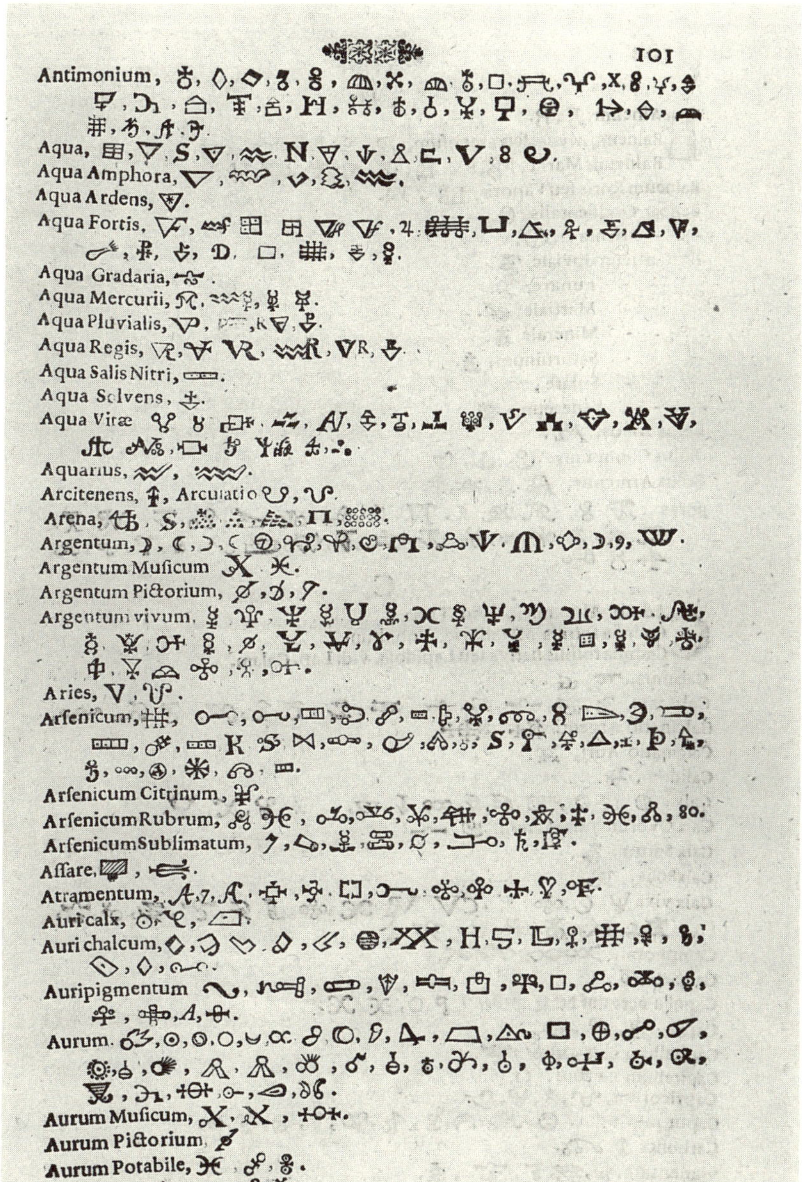

101

Antimonium,

Aqua,

Aqua Amphora,

Aqua Ardens,

Aqua Fortis,

Aqua Gradaria,

Aqua Mercurii,

Aqua Pluvialis,

Aqua Regis,

Aqua Salis Nitri,

Aqua Solvens,

Aqua Vitæ

Aquarius,

Arcitenens, Arcuatio

Arena,

Argentum,

Argentum Musicum

Argentum Pictorium,

Argentum vivum,

Aries,

Arsenicum,

Arsenicum Citrinum,

Arsenicum Rubrum,

Arsenicum Sublimatum,

Assare,

Atramentum,

Auri calx,

Auri chalcum,

Auripigmentum

Aurum.

Aurum Musicum,

Aurum Pictorium

Aurum Potabile,

Autumnus,

N 3 Balneum

FIGURE 29 A page of an explanatory dictionary of alchemical and chemical symbols. From J.C. Sommerhoff, *Lexicon Pharmaceutico-chymicum* (Nuremberg, 1701).

FIGURE 30 The "characters" of chemistry. From D. Diderot (ed.), *Recueil de Planches sur les sciences et les Arts*, vol. 3 (Paris, 1768).

one research field during the eighteenth century in which symbolism was widely employed: the study of chemical affinities. When the French physician Etienne Geoffroy published a memoir on chemical affinities in 1719,[36] many European naturalists began to take an interest in this field. The investigation of chemical attraction was basically intended to determine the *chemical* relations between all known substances. To provide a descriptive and effective scheme of the different degrees of affinity between so many substances a system of abbreviations was needed. It was at this stage in the history of chemistry that symbols began to be regarded as an effective form of abbreviation. In the specific case of chemical attractions, the use of symbols instead of words allowed chemists to display the whole system of relations of an entire class of substances in a single table.

In 1775 the Swedish chemist Torbern Bergman published *A Dissertation on Elective Attractions*, a work that marked a decisive step forward in the investigation of chemical affinities.[37] Bergman, who was a convinced Newtonian, tried to determine the order of chemical attraction of different substances from the intensity of their respective forces.[38] As we can see from the preparatory manuscript to the English edition (Fig. 31), Bergman displayed in the first horizontal row the affinities between vitriolic acid (the first on the left) and other substances, which in order of attraction were phlogiston, calx, platinum, iron, tin, copper, mercury, silver, etc. The second row illustrates the affinities between nitric acid and other substances, and so on down to the last row, which indicates the affinity among aether, oil, and water.

Bergman frequently used this kind of symbolism and justified it as follows:

> To make the present work more useful to my readers, I have had two copper etchings made. One represents the most important chemical symbols and the other the attractions between the simplest substances used in chemistry. It is quite usual to go from one extreme to another and this has also happened to chemical symbols. In the past, sciences were enveloped in darkness and superstition and the most absurd symbols were used to represent things that it was considered best to hide from common knowledge and, perhaps more often, to cover up ignorance. In our times all the signs and symbols of the science have almost

---

[36]E.F. Geoffroy, "Table des différents Rapports observés en Chimie entre différentes substances," in *MARS*, 1718 (published in 1719), 202–212.

[37]Torbern Bergman, "Dissertatio de attractionibus electivis," in *Nova Acta Societatis Regiae Scientiarum Upsaliensis*, 2 (1775), 161–250, translated into English under the title, *A Dissertation on Elective Attractions* (London, 1785).

[38]*Ibid.*, 4.

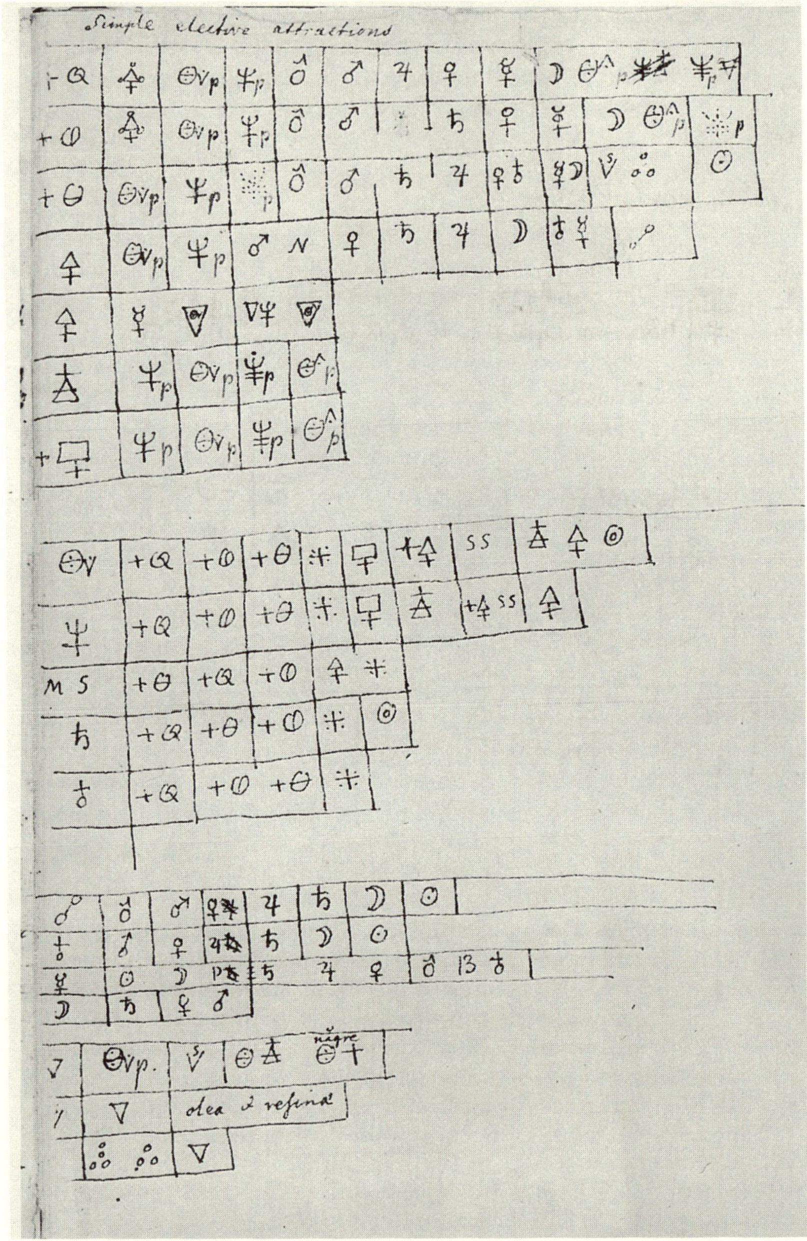

FIGURE 31 Bergman's chemical symbols to denote the chemical attraction between different substances. Courtesy of UUB, Bergmans Manuscripta, D 1459 i.

> been eradicated but here as elsewhere a moderate position is considered best. Such symbols are particularly useful when something has to be written down because by using them one saves time and space.[39]

Exploitation of the utilitarian aspect of symbolism in chemistry increased after Bergman and his unequivocal view of the role of symbols explained the great success of his tables of affinities. Even such a cautious and careful experimenter as the Scottish chemist Joseph Black used Bergman's symbolism in his celebrated lectures on chemistry, as may be seen from a copy of his manuscript (Fig. 32).

Another research field in which symbols proved extremely useful was the classification of salts. The increasing number of salts discovered during the first half of the eighteenth century posed the immediate and urgent problem of their classification, naming, and pictorial representation. Since alchemy had provided a large number of symbols for a small number of substances, it was now necessary either to create new symbols or simply to restrict the range of the existing ones.

Bergman and the French chemist Guillaume-François Rouelle, the most important contributors in this field, carried out an extensive reform of the use of chemical symbols. In the classification of salts, symbols were basically used to solve the problem of squeezing a large quantity of words into a single table. For the representation of double and triple salts, the use of symbols facilitated the understanding of the rather complicated systematics for the student. We can therefore say that the increasing need for clarity and the growing number of chemical objects were the main reasons for the first moves towards a rationalization and reform of alchemical symbolism by Bergman and Rouelle (Fig. 33). However, these moves were made within the limits of alchemical iconography, and although it was now clear that chemistry had nothing to do with alchemical theory or practice, the use of such symbols was seen by an increasing number of eighteenth-century chemists as an ambiguous bridge of continuity between a purely experimental tradition and a metaphysical esotericism. By the end of the 1760s, many naturalists realized that chemistry could not be considered a science until its language was entirely based on the new experimental philosophy. The consistent progress that had been achieved in chemical laboratories during the first half of the eighteenth century began to be regarded as insufficient as long as it could not be supported by a radical reform of chemical nomenclature and symbolism. Torbern Bergman certainly

---

[39]H.T. Scheffer, *Chemiske föreläsningar,* Torbern Bergman (ed.) (Uppsala, 1775), iv–v (my own translation).

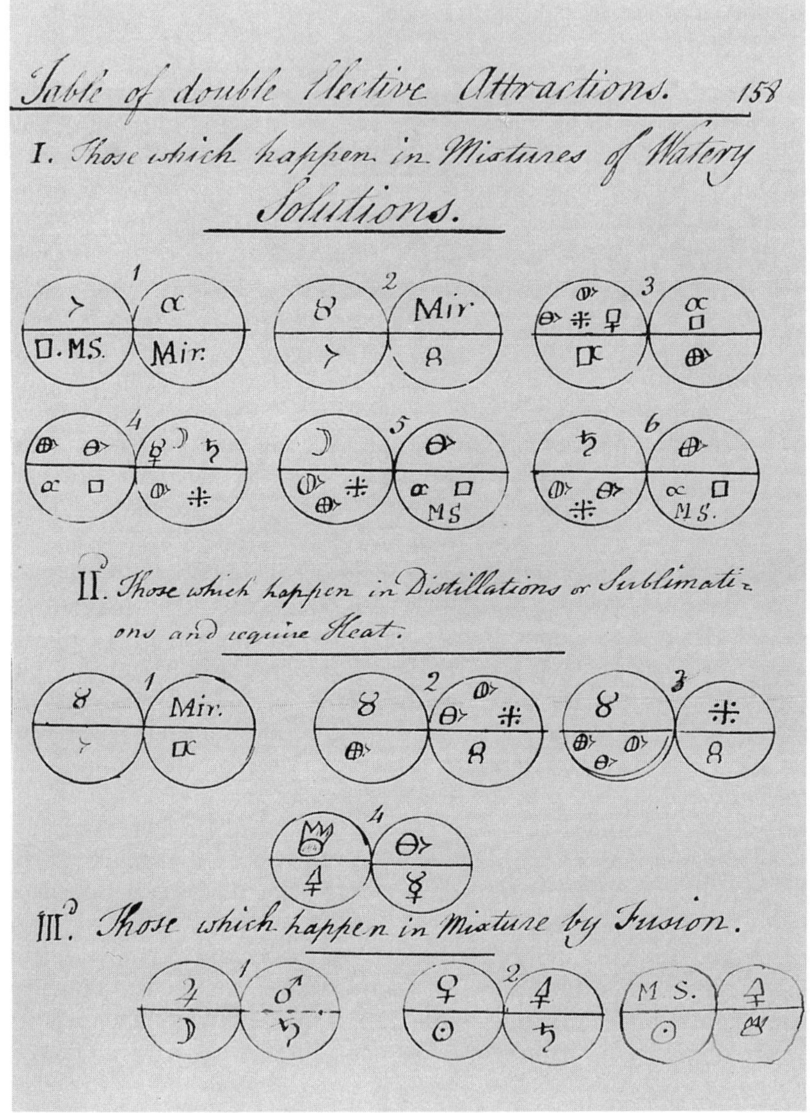

FIGURE 32 Joseph Black's symbols for double chemical attractions. Courtesy of UUB, Wallers samling.

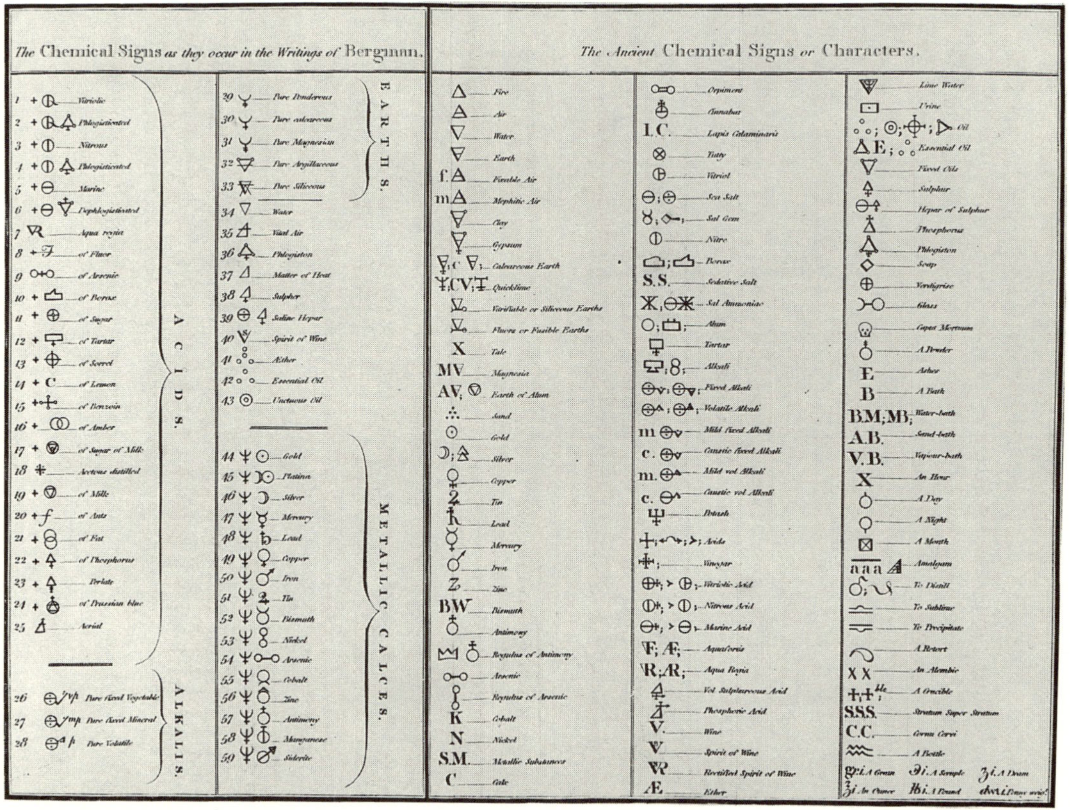

FIGURE 33 The chemical symbolism in use during the second half of the eighteenth century after the reform introduced by Bergman and Rouelle. From William Nicholson, *A Dictionary of Practical and Theoretical Chemistry* (London, 1808).

shared this view of the need for reform and he contributed to the dismantling of most of the alchemical lexicon, but with respect to symbolism, he still believed that a restriction of the meaning of the alchemical iconography would be enough to give chemistry an effective and *scientific* means of expression. Most eighteenth-century chemists in Germany and in the British Isles agreed with Bergman and continued to use alchemical symbols even when the nomenclature of chemistry had been entirely changed. It is not unusual to find alchemical symbolism in German and British textbooks and dictionaries of chemistry printed at the beginning of the nineteenth century.[40]

In France the situation was different, and several authoritative chemists took a negative view of the use of traditional symbolism. Pierre Joseph Macquer was among the first to draw a connection between the progress of chemistry and the clarity and rigor of its language; he suggested several changes in the nomenclature of salts that were similar to those proposed by his friend Torbern Bergman at about the same time.[41] However, Macquer disagreed with his Swedish colleague on symbolism, and he believed that the explanation of chemistry's continued failure to attain recognition as a science lay in the linguistic and iconographic obscurity inherited from alchemy.

In 1749 Macquer remarked that because of the influence of alchemy,

> . . . chymistry became an occult and mysterious Science; its expressions were all tropes and figures, its phrases metaphorical, and its axioms so many enigmas: in short, an obscure unintelligible jargon is the justest character of the Alchymistic Language. Thus, by endeavouring to conceal their secrets, those gentlemen rendered their Art useless to mankind, and brought it into deserved contempt.[42]

Macquer consequently tried to keep alchemical language out of the nomenclature he used, and he eliminated alchemical symbols from his own works.

Macquer's writings were the first in the eighteenth century to omit all symbolism, and they were extremely innovative in their attitude to the role of language in the foundation of a science. Guyton's reform of

---

[40]See, for instance, the article "Characters," in William Nicholson's *A Dictionary of Practical and Theoretical Chemistry* (1795) (London, 1808), where Bergman's table of alchemical and chemical symbols was still considered the best example.

[41]On Macquer's, Bergman's, and Guyton's reforms of chemical nomenclature, see the essay by W.A. Smeaton (1954) and Crosland (1978), 134–152.

[42]P.J. Macquer, *Élémens de chymie Théorique* (1749), translated into English with the title *Elements of the Theory and Practice of Chymistry* (London, 1758), vol. 1, viii–ix.

chemical nomenclature concerned only the verbal language of chemistry, whereas the symbols were totally neglected. It is likely that Guyton shared Macquer's misgivings concerning the ambiguity of traditional chemical symbols but for this reason preferred to eradicate them from the science rather than reform them. This preference was evident a few years later in 1786 when Guyton edited Bergman's tables of elective attractions and replaced the symbols used by the Swedish chemist with the names of the corresponding substance[43] (see Fig. 34). Although the space-saving advantages of symbols were obviously lost, the linguistic classification of affinities gave a clearer picture to those chemists who were no longer acquainted with the alchemical lexicon. In any case a closer look at Guyton's tables of affinities reveals that he did not renounce the use of symbols totally. To indicate the force of adherence between different chemical substances quantitatively, Guyton used an arbitrary number indicating the degree of force. If we look at Fig. 34, we see, for instance, that the attraction between vitriolic acid and potash was represented by the number 9, while the attraction between nitric acid and silver oxide, which was of considerably less force, was represented by the number 2. Similarly numerical relations represented the affinities between potash and nitric acid and between vitriolic acid and silver oxide.

In reality quantification of the force of adhesion was still a "*desideratum*,"[44] but the use of arbitrary numbers began to satisfy the emerging need for expressive rigor in chemical language. Guyton himself declared that the numbers representing degree of affinity were the most important symbols used in chemistry.[45] Although the enthusiasm of Macquer and Guyton for mathematics was confronted with the qualitative structure of eighteenth-century chemistry and the practical impossibility of determining chemical reactions numerically, this highly theoretical commitment was to have important consequences for the

---

[43]Guyton, article "Affinité," in *Encyclopédie Méthodique—Chimie, Pharmacie, Métallurgie*, vol. 1, part 1 (Paris, 1786), 535–613.

[44]Bergman, *A Dissertation* . . . (1785), 4.

[45]"Les *nombres* qui représentent les rapports d'affinité, sont, comme l'on voit, la partie la plus importante de ces symboles, mais aussi la plus difficile. Avant de terminer cet article, je m'en occuperai dans des vues plus étendues & avec tous les détails qu'elles exigent; il suffira maintenant d'être bien averti que les nombres que j'ai employés dans les exemples précédens, n'ont, dans le fait, aucune base certaine; mais puisqu'ils quadrent déjà avec un assez grand nombre d'observations les plus familières, on peut en faire usage sans inconvénient, jusqu'à ce qu'on ait reconnu la necessité de les changer pour les accorder avec d'autres résultats," from Guyton (1786), 557. On this theme, see Goupil (1991), 179–187.

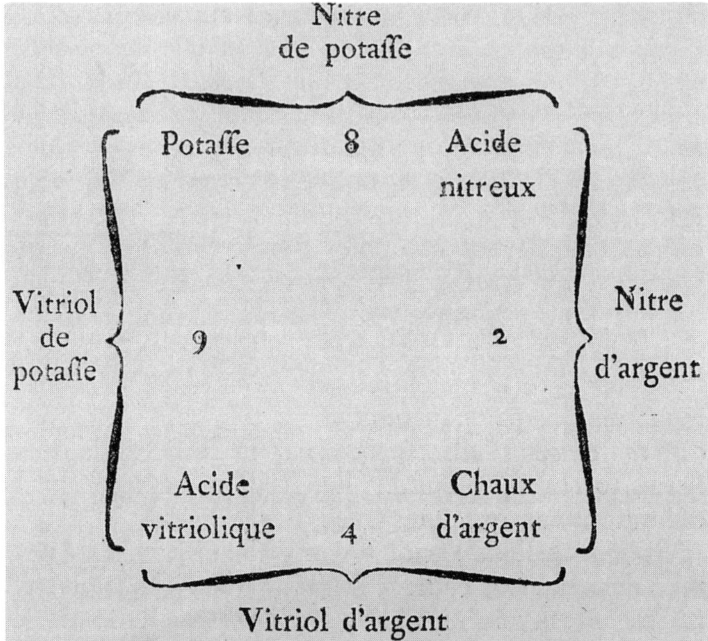

FIGURE 34 Numeric symbols used by Guyton de Morveau to denote the degree of intensity of the force of attraction between different chemical substances. From *Encyclopédie Méthodique—Chimie, Pharmacie, Métallurgie*, vol. 1, part 1 (Paris, 1786).

change of method and the experimental status of chemistry. Furthermore, after Macquer's and Guyton's explicit criticism of alchemical symbolism, a few authoritative chemists, including Kirwan, Priestley, Landriani, Cavendish, and others involved in the investigation of the controversial nature of gases, followed the recommendation to omit all symbolic representations from their works.

The Mathematicization of Chemical Symbols

By the end of the eighteenth century, the role of chemical symbolism had diminished considerably and in an increasing number of cases, all forms of chemical or alchemical symbolism had been abandoned in favor of a purely verbal presentation of the science.

Antoine Laurent Lavoisier approached the problem in an interesting and original way and became the main advocate of the reform of chemical symbolism in the eighteenth century. From the beginning of his scientific career, Lavoisier had consistently tried to modify the traditional chemical symbols. In 1766, when he was only 23, he invented new symbols to denote the most common mineralogical substances and compounds,[46] placing them beside his handwritten annotations on the geological composition of the soil he surveyed with his mentor Jean-Etienne Guettard. His symbol for sand, a triangle, was derived from an Egyptian hieroglyphic; others, such as those for calcareous earth (*pierre calcareuse*), plaster (*craie*), and tufa (*tuf*), were composed of the initial letter inside a square or a circle; sandstone (*gres*) was depicted by a circle with the sand symbol inside.

Although it seems logical nowadays, following the reform introduced by Berzelius, to denote a substance with the initial letter of its name, it is important to remember that when Lavoisier wrote this manuscript it was a new way of logically and effectively indicating substances while avoiding the ambiguities of alchemical and pictorial symbols.

Unfortunately there is no other evidence of Lavoisier using chemical symbols before 1782, when he once again took up a position different from that of his predecessors and contemporaries on this controversial subject. In a memoir attacking the phlogiston theory that he presented at the Academy of Sciences in Paris in that year Lavoisier outlined his theory of acidity and stressed the central role played by oxygen in the calcination of metals.[47] He claimed that a metal that was dissolved in different acids liberated the same quantity of oxygen, through both a humid and an ordinary calcination. To express the quantity of oxygen determined by his measurements with the balance, Lavoisier invented a system of symbols "of a sort which could at first be taken for algebraic formulas"[48] (see Fig. 35). By projecting quantitative parameters and numerical expressions onto his laboratory experiments on the dissolution of metals in acids,

[46]Antoine Laurent Lavoisier, *Journal d'observations minéralogiques faites dans le Brie en Octobre et Novembre 1766*, published for the first time in LO, vol. 5, 109–150. Edouard Grimaux, who after J.B. Dumas became the editor of the last two volumes of Lavoisier's works, decided not to reprint the symbols used by Lavoisier in his manuscript version of the *Journal.* The original manuscript is kept at Arch. Acad., shelfmark: Dossier Lavoisier n. 608.

[47]Lavoisier, "Considérations générales sur la dissolution des métaux dans les acides," in *MARS* (1782) (published in 1783), reprinted in LO, vol. 2, 509–527.

[48]LO, vol. 2, 515, translated into English by C.C. Gillispie, *The Edge of Objectivity. An Essay in the History of Scientific Ideas* (Princeton, 1960), 243 (see the following note).

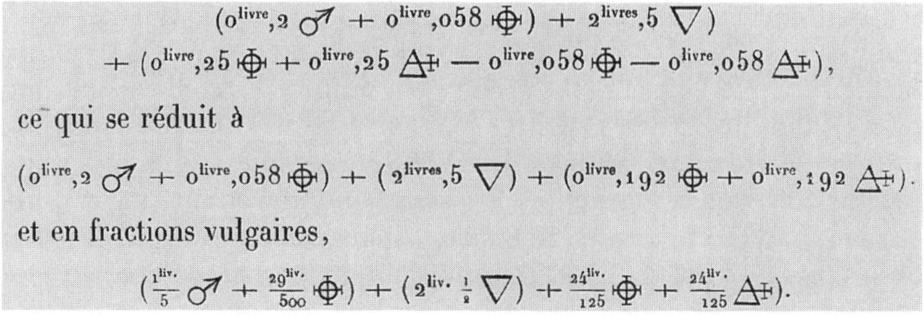

(0^{livre},2 ♂ + 0^{livre},058 ⊕) + 2^{livres},5 ▽)

+ (0^{livre},25 ⊕ + 0^{livre},25 🜁 — 0^{livre},058 ⊕ — 0^{livre},058 🜁),

ce qui se réduit à

(0^{livre},2 ♂ + 0^{livre},058 ⊕) + (2^{livres},5 ▽) + (0^{livre},192 ⊕ + 0^{livre},192 🜁).

et en fractions vulgaires,

($\frac{1^{\text{liv.}}}{5}$ ♂ + $\frac{29^{\text{liv.}}}{500}$ ⊕) + (2$^{\text{liv.}}$ $\frac{1}{2}$ ▽) + $\frac{24^{\text{liv.}}}{125}$ ⊕ + $\frac{24^{\text{liv.}}}{125}$ 🜁).

FIGURE 35 Lavoisier's chemical symbolism used in a memoir on the dissolution of acids published in 1783. From LO, vol. 2.

Lavoisier presented the first chemical equation ever to appear in print. The basic assumption behind the equation was that the total weight of the substances involved in the experiment did not change during the chemical reaction.

However, this revolutionary expression of a chemical reaction was cautiously qualified by Lavoisier himself. Aware that his ideas were bold, Lavoisier remarked that chemists were "still very far from being able to introduce mathematical precision into chemistry," and that the formulas he presented in his memoir had to be considered merely as "simple annotations of which the object is to ease the labors of the mind."[49] However, this initial caution was abandoned in 1787, when Lavoisier, together with Guyton de Morveau, Fourcroy, Berthollet, Hassenfratz, and Adet, published the *Méthode de nomenclature chimique*, a work in which the chemical nomenclature was entirely recreated. Lavoisier, who wrote the two most theoretical memoirs of the *Méthode*, declared that in order to achieve scientific legitimacy, chemistry had to acquire the expressive rigor of algebra. The influence of Condillac is also detectable in the reform of chemical symbols undertaken by Lavoisier's young pupils Hassenfratz and Adet. As with the linguistic nomenclature, Hassenfratz and Adet aimed to construct a methodical system and at the same time to leave no sign of the alchemical tradition, declaring:

[49]"Nous sommes encore bien loin de pouvoir porter dans la chimie la précision mathématique, et je prie, en conséquence, de ne considérer les formules que je vais donner que comme de simples annotations, dont l'objet est de soulager les opérations de l'esprit," Lavoisier (1782), 515, translated by Gillispie (1960), 245. It is significant in that in the same memoir Lavoisier contradicted himself and declared: "J'espère que la lecture de ce mémoire fera entrevoir la possibilité d'appliquer l'exactitude du calcul à la chimie," *ibid.*, 526.

> In our reformation of the chemical characters, we are far from having the same design with the ancient chymists. They endeavoured by every means to screen their science with a mysterious veil from the eyes of the vulgar; we ought on the contrary to use our utmost endeavors to render our knowledge as communicative as possible. Should the chymical characters become uniform among the chymists of every nation, they will resemble the writing of the inhabitants of China, of Tongking, of Japan. Although these people have different languages, they notwithstanding have common characters to represent them. . . .
>
> The chymical characters will resemble the figures of algebra, which expressing the operations of the judgment necessary in that science, afford an easy method to the geometricians of every country to understand one another.[50]

The symbols created by Hassenfratz and Adet developed the general ideas already outlined by Lavoisier during his geological excursions in 1766.

Simple substances, such as oxygen, hydrogen, caloric, and azote, were represented by straight lines (see Fig. 36); semicircles depicted "acidifiable or combustible substances," such as sulphur, charcoal, and phosphorus; circles containing the initial letter of the substance represented the metals; compounds with oxygen were represented by a square with the initial letter of the Latin name inside that referred to the base combining with oxygen; lozenges represented compounds, such as aether and alcohol, that were non-acidifiable.

In addition to the symbols for specific substances, Hassenfratz and Adet created three signs made up of straight lines in different geometrical configurations to represent the solid, liquid, and aeriform states of matter.

The symbols chosen by the Lavoisierian *cotérie* were thus all taken from the elementary figures of geometry and were also an expression of Lavoisier's view of the utility of a mathematicization of chemistry.

The table of the 54 symbols depicting the simple substances in their three possible states gives 24,804 possible three-substance combinations for each substance, which number, "multiplied by thirteen, which expresses the number of positions that these three characters can have, gives 322,452 different combinations."[51] When we recall the small number of chemical symbols, rarely more than 100, used by

---

[50] *Méthode* (1787), 254–255; translated into English by James St. John with the title, *Method of chymical nomenclature* (London, 1788), 192.

[51] *Ibid.*, 285, English translation (1788), 213.

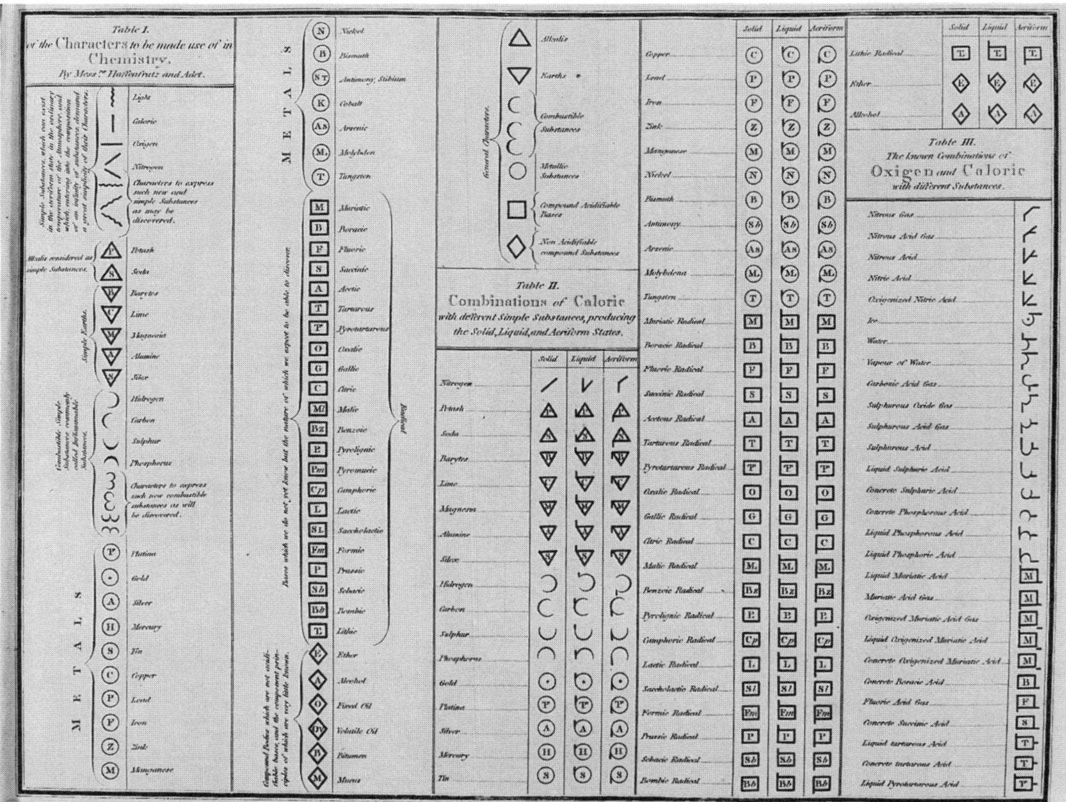

FIGURE 36 The symbols adopted by Lavoisier and his colleagues in the *Méthode de nomenclature chimique* (Paris, 1787). From W. Nicholson, *A Dictionary of Practical and Theoretical Chemistry* (London, 1808).

Lemery, Rouelle, Bergman, and most eighteenth-century chemists, we cannot but be astonished by the revolutionary implications of Lavoisier's vision of matter, which was now realized to be complex and not reducible to a small hierarchy of principles or elements. Lavoisier concluded:

> We think that (the characters) proposed by Messrs. Hassenfratz and Adet are preferable to the ancient; that they have the advantage of painting to the sight, not words, but facts, and giving just ideas of the combinations which they represent. This method seems to possess also another advantage; it determines beforehand the characters of such substances as may be discovered in future, so that there can be nothing arbitrary in the formation of their characters, and a complete table of these characters represents, at the same time all that has been done in the science by former chymists, and all that remains to be discovered.[52]

Unlike all earlier attempts to employ or reform chemical symbolism, that of Lavoisier and his collaborators permitted the representation of those substances or compounds that had not yet been chemically analysed. To use Lavoisier's concept, the new system of symbols was a *method* of scientific discovery and classification.

In referring to algebra as the ideal of scientific language, Lavoisier was not merely making a rhetorical statement; on the contrary, he was committing himself to providing chemistry with a new language and methodology. These apparently abstract theorizations had concrete effects on chemistry at the end of the eighteenth century, when enormous changes were taking place.

However, to do full justice to Lavoisier's ideas on chemical language, it must be remembered that after promoting and supporting a general reform of symbols, the French chemist never used the systematic symbolism published in 1787 again and that like Macquer and Guyton de Morveau before him he preferred to adopt a purely linguistic approach to chemistry.

This prudence may be explained by the fact that it was not until the introduction of the notion of the atom, enriched by the quantitative background of the theories of Dalton, Berzelius, and Avogadro, that the use of symbols could begin to be based on the mathematical principles outlined by Lavoisier.

---

[52] *Ibid.*, 312, English translation (1788), 236

Select Bibliography

The bibliography does not list all the works cited in the footnotes. Articles and books of minor significance have been omitted. On the other hand a number of sources not cited in the footnotes are listed because they are relevant to the central theme discussed in the text.

MANUSCRIPTS

Ithaca, New York: Cornell University, Olin Library-History of Science Collection.

Mss. Bd. Lavoisier QD L41 G930 (1–11); H. Geurlac and C. Perrin, notebooks, 11 vols. These notebooks contain extremely valuable information because they record transcriptions of Lavoisier's letters and manuscripts that have not been published. They also contain useful information on manuscript sources of eighteenth-century chemistry that are kept in the most important European libraries.

Paris: Arch. Acad.

Dossier Lavoisier 102; Cours de chymie, contenant la manière de faire les opérations les plus curieuses et les plus nécessaires pour faciliter l'intelligence de cette science. . . .
Undated manuscript not in Lavoisier's hand.

Paris: Arch. Acad.

Dossier Lavoisier 152; translation by Madame Lavoisier of the notes made by the Italian chemist Vincenzo Dandolo to the preliminary discourse of Lavoisier's *Traité élémentaire de chimie* (1789).

Paris: Arch. Acad.
Dossier Lavoisier 154; A.L. Lavoisier, Tableaux de chimie (1788–1789).

Paris: Arch. Acad.
Dossier Lavoisier 155; A.L. Lavoisier, et al., Nomenclature méthodique française et latine.

Paris: Arch. Acad.
Dossier Lavoisier 380; A.L. Lavoisier, Commencement d'un Traité de chimie. (c. 1764–1766).

Paris: Arch. Acad.
Dossier Lavoisier 608; A.L. Lavoisier, Journal d'observations minéralogiques faites dans le Brie en Octobre et Novembre 1766.

Paris: BI.
Ms.1530; G.E. Sthahl [sic], Cogitationes et utiles reflexiones super contentione de sic dicto Sulphure, et quidem tam de Sulphure communi combustibili aut volatili quam de incombustibili aut fixo. Hallae 1718. With marginalia by Lavoisier and Hellot. A critical edition of the manuscript is in preparation.

Paris: BIP.
Ms. 15; Programme des leçons de Guyton de Morveau ecrit par lui meme. Bound with: Manuscript des leçons faites par Guyton de Morveau en 1805 et 1808.

Paris: BIP.
Ms. 17; [G.F. Rouelle], Introduction to Chymistry (c. 1760).

Paris: BIP.
Reg. 22-38; Régistre[s] des Merchands Apothicaires de la ville de Paris servant aux immatricules, examens, et chef d'ouvres des aspirans a la marchandise et maitrise d'Apothiquaires (1712–1776).

Paris: Bibliothèque Mazarine.
Ms. 4441; Régistre des Procès-verbaux des séances de la Société Linnéenne de Paris (1787–1827).

Paris: BMHN.
Ms.1202; Cours de Chymie Recueilli des leçons de M. Rouelle Apothicaire, Demonstrateur en Chymie au Jardin du Roy, de l'Academie Royale des Sciences de Paris. Par A.L. de Jussieu (Paris, 1767).

Paris: BN.
Ms. fr. 9133; P.J. Macquer, Notes pour son cours de chimie (1771–1779).

Uppsala: UUB.
Bergmans Manuscripta, vol. IV, D 1459 d; manuscript version of Bergman's histories of chemistry.

Uppsala: UUB.
Bergmans Manuscripta, vol. VII, D 1459 g; Bergman, Observations esperses sur les sels (c. 1769).

Uppsala: UUB.
Bergmans Manuscripta, vol. XIII, D 1459 n; Swedish manuscript version of Bergman's Latin nomenclature published in 1784.

Uppsala: UUB.
Wallers Samling, 653 D:3; [Joseph Black], Lectures upon chemistry by Dr Black in 3 volumes (Edinburgh, 1768–1769).

PRIMARY SOURCES

Adanson, Michel. *Famille des Plantes.* 2 vols. (Paris, 1763).

Agricola, Georgius. [*Opera*] (Basel, 1546).

———. *De re metallica libri XII* (Basel, 1556).

———. *Vom Bergkwerck xij Bücher* (Basel, 1557).

———. *Del'arte dei metalli partita in XII Libris* (Basel, 1563).

———. *De re metallica.* Trans. H.C. and L.H. Hoover (1912), reprint (New York, 1950).

———. *De Natura Fossilium (Textbook of Mineralogy).* Trans. M.C. and J.A. Bandy (New York, 1955).

———. *Ausgewählte Werke,* 12 vols. (Berlin, 1955–1974).

———. *Bermannus.* Trans. R. Halleux and A. Yans (Paris, 1990).

Aikin, A. and C.R. *A Dictionary of Chemistry and Mineralogy.* 2 vols. (London, 1807).

d'Alembert, Jean Le Rond. *Discours préliminaire de l'Encyclopédie,* E. Picavet (ed.). Second Ed. (Paris, 1984).

———. *Essai sur les éléments de philosophie* (1759) (Paris, 1986).

———. *Preliminary Discourse to the Encyclopedia of Diderot.* Trans. R.N. Schwab (New York, 1963).

Aréjula, Juan Manuel. "Réflexions sur la nouvelle nomenclature." *Observations sur la physique,* 33 (1788), 262–286.

———. *Reflexîones sobre la nueva nomenclatura quìmica* (1788). Reprint edited by Ramon Cago and Juan L. Carrillo (Malaga, 1979).

Babington, William. *Syllabus of a Course of Chemical Lectures Read at Guy's Hospital* (London, 1797).

Bailly, Jean-Sylvain. *Histoire de l'astronomie ancienne depuis son origine jusqu'à l'établissement de l'école d'Alexandrie* (Paris, 1775).

———. *Histoire de l'astronomie moderne depuis la fondation de l'école d'Alexandrie jusqu'à l'époque de MDCCXXX.* 3 vols. (Paris, 1779–1783).

Barlet, Annibal. *Le vray et methodique cours de la physique resolutive* (Paris, 1657).

Baumé, Antoine. *Manuel de Chymie* (Paris, 1766).

———. *Chymie Expérimentale et Raisonnée.* 3 vols. (Paris, 1773).

———. *Opuscules Chimiques* (Paris, 1798).
Becher, Johann Joachim. *Parnassus illustratus medicinalis* (Ulm, 1663).
———. *Tripus Hermeticus Fatidicus, Pandens Oracula Chymica* (Frankfurt, 1689).
———. *Physica Subterranea* (Leipzig, 1703).
Bergman, Torbern. *Dissertatio gradualis de primordiis chemiae* (Uppsala, 1779).
———. *Opuscula physica et chemica.* 6 vols. (Stockholm, Uppsala, and Leipzig, 1779–1790).
———. *Opuscules chymiques et physiques.* 2 vols. (Dijon, 1780–1785).
———. *Dissertatio gradualis sistens chemiae progressus a medio saec. VII ad medium saec. XVII* (Uppsala, 1782).
———. *An Essay on the Usefulness of Chemistry, and Its Application to the Various Occasions of Life* (London, 1783).
———. *Outlines of Mineralogy* (Birmingham, 1783).
———. "Meditationes de systemate fossilium naturali." *Nova acta Regiae Societatis Scientiarum Upsaliensis,* 4 (1784), 63–128.
———. *Physical and Chemical Essays.* 3 vols. (London and Edinburgh, 1784–1790).
———. *A Dissertation on Elective Attractions* (London, 1785).
———. *Torbern Bergman's Foreign Correspondence.* Göte Carlid and Johan Nordström (eds.) (Stockholm, 1965).
Bermännisches Wörterbuch (Chemnitz, 1778).
Berthollet, Claude Louis. "Observations sur quelques combinaisons de l'acide muriatique oxigéné." *Observations sur la physique,* 33 (1788), 217–224.
———. *Essai de statique chimique.* 2 vols. (Paris, 1803).
———. *An Essay on Chemical Statics.* 2 vols. (London, 1804).
Berzelius, Jöns Jacob. *Traité de Chimie.* 8 vols. (Paris, 1829–1833).
Bibliotheca Metallica (Leipzig, 1728).
Biringuccio, Vannoccio. *De la Pirotechnia* (Venice, 1540).
———. *The Pirotechnia of Vanoccio Biringuccio.* Trans. C.S. Smith and M.T. Gnudi (New York, 1942).
Black, Joseph. *Experiments upon Magnesia Alba, Quick-lime, and other Alcaline Substances* (1756). Reprint (Edinburgh, 1910).
———. *Lectures on the Elements of Chemistry.* 2 vols. (Edinburgh, 1803).
Boerhaave, Herman. *A Method of Studying Physick* (London, 1719).
———. *Institutiones et Experimenta Chemiae.* 2 vols. (Paris, 1724).
———. *A New Method of Chemistry; including the Theory and Practice of that Art* (London, 1727).
———. *Elementa Chemiae.* 2 vols. (Leyden, 1732).
———. *Elements of Chemistry.* 2 vols. (London, 1735).
———. *Élémens de Chymie.* 6 vols. (Paris, 1754).
Borch, Oluf. *De ortu et progressu chemiae dissertatio* (Copenhagen, 1668).
———. *Hermetis Aegyptiorum et chemicorum sapientia ab Hermanni Conringii animadversionibus vindicata* (Copenhagen, 1674).
———. *Conspectus scriptorum illustrorum* (Copenhagen, 1697).
Bostock, John. *Remarks on the Reform of the Pharmaceutical Nomenclature* (Liverpool, 1807).

Bourguet, Dav. Ludwig. *Chemisches Handwöterbuch nach den neuesten Entdeckungen entworfen.* 6 vols. (Berlin, 1798–1805).

Boyle, Robert. *The Works.* 6 vols. (London, 1772).

Brisson, Mathurin Jacques. *Élémens ou Principes Physico-Chymiques* (Paris, 1800).

———. *The Physical Principles of Chemistry* (London, 1801).

Brugnatelli, Luigi Valentino. "Prospetto di riforma alla nuova nomenclatura chimica proposta dai Sigg. Morveau, Lavoisier, Berthollet, e Fourcroy." *Annali di chimica,* 8 (1795), 149–173.

———. *Elementi di Chimica Appoggiati alle più recenti scoperte Chimiche e Farmaceutiche.* 3 vols. (Pavia, 1795–1798).

———. *Tavola della nomenclatura moderna di chimica* (Pavia, 1800).

———. *Synonimie des nomenclatures chimiques modernes* (Brussels, 1802).

———. *Diario del viaggio compiuto in Svizzera e in Francia con Alessandro Volta nel 1801* (Pavia, 1953).

Buffon, Georges-Louis Leclerc, Comte de. *Histoire naturelle des minéraux.* 5 vols. (Paris, 1783–1788).

———. *Oeuvres philosophiques* (Paris, 1954).

Cabanis, Pierre. *Du degré de certitude de la médicine* (Paris, 1798).

———. *Coup d'oeil sur les révolutions et sur la réforme de la médicine* (Paris, 1804).

Carrillo, Juan L., and Ramon Cago (eds.). *Memoria sobre una nueva y metodica clasificacion de los fluidos elasticos permanentes y gaseosos de Juan Manuel de Arejula (1755-1830)* (Malaga, 1980).

Cavendish, Henry. *The Scientific Papers.* 2 vols. (Cambridge, 1921).

Caventou, Jean Bienaimé. *Nouvelle Nomenclature Chimique* (Paris, 1816).

Chaptal, Jean Antoine Claude. *Élémens de Chimie.* 3 vols. (Montpellier, 1790).

———. *Elements of Chemistry.* 3 vols. (London, 1801).

Chenevix, Richard. *Remarks Upon Chemical Nomenclature* (London, 1802).

Chénier, Marie-Joseph. *Litterature Française. Rapports à l'Empereur sur les progrès des sciences, des lettres et des arts depuis 1789. Vol. III.* Reprint (Paris, 1989).

Coelho de Seabra, Vincente. *Elementos de Chimica.* 2 vols. (Coimbra, 1780–1788).

Compagnoni, Giovanni. *La Chimica per le Donne.* 2 vols. (Venice, 1796).

Condillac, Etienne Bonnot de. *Essay on the Origin of Human Knowledge.* Trans. T. Nugent (London, 1756).

———. *Cours d'Etude pour l'instruction du Prince de Parme.* 16 vols. (Parma, 1775).

———. *Oeuvres Philosophiques.* 3 vols. (Paris, 1947–1951).

———. *La Logique. Logic.* Trans. W.R. Albury (New York, 1980).

Condorcet, Antoine-Nicolas Caritat, Marquis de. *Esquisse d'un tableau historique des progrès de l'esprit humain* (Paris, 1795).

———. *Correspondance inédite de Condorcet et de Turgot (1770–1779)* (Paris, 1883).

———. *Sketch of a Historical Picture of the Progress of the Human Mind* (London, 1955).

Conring, Hermann. *De Hermetica medicina* (Helmstadt, 1669).

Critical Examination of the First Part of Lavoisier's Elements of Chemistry (London, 1797).

Cronstedt, Axel Fredrik. *An Essay towards a System of Mineralogy* (London, 1772).

Cuvier, Georges. *Chimie et sciences naturelles: Rapports à l'Empereur sur le progrès des sciences, des lettres et des arts depuis 1789. Vol. II.* Reprint (Paris, 1989).
———. *Histoire des sciences naturelles.* 5 vols. (Paris, 1841).
Dandolo, Vincenzo. *Fondamenti della scienza chimico-fisica* (Venice, 1795).
Davy, Sir Humphry. *Collected Works.* 9 vols. (London, 1839–1840).
Delambre, Jean-Baptiste. *Sciences mathématiques: Rapports à l'Empereur sur les progrès des sciences, des lettres et des arts depuis 1789. Vol. I.* Reprint (Paris, 1989).
Descartes, René. *Oeuvres.* 12 vols. (Paris, 1897–1910).
———. *Philosophical Writings.* Trans. N. Kemp Smith (London, 1952).
Dickson, Stephen. *An Essay on Chemical Nomenclature* (Dublin, 1796).
Diderot, Denis. *Oeuvres complètes.* 19 vols. (Paris, 1976–).
Dumas, Claude-Louis. *Système Méthodique de nomenclature et de classification des muscles du corps humain* (Montpellier, 1797).
———. *Principes de physiologie ou introduction à la science expérimentale, philosophique et médicale de l'homme vivant.* 4 vols. (Paris, 1800).
Eimbke, Georg. *Versuch einer systematischen Nomenklatur für die phlogistische und antiphlogistische Chemie* (Halle, 1793).
Ekeberg, Anders Gustaf. *Försök till Svensk Nomenklatur för Chemien lämpad efter de sednaste uptäckterne* (Stockholm and Uppsala, 1795).
Encyclopaedia Britannica. Third Ed. 18 vols. (Edinburgh, 1790).
Encyclopédie, ou Dictionnaire raisonnée des sciences, des arts et métiers. 17 vols. (Paris, 1751–1765).
———. *Supplément à l'Encyclopédie.* 4 vols. (Amsterdam, 1776–1777).
Encyclopédie Méthodique: Botanique. 13 vols. (Paris, 1783–1817).
Encyclopédie Méthodique: Chymie, Pharmacie et Métallurgie. 6 vols. (Paris, 1786–1815).
Encyclopédie Méthodique. Philosophie. 3 vols. (Paris, 1791–1792).
Ercker, Lazarus. *Beschreibung Aller fürnemisten Mineralischen Ertzt unnd Bergkwercksarten.* Second Ed. (Frankfurt, 1580).
Fabricius, Georgius. *De metallicis rebus ac nominibus* (Zurich, 1565).
Fontana, Felice. *Opuscules Physiques et Chymiques* (Paris, 1784).
Fougeron, J.-B. *Nouvelle Synonymie Chimique* (Paris, 1820).
Fourcroy, Antoine-François. *Leçons Élémentaire d'Histoire Naturelle et de Chimie* Third Ed. 5 vols. (Paris, 1789).
———. *Philosophie chimique* (Paris, 1792).
———. *Philosophia chemica* (Stockholm, 1795).
———. *Notice sur la vie et les travaux de Lavoisier* (Paris, 1796).
———. *A General System of Chemical Knowledge.* 11 vols. (London, 1804).
Gadolin, Johan. *Animadversiones in novam nomenclaturae chemicae* (Turku, 1788).
———. *Wissenschaftliche Abhandlungen* (Helsinki, 1910).
Garat, Dominique-Joseph. *Mémoires historiques sur la vie de M. Suard.* 2 vols. (Paris, 1820).
Gérando, J.M. de. *Des signes et de l'art de penser.* 4 vols. (Paris, 1800).
Girtanner, Christoph. *Neue chemische Nomenklatur für die Deutsche Sprache* (Berlin, 1791).

———. *Anfangsgründe der antiphlogistischen Chemie* (Berlin, 1795).

Glaser, Christophe. *Traite de la Chymie.* Second Ed. (Paris, 1676).

Gmelin, Johann Friedrich. *Geschichte der Chemie.* 3 vols. (Göttingen, 1797–1799).

Gobet, Nicolas. *Les anciens minéralogistes du royaume de France.* 2 vols. (Paris, 1779).

Gren, Friedrich Albrecht Carl. *Systematisches Handbuch der gesammten Chemie.* 4 vols. (Halle, 1794–1796).

———. *Grundriss der Chemie.* Second Ed. 2 vols. (Halle, 1800).

———. *Principles of Modern Chemistry.* 2 vols. (London, 1800).

Guyton de Morveau, Louis Bernard. *Digressions académiques* (Dijon, 1772).

———. *Élémens de Chymie, théorique et pratique* (in collaboration with Maret and Durande). 3 vols. (Dijon, 1777–1778).

———. "Mémoire sur les dénominations chimiques." *Observations sur la physique,* 19 (1782), 370–383.

Haüy, René Just. *Essai d'une théorie de la structure des cristaux* (Paris, 1784).

Helmont, Jean Baptiste Van. *Les Oeuvres* (Lyon, 1670).

Hermbstädt, Sigismund Friedrich. *Systematischer Grundriss der allgemeinen Experimentalchemie.* 4 vols. (Berlin, 1791–1793).

The Hermetic Museum. Reprint (Yorkbeach, 1991).

Higgins, William. *A Comparative View of the Phlogistic and Antiphlogistic Theories* (London, 1789).

Hoefer, Ferdinand. *Nomenclature et Classifications Chimiques* (Paris, 1845).

Hulsius, Levinus. *Dictionnaire François-Allemand & Allemand-François.* Third Ed. (Frankfurt, 1607).

Jungken, Johan Hefric. *Lexicon chimico-pharmaceuticum* (Nuremberg, 1699).

Keir, James. *The First Part of a Dictionary of Chemistry* (London, 1789).

Kirwan, Richard. *Elements of Mineralogy.* Second Ed. 2 vols. (London, 1784).

———. *An Essay on Phlogiston and the Constitution of Acids* (London, 1787).

———. *Essai sur le Phlogistique* (Paris, 1788).

Klaproth, Martin Heinrich. *Chemisches Wörterbuch.* 5 vols. (Berlin, 1807–1810).

Lamarck, Jean Baptiste. *Flore Française.* 3 vols. (Paris, 1779).

———. *Recherches sur les causes des principaux faits physiques* (Paris, 1794).

———. *Réfutation de la théorie pneumatique* (Paris, 1796).

Landriani, Marsilio. *Relazioni di Marsilio Landriani sui progressi delle manifatture in Europa alla fine del Settecento* (Milan, 1981).

La Planche, Laurent Charles de. *Plan d'un cours d'Operations de chimie & pharmacie* (Paris, 1750).

———. *Plan d'un cours de chymie* (Paris, 1764).

Lavoisier, Antoine Laurent. *Opuscules Physiques et Chymiques* (Paris, 1774).

———. *Essays Physical and Chemical.* Trans. T. Henry (London, 1776).

———. *Méthode de nomenclature chimique* (Paris, 1787).

———. *Method of Chymical Nomenclature.* Trans. J. St. John (London, 1788).

———. *Metodo de la nueva Nomenclatura química.* Trans. G. Bueno (Madrid, 1788).

———. *Metodo di Nomenclatura Chimica.* Trans. P. Calloud (Venice, 1790).

———. *Dizionarii Vecchio e Nuovo, Nuovo e Vecchio di Nomenclatura Chimica.* Trans. V. Dandolo (Venice, 1791).

———. *Methode der chemischen Nomenklatur für das antiphlogistische System.* Trans. K.F. Meidinger (Vienna, 1793).

———. *Traité élémentaire de chimie.* 2 vols. (Paris, 1789).

———. *Elements of Chemistry.* Trans. R. Kerr (Edinburgh, 1790).

———. *System der Antiphlogistischen Chemie.* 2 vols. (Berlin and Stettin, 1792).

———. *Oeuvres.* 6 vols. (Paris, 1862–1893).

———. *Correspondance.* 4 vols. (Paris, 1955–1986).

Lemery, Nicolas. *Cours de Chymie.* Eleventh Ed. (Paris, 1756).

———. *Pharmacopée universelle.* Third Ed. (Paris, 1734).

Lichtenberg, Georg Christoph. *Vermischte Schriften.* 8 vols. (Göttingen, 1846–1847).

Linnaeus, Carl. *Fundamenta Botanica* (1737) (Stockholm, 1747).

———. *Critica Botanica* (Stockholm, 1737).

———. *Species plantarum.* 2 vols. (Stockholm, 1753).

———. *Vita Caroli Linnaei* (Stockholm, 1957).

Löhneyss, Georg Engelhardt. *Bericht vom Bergkwerck* (Zellerfeldt, 1617).

Locke, John. *The Works.* Sixth Ed. 3 vols. (London, 1759).

Macquer, Pierre-Jospeh. *Élémens de Chymie-Théorique* (Paris, 1749).

———. *Élémens de Chymie-Pratique.* (Paris, 1751).

———. *Elements of the Theory and Practice of Chymistry.* 2 vols. (London, 1758).

———. *Dictionnaire de chymie.* 2 vols. (Paris, 1766).

———. *A Dictionary of Chemistry.* 2 vols. (London, 1771).

Maier, Michael. *Atalanta fugiens* (Oppenheim, 1618).

Malouin, Paul-Jacques. *Traité de chimie* (Paris, 1734).

Manget, Jean-Jacques. *Bibliotheca Chemica Curiosa.* 2 vols. (Geneva, 1702).

Marum, Martinus van. *Martinus Van Marum: Life and Work.* 6 vols. (Haarlem, 1969–1976).

Mercati, Michele. *Metallotheca* (Rome, 1717–1719).

Mitchill, Samuel Latham. *Explanation of the Synopsis of Chemical Nomenclature* (New York, 1801).

Monnet, Antoine Grimoald. *Demonstration de la fausseté des principes des nouveaux chymistes* (Paris, 1798).

Moscati, Pietro. *Discorso accademico dei vantaggi dell'educazione filosofica nello studio della chimica* (Milan, 1784).

Nazari, G.B. *Della trasmutatione metallica, sogni tre* (Brescia, 1599).

Newton, Isaac. *Opticks* (1717). Reprint (New York, 1952).

Nicholson, William. *A Dictionary of Practical and Theoretical Chemistry.* Second Ed. (London, 1808).

Oersted, J.C. *Tentamen nomenclaturae chemicae omnibus linguis Scandinavico-Germanicis communis* (Copenhagen, 1814).

Paracelsus. *Selected Writings* (Princeton, 1988).

Pearson, George. *A Translation of the Table of Chemical Nomenclature.* Second Ed. (London, 1799).

Peart, Edward. *The Anti-phlogistic Doctrine of M. Lavoisier Critically Examined* (London, 1795).
Pinel, Philippe. *Nosographie philosophique ou la méthode de l'analyse appliquée à la médicine.* 2 vols. (Paris, 1796).
———. *La médicine clinique rendue plus précise et plus exacte par l'application de l'analyse.* Second Ed. (Paris, 1804).
Pini, Ermenegildo. *Sulla metachimica ossia sulla nuova teoria e nomenclatura chimica* (Milan, 1793).
Priestley, Joseph. *The History and Present State of Electricity.* Third Ed. (London, 1775).
———. *The History and Present State of Discoveries Relating to Vision, Light, and Colours* (London, 1772).
———. *Experiments and Observations on Different Kinds of Air.* 3 vols. (London, 1775–1777).
———. *Experiments and Observations Relating to Various Branches of Natural Philosophy* (London, 1779).
———. *Considerations on the Doctrine of Phlogiston and the Decomposition of Water* (Philadelphia, 1796).
———. *Scientific Correspondence of Joseph Priestley.* H.C. Bolton (ed.) (New York, 1892).
———. *Scientific Autobiography of Joseph Priestley (1733–1804).* Edited with commentary by R.E. Schofield (Cambridge, Mass., 1966).
Prieur, C.A. *Note instructive sur les poids et mesures* (Paris, 1794).
Pye, C. *The New Chemical Nomenclature* (London, 1802).
Remler, J.C. *Neues chemisches Wörterbuch oder Handlexicon* (Erfurt, 1793).
Rinman, Sven. *Bergwerks-Lexicon.* 2 vols. (Stockholm, 1788–1789).
Romé de L'Isle, J.B. *Essai de cristallographie* (Paris, 1772).
———. *Description méthodique d'une collection des minéraux* (Paris, 1773).
Rössler, Balthasar. *Speculum Metallurgiae Politissimum* (Dresden, 1700).
Ruland, Martin. *Lexicon Alchemiae* (Frankfurt, 1612).
Sage, Balthasar-Georges. *Exposé des effets de la contagion nomenclative* (Paris, 1810).
Shaw, Peter. *Chemical Lectures* (London, 1734).
Scheffer, Henrik Theophilus. *Chemiske Föreläsningar, Rörande Salter, Jordarter, Vatten* (Uppsala, 1775).
Scheele, Carl Wilhelm. *The Collected Papers of Carl Wilhelm Scheele* (London, 1931).
———. *The Brown Book,* 2 vols. (Stockholm, 1968).
Scherer, Alexander Nicolaus. *Nachträge zu Grundzügen der neuern chemischen Theorie* (Jena, 1796).
———. *Grundriss der Chemie* (Tübingen, 1800).
Scherer, J.A. *Versuch einer neuen Nomenclatur für Deutsche Chymisten* (Vienna, 1792).
Schlüter, Christoph Andreas. *Gründlicher Unterricht von Hütten-Werken* (Brunswick, 1738).
Senac, Jean-Baptiste. *Nouveau cours de chymie suivant les principes de Newton et de Sthall.* Second Ed. (Paris, 1737).

Smith, Thomas P. *A Sketch of the Revolutions in Chemistry* (Philadelphia, 1798).
Sommerhoff, J.C. *Lexicon Pharmaceutico-chymicum* (Nuremberg, 1701).
Spalding, Lyman. *A New Nomenclature of Chemistry* (London, 1794).
Spallanzani, Lazzaro. *Chimico esame degli esperimenti del Professor Göttling* (Modena, 1796).
———. *Epistolario.* 5 vols. (Florence, 1958–1964).
Stahl, Georg Ernst. *Fundamenta chymiae dogmatico-rationalis et experimentalis* (Nuremberg, 1718).
———. *Philosophical Principles of Universal Chemistry* (London, 1730).
———. *Traité du Soufre* (Paris, 1766).
———. *Traité des sels* (Paris, 1771).
Stromeyer, Friedrich. *Tabellarische Uebersicht der chemisch einfachen und zusammengesetzen Stoffe* (Göttingen, 1806).
Turgot, Anne-Robert-Jacques. *Turgot on Progress, Sociology and Economics.* Trans. Ronald Meek (Cambridge, 1973).
Tychsen, N. *Fransk Chemisk Nomenklatur* (Copenhagen, 1794).
———. *Chemisk Haandbog.* 3 vols. (Copenhagen, 1794).
Vicq d'Azyr, Félix. *Oeuvres.* 6 vols. (Paris, 1805).
Volta, Alessandro. *Epistolario.* 5 vols. (Bologna, 1949–1955).
Wall, Martin. *Dissertations on Select Subjects in Chemistry and Medicine* (Oxford, 1783).
Wallerius, Johann Gottschlak. *Minéralogie, ou description générale du regne minéral.* 2 vols. (Paris, 1753).
Waltersdorff, Johan Lucas. *Systema Minerale* (Berlin, 1748).
Watson, Richard. *Chemical Essays.* 5 vols. (Cambridge and London, 1781–1787).
Werner, Abraham Gottlob. *On the External Characters of Minerals.* Trans A.V. Carozzi (Urbana, Ill., 1962).
Westrumb, Johann Friedrich. *Keline physikalisch-chemische Abhandlungen.* 4 vols. (Leipzig and Hannover, 1786–1795).
White, Robert. *An Analysis of the New London Pharmacopoeia* (New Market, 1792).
Wiegleb, Johann Christian. *Geschichte des Wachsthums und der Erfindungen in der Chemie in der neueren Zeit.* 2 vols. (Berlin and Stettin, 1790–1791).
Zeisig, Johann Caspar. *Neues und wohlein-gerichtetes Mineral- und Bergwercks-Lexicon.* Second Ed. (Chemnitz, 1743).

SECONDARY SOURCES

Aarsleff, Hans. *From Locke to Saussure: Essays on the Study of Language and Intellectual History* (London, 1982).
Abbri, Ferdinando. *La chimica del Settecento* (Turin, 1978).
———. *Elementi, principi e particelle: Le teorie chimiche da Paracelso a Stahl* (Turin, 1980).
———. "Spallanzani e la diffusione delle teorie chimiche di Lavoisier in Italia." In G. Montalenti and P. Rossi (eds.). *Lazzaro Spallanzani e la biologia del Settecento* (Florence, 1983), 121–136.

———. "Lavoisier e Dandolo: Le edizioni italiane del *Traité élémentaire de chimie.*" In *Annali dell'Istituto di Filosofia* (Università di Firenze), VI (1984), 163–182.

———. *Le terre, l'acqua, le arie: La rivoluzione chimica del Settecento* (Bologna, 1984).

———. "The Chemical Revolution: A Critical Assessment." *Nuncius,* IV:2 (1989), 303–315.

———. "Immagini, teorie e strumenti: Lavoisier e la rivoluzione chimica." In S. Poggi and M. Mugnai (eds.). *Tradizioni filosofiche e mutamenti scientifici* (Bologna, 1990), 69–89.

———. "Tradizioni chimiche e meccanismi di difesa: G.A. Scopoli e la 'Chimie Nouvelle'." *Archivio di storia della Cultura,* IV (1991), 75–92.

———. *Science de l'air: Studi su Felice Fontana* (Cosenza, 1991).

Acton, H.B. "The Philosophy of Language in Revolutionary France." *Proceedings of the British Academy* (1959), 199–219.

Ahlers, Willem C. *Un chimiste du XVIIIe siècle: Pierre-Jospeh Macquer (1718–1784).* Thèse présentée en vue de l'obtention du Doctorat de Troisième Cycle. École Pratique des Hautes Études VIe Section (Paris, 1969).

Albury, William Randall. *The Logic of Condillac and the Structure of French Chemical and Biological Theory (1780–1800).* Ph. D. thesis, The Johns Hopkins University (Baltimore, Md., 1972).

———. The Order of Ideas: Condillac's Method of Analysis as a Political Instrument in the French Revolution." In J.A. Schuster and R.R. Yeo (eds.). *The Politics and Rhetoric of Scientific Method* (Dordrecht and Boston, 1986), 203–225.

Altieri Biagi, Maria Luisa. "Lingua della scienza tra Seicento e Settecento." *Lettere Italiane,* XXVIII (1976), 410–461.

Anderson, R. and C. Lawrence (eds.). *Science, Medicine, and Dissent: Joseph Priestley* (London, 1988).

Anderson, W.G. *Between the Library and the Laboratory* (Baltimore and London, 1984).

Baker, Keith Michael. *Condorcet: From Natural Philosophy to Social Mathematics* (Chicago, 1975).

Barsanti, Giulio. "Linné et Buffon: deux visions différentes de l'histoire naturelle." *Revue de Synthèse,* 113–114 (1984), 83–107.

———. *La scala, la mappa, l'albero: Immagini e classificazioni della natura fra Sei e Ottocento* (Florence, 1992).

Bensaude-Vincent, Bernadette. *A propos de Méthode de nomenclature chimique* (Paris, 1983).

———. "A View of the Chemical Revolution through Contemporary Textbooks: Lavoisier, Fourcroy and Chaptal." *BJHS,* 23 (1990), 435–460.

Beretta, Marco. "Luigi Valentino Brugnatelli e la chimica in Italia alla fine del Settecento." *Storia in Lombardia,* 2 (1988), 3–31.

———. "T.O. Bergman and the Definition of Chemistry." *Lychnos* (1988), 37–67.

———. "A.L. Lavoisier en Italie (1774–1800)." In *Échanges d'influences scientifiques et techniques entre pays européens de 1780 à 1830* (Paris, 1990), 125–144.

———. "The *Société linnéenne de Paris* (1787–1827)." *Svenska Linnésällskapets Årsskrift* (1990–1991), 151–175.

———. "The Historiography of Chemistry in the Eighteenth Century: A Preliminary Survey and Bibliography." *Ambix*, 39 (1992), 1–10.

———. "Torbern Bergman in France: An Unpublished Letter by Lavoisier to Guyton de Morveau." *Lychnos* (1992), 167–170.

———. "The Library of Lavoisier." In E. Canone (ed.). *Filosofi, suenziati e i libri: Da Cusano a despardi* Forthcoming.

———. "Chemists in the Storm: Lavoisier and Priestley and the French Revolution." In *Nuncius* (1993), forthcoming.

———. *A New Course in Chemistry: Lavoisier's First Chemical Paper.* Forthcoming.

Berthelot, Marcelin. *Les origines de l'alchimie* (Paris, 1885).

———. *Introduction à l'étude de la chimie des anciens et du Moyen Age* (Paris, 1889).

———. *La révolution chimique: Lavoisier* (Paris, 1890).

Bouchard, Georges. *Guyton Morveau, chimiste et conventionnel* (Paris, 1938).

Brunschvicg, Léon. *De la connaissance de soi* (Paris, 1931).

Cassirer, Ernst. *The Problem of Knowledge: Philosophy, Science, and History since Hegel* (New Haven, 1950).

———. *The Philosophy of the Enlightenment* (Princeton, 1953).

———. *Philosophie der symbolischen Formen.* Third Ed., vol. 2 (Oxford, 1958).

Caven, R.M. and J.A. Cranston. *Symbols and Formulae in Chemistry* (London, 1928).

Clericuzio, Antonio. "A Redefinition of Boyle's Chemistry and Corpuscolar Philosophy." *Annals of Science*, 47 (1990), 561–589.

Cohen, I. Bernard. *Revolution in Science* (Cambridge, Mass., 1985).

Corsi, Pietro. *The Age of Lamarck: Evolutionary Theories in France 1790–1830* (Berkeley, 1989).

Crosland, Maurice P. *Historical Studies in the Language of Chemistry.* Second Ed. (New York, 1978).

———. "The Development of Chemistry in the Eighteenth Century." *Studies on Voltaire and the Eighteenth Century*, 24 (1963), 369–441.

———. *Les héritiers de Lavoisier* (Paris, 1968).

———. "Nature and Measurement in Eighteenth-Century France." *Studies on Voltaire and the Eighteenth Century*, 87 (1972), 277–309.

———. "Chemistry and the Chemical Revolution." In G.S. Rousseau and R. Porter (eds.). *The Ferment of Knowledge* (Cambridge, 1980), 389–416.

Dagognet, François. *Tableaux et langages de la chimie* (Paris, 1969).

———. *Le catalogue de la vie* (Paris, 1970).

Dal Pra, Mario. *Condillac* (Milan, 1942).

Daudin, Henri. *De Linné à Jussieu—Méthodes de la classification et idée de série en botanique et en zoologie. (1740–1790)* (Paris, 1926).

Daumas, Maurice. *Lavoisier théoricien et expérimentateur* (Paris, 1955).

———. *Scientific Instruments of the 17th & 18th Centuries and Their Makers* (London, 1989).

Debus, Allen G. *The Chemical Philosophy.* 2 vols. (New York, 1977).

———. *The French Paracelsians* (Cambridge, 1991).

Delhez, R. "Révolution chimique et révolution française: le discours préliminaire au Traité élémentaire de chimie de Lavoisier." *Revue des questions scientifiques*, 143 (1972), 3–26.

Dhombres, Nicole and Jean Dhombres. *Naissance d'un nouveau pouvoir: sciences et savants en France 1793–1824* (Paris, 1989).

Donovan, Arthur. *Philosophical Chemistry in the Scottish Enlightenment* (Edinburgh, 1975).

——— (ed.). *The Chemical Revolution: Essays in Reinterpretation. Osiris.* Second series, vol. 4 (Philadelphia, 1988).

Duhem, Pierre. *Le miste et la combinaison chimique* (Paris, 1902).

———. *La chimie est-elle une science française?* (Paris, 1916).

Duveen, Denis I. and H. Klickstein. "The Introduction of Lavoisier's Chemical Nomenclature in America." *Isis* (1954), 278–292 and 368–382.

———. *A Bibliography of the Works of Antoine Laurent Lavoisier (1743–1794)* (London, 1954).

Duveen, Denis I. *Bibliotheca Alchemica et Chemica* (London, 1949).

———. *Supplement to a Bibliography of the Works of Antoine Laurent Lavoisier (1743–1794)* (London, 1965).

Ehrard, Jean. *L'idée de Nature en France dans la première moitié du XVIIIe siècle.* 2 vols. (Paris, 1963).

Eklund, Jon. *The Incompleat Chymist* (Washington, 1975).

Engelhardt, Dietrich von. *Historisches Bewusstsein in der Naturwissenschaft von der Aufklärung bis zum Positivismus* (Freiburg and Munich, 1979).

Fayet, J. *La Révolution française et la science: 1789–1795* (Paris, 1960).

Formigari, Lia. *Linguistica ed empirismo nel Seicento inglese* (Bari, 1970).

Frängsmyr, Tore (ed.). *Linnaeus: The Man and His Work* (Berkeley, 1983).

———, *et al.* (eds.). *The Quantifying Spirit in the 18th Century* (Berkeley, 1990).

Freudenthal, Gad (ed.). *Études sur Hélène Metzger* (Leiden and New York, 1990).

Gille, Bertrand. *Les origines de la grande industrie métallurgique en France* (Paris, 1947).

Gillispie, Charles C. *The Edge of Objectivity* (Princeton, 1960).

———. *Science and Polity in France at the End of the Old Regime* (Princeton, 1980).

Givry, Grillot de. *Witchcraft, Magic & Alchemy* (1931). Reprint (New York, 1971).

Golinski, Jan. "Chemistry in the Scientific Revolution: Problems of Language and Communication." In David Lindberg, Robert S. Westman (eds.) *Reappraisals of the Scientific Revolution.* (Cambridge, 1990), 367–396.

Goltz, Dietlinde. *Studien zur Geschichte der Mineralnamen in Pharmazie, Chemie, und Medizin von den Anfängen bis Pracelsus* (Wiesbaden, 1972).

Gordon, Cosmo Alexander. *A Bibliography of Lucretius* (London, 1962).

Gough, J.B. "Lavoisier's Early Career in Science: an Examination of Some New Evidence." *BJHS*, 4 (1968), 52–57.

Goupil, Michelle. *Le chimiste C.L. Berthollet (1748–1822)* (Paris, 1977).

———. "Les tentatives de mathématisation de la chimie au XVIIIe siècle. Echecs et oppositions." *Sciences et techniques en perspective* (1981–1982), vol. 1, 1–20.

———. *Du flou au clair? Histoire de l'affinité chimique* (Paris, 1991).

Guareschi, Icilio. "Lavoisier, sua vita e sue opere." *Supplemento annuale all'Enciclopedia chimica*, 19 (1903), 307–460.

———. "La chimica in Italia dal 1750 al 1800." *Supplemento annuale alla Enciclopedia chimica,* 25 and 26 (1909–1910).

Grimaux, Edouard. *Lavoisier, 1743–1794* (Paris, 1888).

Guerlac, Henry. *Lavoisier—The Crucial Year* (Ithaca, N.Y., 1961).

———. *A.L. Lavoisier, Chemist and Revolutionary* (New York, 1975).

———. *Essays and Papers in the History of Modern Science* (Baltimore and London, 1977).

Hall, A.R. and M. Boas Hall. "Philosophy and Natural Philosophy." In R. Taton and F. Braudel (eds.). *Mélanges Alexandre Koyré* (Paris, 1964). Vol. 2, 241–256.

Halleux, Robert. *Le problème des métaux dans la science antique* (Paris, 1974).

———. "La controverse sur les origines de la chimie de Paracelse à Borrichius." In *Acta conventus neo-latini Touronensis* (Paris, 1980). Vol. 2, 807–819.

———. "Pratique industrielle et chimie philosophique de l'Antiquité au XVIIe siècle." *L'actualité chimique* (Janvier-Févier, 1987), 16–20.

Hannaway, Owen. *The Chemists and the Word: The Didactic Origins of Chemistry* (Baltimore and London, 1975).

Haupt, Bettina. *Deutschsprachige Chemielehrbücher (1775–1850)* (Stuttgart, 1987).

Heilbron, John L. *Electricity in the 17th and 18th Centuries* (Berkeley, 1979).

Heller, John Lewis. *Studies in Linnean Method and Nomenclature* (Frankfurt, 1983).

Holmes, Fredric Lawrence. *Lavoisier and the Chemistry of Life* (Madison, 1985).

———. *Eighteenth-Century Chemistry as an Investigative Enterprise* (Berkeley, 1989).

———. "Lavoisier the Experimentalist." *Bulletin of History of Chemistry,* 5 (1989), 24–31.

Holmyard, E.J. *Alchemy.* Reprint (New York, 1990).

Hufbauer, Karl. *The Formation of the German Chemical Community (1725–1795)* (Berkeley, 1982).

Jones, Peter (ed.). *Philosophy and Science in the Scottish Enlightenment* (Edinburgh, 1988).

Jordanova, Ludmilla (ed.). *Languages of Nature* (London, 1986).

Julia, Dominique. *Les trois couleurs du tableau noir: La Révolution* (Paris, 1981).

Jullien, Charles Edouard. *La chimie nouvelle ou le crassier de la nomenclature chimique de Lavoisier* (Paris, 1870).

Jung, Carl Gustav. *Psychology and Alchemy* (London, 1953).

———. "Paracelsus as a Spiritual Phenomenon." In *idem, Collected Works* (New York, 1967). Vol. 13, 109–189.

Kahlbaum, G.W. and A. Hoffmann. *Die Einführung der Lavoisier'schen Theorie im besonderen in Deutschland* (Leipzig, 1897).

Knight, David. "Revolutions in Science: Chemistry and the Romantic Reaction to Science." In William R. Shea (ed.). *Revolutions in Science* (Canton, Mass., 1988), 49–69.

Knight, Isabel. *The Geometric Spirit: The Abbé Condillac and the French Enlightenment* (New Haven and London, 1968).

Knowlson, James. *Universal Language Schemes in England and France, 1600–1800* (Toronto and Buffalo, 1975).

Koyré, Alexandre. *Mystiques, spirituels, alchimistes du XVIe siècle allemand* (Paris, 1971).

———. *Études d'histoire de la pensée philosophique.* Second Ed. (Paris, 1971).

———. *Études d'histoire de la pensée scientifique.* Second Ed. (Paris, 1973).

Kristeller, Paul Oskar. *Renaissance Thought and Its Sources* (New York, 1979).

Kuhn, Thomas S. "Mathematical versus Experimental Traditions in the Development of Physical Science." In *idem, The Essential Tension* (Chicago, 1977), 31–65.

Larson, James L. *Reason and Experience: The Representation of Natural Order in the Work of Carl von Linné* (Berkeley, 1971).

Lenoble, Robert. *Esquisse d'une histoire de l'idée de nature* (Paris, 1969).

Levere, Trevor H. "Lavoisier: Language, Instruments, and the Chemical Revolution." In T.H. Levere and W. Shea (eds.). *Nature, Experiment, and the Sciences* (Dordrecht and Boston, 1990), 207–223.

Lindroth, Sten. *Gruvbrytning och Kopparhantering vid Stora Kopparberget intill 1800-talets början.* 2 vols. (Uppsala, 1955).

Long, Pamela O. "The Openness of Knowledge: An Ideal and Its Context in the 16th-Century Writings on Mining and Metallurgy." *Technology and Culture,* 32 (1991), 318–355.

Lüdy-Tenger, Fritz. *Alchemistische und Chemische Zeichnen.* Reprint (Vaduz, 1981).

Lundgren, Anders. "The New Chemistry in Sweden: the Debate that Wasn't." In A. Donovan (ed.). *The Chemical Revolution: Essays in Reinterpretation, Osiris,* 4 (1988), 147–155.

Manuel, Frank E. *The Prophets of Paris* (Cambridge, Mass., 1962).

McEvoy, J.G. and J.E. McGuire. "God and Nature: Priestley's Way of Rational Dissent." *Historical Studies in Physical Sciences,* 6 (1975), 325–404.

McKie, Douglas. *Antoine Lavoisier: Scientist, Economist, Social Reformer.* Reprint (New York, 1990).

Meinel, Cristoph (ed.). *Die Alchemie* (Wiesbaden, 1986).

———. "Artibus Academicis Inserenda: Chemistry's Place in the Eighteenth-Century and Early Nineteenth-Century Universities." *History of Universities,* 7 (1988), 89–115.

Meldrum, Andrew Norman. *The Eighteenth-Century Revolution in Science—The First Phase* (Calcutta, 1930).

Melhado, Evan M. *Jacob Berzelius: The Emergence of His Chemical System* (Uppsala, 1981).

———. "Chemistry, Physics, and the Chemical Revolution." *Isis,* 76 (1985), 195–211.

Metzger, Hélène. *La genèse de la science des cristaux* (1918). Reprint (Paris, 1969).

———. "L'évolution du règne métallique d'après les alchimistes du XVIIe siècle." *Isis,* 4 (1922), 466–482.

———. *Les doctrines chimiques en France du début du XVIIe siècle à la fin du XVIIIe siècle.* Second Ed. (Paris, 1969).

———. *Newton, Stahl, Boerhaave e la doctrine chimique* (1930) (Paris, 1974).

———. *La chimie* (Paris, 1930).

———. "Introduction à l'étude du rôle de Lavoisier dans l'histoire de la chimie." *Archeion*, 14 (1932), 31–50.
———. *La philosophie de la matière chez Lavoisier* (Paris, 1935).
———. "Alchimie." *Revue de Synthèse*, 16 (1938), 43–53.
———. *La méthode philosophique dans l'histoire des sciences* (Paris, 1987).
Meyerson, Émile. *Identité et realité* (Paris, 1908).
———. *Explanation in the Sciences* (1921) (Dordrecht, 1991).
———. *Du cheminement de la pensée.* 3 vols. (Paris, 1931).
Micheli, Gianni. "Ulteriori sviluppi dell'illuminismo francese." In L. Geymonat (ed.), *Storia del pensiero filosofico e scientifico* (Milan, 1979). Vol.3, 328–332.
———. "Natura." In *Enciclopedia. Einaudi* (Turin, 1980). Vol. 9, 715–756.
Mieli, Aldo. *Lavoisier.* Second Ed. (Rome, 1926).
———. *Pagine di storia della chimica* (Rome, 1922).
———. "Le rôle de Lavoisier dans l'histoire des sciences." *Archeion*, 14 (1932), 51–56.
Moravia, Sergio. *Il pensiero degli Idéologues: Scienza e filosofia in Francia (1780–1815)* (Florence, 1974).
———. *Filosofia e scienze umane nell'età dei Lumi* (Florence, 1982).
Naville, Pierre. *D'Holbach et la philosophie scientifique du XVIIIe siècle* (Paris, 1967).
Olschki, Leo. *Geschichte der neuespralichen wissenschaftlichen Literatur.* 2 vols. (Heidelberg-Leipzig, 1919–1922).
Olsson, Hugo. *Kemiens historia i Sverige intill år 1800* (Uppsala, 1971).
Pagel, Walter. *Paracelsus: An Introduction to Philosophical Medicine in the Era of Renaissance* (Basel, 1982).
———. *Joan Baptista Van Helmont: Reformer of Science and Medicine* (Cambridge, 1982).
Partington, J.R. *A History of Chemistry.* 4 vols. (London, 1961–1970).
——— and D. McKie. *Historical Studies on the Phlogiston Theory.* Reprint (New York, 1981).
Perrin, C.E. "The Triumph of the Antiphlogistians." In H. Woolf (ed.). *The Analytic Spirit,* (Ithaca, N.Y., 1981), 40–63.
Pioger, Léger-Marie. *La clef de la chimie, ou la nomenclature chimique mise a la portée de toutes les intelligences* (Le Mans, 1853).
Ramsay, William. *Life and Letters of Joseph Black* (London, 1918).
Rappaport, Rhoda. "Rouelle and Stahl—The Phlogistic Revolution in France." *Chymia,* 7 (1961), 73–102.
Read, John. *Prelude to Chemistry.* Reprint (London, 1969).
Robinet, André. *Le langage à l'âge classique* (Paris, 1978).
Rossi, Paolo. *Clavis Universalis: Arti della memoria e logica combinatoria da Lullo a Leibniz.* Second Ed. (Bologna, 1983).
———. Lingue artificiali, classificazioni, nomenclature." In *idem, Aspetti della rivoluzione scientifica* (Naples, 1971), 295–367.
———. *Immagini della scienza* (Rome, 1977).
Rousseau, Nicolas. *Connaissance et langage chez Condillac* (Geneva, 1986).
Sabbah, Guy (ed.). *Le latin médical: La constitution d'un langage scientifique* (Saint-Étienne, 1991).

Shea, William (ed.). *Revolutions in Science* (Canton, Mass., 1988).

Schneider, Hans-Georg. "The Fatherland of Chemistry: Early Nationalistic Currents in the Late Eighteenth-Century German Chemistry." *Ambix*, 36 (1989), 14–21.

Schneider, Wolfgang. *Lexicon Alchemistisch-pharmazeutischer Symbole* (Weinbeim, 1962).

Sgard, Jean. *Corpus Condillac* (Geneva, 1981).

——— (ed.). *Condillac et les problèmes du langage* (Geneva, 1982).

Siegfried, Robert. "Lavoisier's View of Gaseous State and Its Early Application to Pneumatic Chemistry." *Isis*, 63 (1972), 59–78.

Slaughter, M.M. *Universal Languages and Scientific Taxonomy in the Seventeenth Century* (Cambridge, 1972).

Smeaton, William A. "The Contributions of P.-J. Macquer, T.O. Bergman and L.B. Guyton de Morveau to the Reform of Chemical Nomenclature." *Annals of Science*, 10 (1954), 87–103.

———. "L.B. Guyton de Morveau (1737–1816)." *Ambix*, 6 (1957), 18–34.

———. *Fourcroy, Chemist and Revolutionary, 1755–1809* (London, 1962).

Stafleu, Frans A. *Linnaeus and the Linneans* (Utrecht, 1971).

Stearn, William T. "Species Plantarum and the Language of Botany." *Proceedings of the Linnean Society of London*, 165 (1955), 158–164.

Storost, Jurgen. "Zur Herausbildung der Grundsätze der modernen französischen Beachtung des philosophischen Einfluss von Condillac." *Beiträge zur Romanischen Philologie*, 11 (1972), 292–311.

Strube, Irene. "Die Phlogistonlehre Georg Ernst Stahls (1659–1734) in Ihrer historisches Bedeutung." *NTM*, 1 (1960), 27–51.

———. *Georg Ernst Stahl* (Leipzig, 1984).

Strube, Wilhelm. *Die Chemie und ihre Geschichte* (Berlin, 1974).

Taton, René (ed.). *Enseignement et diffusion des sciences en France au XVIIIe siècle.* Second Ed. (Paris, 1986).

Testi, Gino. *Dizionario di alchimia e di chimica antiquaria.* Second Ed. (Rome, 1980).

Tugnoli-Pattaro, Sandra. *La Teoria del flogisto: Alle origini della rivoluzione chimica* (Bologna, 1983).

Vartanian, Aram. *Diderot and Descartes: A Study of Scientific Naturalism in the Enlightenment* (Princeton, 1953).

Weyer, J. *Chemiegeschichtsschreiubung von Wiegleb (1790) bis Partington (1970)* (Hildesheim, 1975).

Woolf, H. (ed.). *The Analytic Spirit* (Ithaca, N.Y., 1981).

Worth Estes, J. *Dictionary of Protopharmacology. Theurapeutic Practices, 1700–1850* (Canton, Mass., 1990).

Yates, F.A. "The Hermetic Tradition in Renaissance Science." In C.E. Singleton (ed.). *Art, Science and History in the Renaissance* (Baltimore, 1968), 255–274.

Index of Names

Aarsleff, Hans, 46*n*
Abbri, Ferdinando, 99*n*, 160, 176*n*–177, 181*n*, 204–206*n*, 223*n*, 234*n*, 241*n*, 250*n*, 254*n*, 309*n*, 311*n*
Acton, H.B., 324*n*
Adanson, Michel, 55–56, 59
Adet, Pierre Auguste, 184–185, 187, 216, 219, 244, 283, 311, 364–365
Afzelius, Johan, 318
Agricola, Georgius, 28, 79–90, 92–95, 106–107, 112, 114, 121–122, 135*n*, 275, 346
Ahlers, William C., 134*n*, 136*n*, 149*n*
Albertus Magnus, 16, 83
Albury, William R., 3*n*, 65–68*n*, 154*n*, 161*n*, 176*n*, 179*n*–180, 188*n*, 194*n*, 198*n*, 202*n*, 210*n*, 219*n*–220*n*, 257*n*, 263*n*, 324*n*–325*n*
Aldrovandi, Ulisse, 93*n*
Alembert, Jean le Rond d', 2, 5–7, 35, 37, 129, 189, 248–249, 252
Altieri Biagi, Maria Luisa, 46*n*
Amoretti, Carlo, 313*n*
Ampère, André Marie, 288, 290
Anderson, G.W., 188*n*
Anderson R. and C. Lawrence, 174*n*
Aréjula, Juan Manuel de, 225–227, 283–284, 291, 315
Argenville, Antoine Joseph Dezallier d', 59
Aristotle, 78, 82, 107, 110, 345
Arnald of Villanova, 16
Avogadro, Amedeo, 367

Babington, William, 254
Bacon, Francis, 2, 5–6, 8, 173, 193*n*, 196
Bacon, Roger, 16
Bailly, Jean Sylvain, 12–13, 38–39*n*, 62, 236
Baker, Keith Michael, 9*n*
Banks, Joseph, 59
Barlet, Annibal, 337–338, 340–341
Barsanti, Giulio, 50*n*, 52–53*n*
Batteux, Charles, 189
Baumé, Antoine, 137, 141, 166, 181, 187, 214–215, 218, 222, 287
Bayen, Pierre, 59, 131
Bayle, Pierre, 33

Becher, Johann Joachim, 18, 22, 125–126, 128, 217, 234, 346–349
Beddoes, Thomas, 229
Bensaude Vincent, Bernadette, 132*n*, 160–163*n*, 166*n*, 168*n*, 279–281, 283*n*
Bentham, Jeremy, 326
Berg, Fredrik, 63*n*
Bergman, Torbern, 15–16*n*, 22*n*, 90, 100–103, 106, 132, 137, 138–150, 153, 155–156, 181, 202–203, 205, 211–212, 227, 229, 234*n*, 239–240, 253, 262, 273, 297, 299, 307, 318, 320–321, 355–357, 359–361, 367
Berkeley, George, 34*n*
Berthelot, Marcelin, 106*n*, 168*n*, 247–248, 255, 259*n*, 263*n*, 330*n*, 332*n*–335*n*
Berthollet, Claude Louis, 151, 184, 186*n*–187, 211, 216, 226–230, 239, 244, 282, 284–286, 288, 292, 309–310*n*, 313, 364
Berzelius, Jöns Jacob, 103, 269, 321, 323, 363, 367
Beyern, August von, 92
Bigourdan, Camille Guillaume, 38*n*
Birembaut, Arthur, 167*n*
Biringuccio, Vannoccio, 28, 81–83, 86, 107
Black, Joseph, 22, 148, 167, 171–172, 174, 231–233, 236, 239, 240, 250, 253, 263, 292–293, 312, 357–358
Blagden, Charles, 224
Boerhaave, Hermann, 13, 16, 22, 61*n*, 122–123, 352
Bolton, H.C., 332*n*
Bonnet, Charles, 189*n*, 205, 256
Boodt, Anselmus Boëtius de, 93*n*, 95
Borch, Oluf, 14–15, 17
Borda, Jean Charles de, 38–40*n*
Bosc, Louis, 59
Bouchard, Georges, 149*n*, 185
Boulduc, Gilles François, 131
Bourdelin, Claude Louis, 15, 131
Bourguet, David Ludwig, 308
Boyle, Robert, 18, 22, 114–119, 121–122, 135*n*, 193, 204
Brahe, Tycho, 4, 6
Brisson, Mathurin Jacques, 282
Brosses, Charles de, 189
Broussonet, Pierre Marie Auguste, 58
Brugnatelli, Luigi Valentino, 283, 286, 310*n*–315
Bruno, Giordano, 107
Brunschvicg, Léon, 1, 267*n*
Bueno, Pedro Gutiérrez, 225, 227, 315–316
Buffon, Georges Louis Leclerc, 36–38*n*, 53–56, 58–60, 70–71, 104, 106*n*, 129, 157, 189
Buhle, J.G., 11*n*
Burlingame, Leslie J., 286*n*

Cabanis, Pierre, 70–71, 257, 288
Cadet de Gassicourt, Charles Louis, 187, 214–215, 218, 222
Cadet de Vaux, Antoine Alexis, 254
Caesalpinus, Andreas, 51, 93*n*
Cage, A.T., 59*n*
Cago, Ramon, 225*n*, 315–316*n*
Cago, Ramon and J.L. Carrillo, 225*n*
Caillet, 250
Calloud, Pietro, 310–311
Calvin, 233
Carburi, Marco, 314–315
Carozzi, Albert V., 104*n*
Carradori, Gioacchino, 314
Cassirer, Ernst, 31*n*, 327, 332*n*
Caven, R.M. and J.A.Cranston, 332*n*
Cavendish, Henry, 22, 181, 208, 224–225, 231, 292, 362
Caventou, Jean Bienaimé, 288
Chaptal, Jean Antoine Claude, 183, 236–237, 252, 256, 258, 282–283, 297, 313
Charpentier, Johann, 92*n*

Chénier, Marie Joseph, 325
Chevreul, Eugène, 288
Cicero, 239
Clairaut, Alexis Claude, 198*n*
Clericuzio, Antonio, 115, 117*n*–118*n*
Clow, Archibald and Nan Clow, 90*n*
Cohen, I. Bernard, 12*n*, 246, 248, 252*n*
Colonna, Francesco, 337
Comte, Auguste, 325–326
Condillac, Etienne Bonnot de, 2–5, 7–8, 30, 36–37, 66, 68, 71, 153–154, 180, 188–192, 194–203, 206, 219–220, 237, 240, 257, 259–262, 279–281, 288, 324–325, 364
Condorcet, Marie Jean Antoine Nicolas Caritat, marquis de, 2, 9, 38–41, 58, 65, 189, 207, 256–257
Conring, Hermann, 14
Copernicus, Nicolas, 4
Corneille, Pierre, 6
Corsi, G. conte di Viano, 314
Corsi, Pietro, 56–57*n*
Coulomb, Charles Augustin, 62, 282
Cousin, Jacques Antoine Joseph, 183*n*, 236
Cramer, Johannes Andreas, 92
Crell, Lorenz, 157, 235, 238*n*, 301, 306–307
Croll, Oswald, 107–109, 111
Cronstedt, Axel Fredrik, 99–101, 318
Crosland, Maurice P., 15*n*, 38, 40*n*, 102*n*, 106*n*, 134*n*, 137*n*, 147*n*, 152*n*–153*n*, 155*n*, 185, 188*n*, 193*n*, 210*n*–211*n*, 229, 231*n*–232*n*, 250*n*, 256*n*, 263*n*, 333*n*, 360*n*
Crowther, J.G., 171*n*
Cullen, Edmund, 142*n*
Cullen, William, 68, 193*n*
Curie, James, 235
Cuvier, Georges, 57*n*–58*n*, 245, 256, 285–286, 323, 329

Dagognet, François, 50*n*, 70*n*, 188*n*, 263*n*
Dal Pra, Mario, 189*n*
Dalton, John, 323, 367
Dandolo, Vincenzo, 311–312, 315
Dante Alighieri, 73
Darcet, Jean, 62, 185, 187, 214–215, 218, 222
Darwin, Erasmus, 289
Daubenton, Louis Jean Marie, 59, 62
Daudin, Henri, 50*n*, 56*n*
Daumas, Maurice, 160*n*, 166*n*, 168*n*, 178*n*–179*n*, 185*n*, 207*n*, 219*n*, 258, 260, 274*n*, 279
Davy, Humphry, 268, 281*n*, 299, 323
Debus, Allen G., 75*n*, 331*n*
Dégerando, Joseph Marie, 324–325
De Jong, H.M.E., 331*n*, 339*n*
Delambre, Jean Baptiste Joseph, 38, 40, 42, 256
Deloche Bernard and J.M. Leniaud, 41*n*, 43*n*
Demachy, Jacques François, 287
Descartes, René, 3, 6, 8, 29, 31–32, 36–37, 47, 111, 117, 143, 190, 192, 194, 196–198, 281
Desfontaines, René Louiche, 59
Destutt de Tracy, Antoine, 324
Dhombres, Jean, 200
Dhombres, Nicole and J. Dhombres, 324
Dickson, Stephen, 252, 283, 286, 293–296, 298, 312
Diderot, Denis, 1, 18–19, 27, 34, 47–49, 71, 130, 152, 188–189, 240, 249, 352, 354
Dietrich, Philippe Frédéric, baron de, 282
Dioscorides, 79
Dobbs, Betty J.T., 331*n*
Donadei, 316

Donovan, Arthur, 166, 171*n*–172*n*, 263*n*
Dorn, Gerard, 107
Drake, Stillman, 276*n*
Duhamel, Jean Pierre François Guillot, 59, 205
Duhem, Pierre, 124*n*, 126
Du Marsais, C., 189
Dumas, Charles Louis, 69–71, 288
Dumas, J.B., 363*n*
Duméril, Constant, 68, 288
Durande, Jean François, 150
Du Pré, 314
Durkheim, Émile, 1
Duveen, I. Denis, 337–338*n*
Duveen, I. Denis and H. Klickstein, 184–186*n*, 220, 235*n*, 251*n*, 258*n*, 301*n*

Ehrard, Jean, 31*n*
Eichorn, Johann Gottfried, 11
Eimbke, Georg, 304
Elvius, Pehr, 11*n*
Engelhardt, Dietrich von, 10*n*
Engels, Friedrich, 159
Engeström, Gustaf von, 99*n*
Ekeberg, Anders Gustaf, 320–321
Entzelt, Christopher, 90
Epicurus, 34
Erasmus, 79, 233
Ercker, Lazarus, 90
Eriksson, Gunnar, 50*n*, 52*n*
Erxleben, Johann Christian Polykarp, 352
Euclid, 135
Euler, Leonard, 189, 198*n*, 200

Fabbroni, Giovanni, 175*n*, 234, 314–315
Faraday, Michael, 323
Faujas de Saint Fond, Barthélémi, 59
Fayet, J., 38*n*
Ferner, Jacob, 92*n*
Ferrone, Vincenzo, 10*n*
Filgueiras, Carlos L., 317*n*
Fontana, Felice, 157, 256, 309, 314–315
Fontenelle, Bernard le Bovier de, 248
Forbes, R.J., 331*n*
Formigari, Lia, 46*n*, 197*n*
Fortin, Nicolas, 274
Foucault, Michel, 188*n*
Fourcroy, Antoine François, 15–16*n*, 18, 22–24, 59, 157, 184, 187, 206, 208, 212–213, 216, 225, 239, 244, 254–255, 258, 282–283, 286, 297, 307, 310*n*, 313, 319, 364
Frängsmyr, Tore, 29*n*, 38*n*, 50*n*
Franklin, Benjamin, 165, 218, 235, 250, 301
Freind, John, 13
Fric, René, 168*n*–169*n*, 251*n*

Gadolin, Johan, 227–228, 283, 318–319, 321
Galen, 79
Galilei, Galileo, 4, 6, 8, 29, 133, 275–276, 278
Gassendi, Pierre, 33
Garat, Dominique J., 257, 325
Gay-Lussac, Louis Joseph, 268, 323
Geber, 86*n*
Gellert, Chriestlieb Ehrgott, 99, 241
Geoffroy, Etienne François, 15–16, 131, 141, 355
Geoffroy, Etienne Louis, 59
Gérard, F., 313
Gesner, Conrad, 51
Geymonat, Ludovico, 7*n*
Gille, Bertrand, 90*n*
Gillispie, Charles C., 62*n*, 178*n*, 188*n*, 209*n*, 363*n*–364*n*
Giobert, Giovanni Antonio, 309, 311–312, 314–315
Giorgi, Ferdinando, 314
Gilson, Etienne, 47*n*
Girtanner, Christoph, 254, 302–304, 319
Glaser, Cristopher, 112–114, 116

Gmelin, Johann Friedrich, 11*n*, 15, 234, 307, 314*n*
Gnudi, Martha T., 81*n*
Goethe, Johann Wolfgang, 27
Goguet, Antoine Yves, 13*n*
Golinski, Jan, 115*n*
Goltz, Dietlinde, 74*n*
Gordon, Cosmo Alexander, 33*n*–34*n*
Göttling, Johann Friedrich August, 301–304, 306, 308, 313
Gottlieb, Johann, 92*n*
Gouan, Antoine, 59
Gough, J.B., 160*n*, 167–168*n*, 247*n*
Goulin, Jean, 11*n*
Goupil, Michelle, 142*n*, 148*n*, 230*n*, 361*n*
Gouveia, A.J.A., 317*n*
Greene, John C., 235*n*
Grégoire, Henri, 43
Gren, Friedrich Albert Carl, 234, 253, 304–305
Grillot de Givry, E.A., 332, 343*n*–344*n*
Grimaux, Edouard, 160–161, 168*n*, 178*n*, 205*n*, 252*n*, 259*n*, 263*n*, 363*n*
Guareschi, Icilio, 309*n*, 314*n*
Guerlac, Henry, 160, 163, 167*n*–169*n*, 171*n*–172, 184, 205*n*, 250*n*, 254*n*, 259*n*, 263*n*
Guerci, L., 4*n*
Guettard, Jean Etienne, 90, 160, 163, 205–206, 363
Guillaume, J., 41*n*
Guntau, Martin, 104*n*
Gusdorf, Georges, 31*n*
Guyton de Morveau, Louis Bernard, 21–22, 41, 59, 102, 106, 142, 147–157, 180–181, 184–187, 202–203, 205–212, 219, 230–231, 238–241, 243*n*–244, 258, 262, 273, 283–286, 310*n*, 313, 360–362, 364

Hadzsits, George D., 33*n*
Hagen, Karl Gottfried, 307
Hales, Stephen, 22, 167
Hall. A.R. and M. Boas Hall, 117*n*
Hall, Marie Boas, 115*n*
Halleux, Robert, 14*n*, 77, 79*n*–80, 87, 332
Hankins, T., 6*n*
Hannaway, Owen, 107–108*n*, 112
Harris, John, 124*n*
Hartsoeker, Nicolas, 122
Hassenfratz, Jean Henri, 59, 184–185, 187, 216, 219, 244, 283, 311, 364–365
Haüy, René Just, 38, 59, 103, 282, 313
Hazard, Paul, 31*n*–32*n*
Heilbron, John, 29*n*, 38*n*, 165*n*
Heilbronner, Johann C., 10*n*
Heller, John Lewis, 50*n*
Hellot, Jean, 93, 131, 133, 166–167
Helmont, Joan Baptista van, 16, 18, 73, 110–111, 115, 266
Henckel, Johann Friedrich, 99, 300
Henry, Thomas, 175*n*
Herder, Johann Gottfried, 30, 189
Hermbstädt, Sigismund Friedrich, 234, 303, 308, 319
Hermes Trismegistus, 14*n*, 76, 86*n*
Heym, Gerard, 333*n*
Hippocrates, 68
Hobbes, Thomas, 192–193*n*, 197
Holbach, Henry Thiry Paul, baron d', 27, 33–34, 37, 131, 167*n*
Holmes, Frederic Lawrence, 15*n*, 132*n*, 168*n*, 247*n*, 278
Holmyard, E.J., 333*n*, 335*n*
Homberg, Wilhelm, 15–16, 122, 131
Hoover, H.C., 79*n*, 85*n*, 87*n*
Hopson, Charles Rivington, 291
Huard, Pierre, 61*n*
Hufbauer, Karl, 234*n*, 300*n*–301*n*
Hulsius, Levinus, 89

Imperato, Ferrante, 93*n*

Jars, Gabriel, 92*n*
Jefferson, Thomas, 235

Jensen, William B., 100*n*
Jones, Peter, 63*n*
Jonson, Ben, 345–346*n*
Jordanova, Ludmilla, 31*n*, 324*n*
Julia, Dominique, 45–46*n*
Juncker, Johann, 300
Jung, Carl Gustav, 77*n*, 106*n*–107*n*, 332–333*n*, 335
Jussieu, Antoine Laurent de, 55, 57–58
Jussieu, Bernard de, 60, 160, 163, 205–206

Kahlbaum, G.W. and A. Hoffmann, 300*n*, 305*n*–306*n*
Kant, Immanuel, 133, 256
Kästner, Abraham Gotthelf, 11*n*, 13*n*
Keill, John, 13
Keir, James, 252, 256, 289–292, 298–299*n*
Kemp Smith, Norman, 133*n*
Kepler, Johann, 4,6, 8, 29
Kerr, Robert, 209*n*, 254
King, Edward, 237–238
Kircher, Athanasius, 46
Kirwan, Richard, 142, 147, 157, 181, 208, 229–231, 239–244, 252, 256, 262, 286, 292–296, 298–299, 362
Knight, Isabel, 4*n*, 189*n*
Knowlson, James, 46*n*
Koyré, Alexandre, 9*n*, 107, 275–276*n*, 332, 335*n*
Kristeller, Paul Oskar, 107
Kuhn, Thomas S., 29*n*, 115*n*
Kula, Witold, 38*n*

Lacaille, Nicolas Louis de, 160, 200
Lacépède, Bernard de la Ville sur Illon, comte de, 59
Lacretelle, Pierre Joseph, 189*n*
La Grange, 34
Lagrange, Joseph Louis de, 38–40*n*, 198*n*–200, 303
Lamarck, Jean Baptiste de Monnet, chevalier de, 55–59, 286–287
Lambert, J., 188*n*
La Métherie, Jean Claude de, 59, 222–223, 282–283, 287, 301
Lampadius, Willhelm August, 307
Landriani, Marsilio, 157, 231–233, 309–312, 315, 362
Laplace, Pierre Simon, 38–39*n*, 62, 183*n*, 208, 236, 244, 256, 313–314
La Planche, Charles Louis de, 161–162
Larson, James L., 50*n*, 52
Lavoisier, Antoine Laurent, 3*n*, 21, 23–24, 38–41*n*, 59, 62, 65, 68, 90, 125, 131–133, 154*n*, 169–188, 194, 200–210, 212–229, 231–244, 246–247, 249–289, 291–302, 306–307, 309, 311–318, 323–327, 363–367
Lavoisier, Marie Anne Pierrette Paulze, Madame, 159, 231, 235, 237, 244, 277
Leclerc, Daniel, 10*n*
Le Febvre, Nicaise, 111, 116
Leibniz, Gottfried Wilhelm, 36, 46, 196
Lemery, Nicolas, 13, 15, 120–122, 218, 348, 350–352, 367
Lenoble, Robert, 1, 31
Leonhardi, Johann Gottfried, 157, 234, 307
Libavius, Andreas, 112–113, 116
Lichtenberg, Georg Christoph, 12, 301
Liebig, Justus von, 330–331
Lindqvist, Svante, 90*n*
Linnaeus, Carolus, 10*n*, 35, 50–56, 58–61, 68, 71, 95–96, 98, 100, 102–104, 106, 148–149, 204–206, 234*n*
Lindroth, Sten, 11*n*, 90*n*
Llana, James W., 205*n*
Locke, John, 3, 5, 8, 47, 188, 192–196, 198, 223, 240
Löhneyss, Georg Engelhard von, 90–91

Long, Pamela O., 79*n*, 89*n*
Lovejoy, Arthur O., 31*n*
Lubbock, Richard, 229
Lucretius, 33–34, 73
Lüdy Tenger, Fritz, 332*n*
Lull, Ramon, 86*n*
Lundgren, Anders, 90*n*, 142*n*, 317*n*
Luther, Martin, 233

Macquer, Pierre Joseph, 13*n*, 15–16*n*, 19–20, 22, 93, 131–139, 141, 145, 149–150, 157, 165–166, 18–181, 212, 239, 273, 289, 299, 360–362, 367
Macri, 314
Madison, James, 235
Maier, Michael, 339, 342–342
Malebranche, R.P. Nicolas, 197
Manget, Jean Jacques, 14*n*
Manuel, Frank E., 7*n*–9*n*
Maret, Huges, 150
Marggraf, Andreas Sigismund, 101, 239
Marino, Luigi, 11*n*
Marivetz, Etienne Clément, 223
Markus, Thomas, 63*n*
Marum, Martinus van, 184
Martyn, Thomas, 59
Mascheroni, Lorenzo, 313
Maupertuis, Pierre Louis Moreau de, 30, 189
Mayow, John, 22
McEvoy, J.G., 176*n*
McEvoy, J.G. and J.E. McGuire, 174*n*
McKie, Douglas, 183*n*, 259*n*, 263*n*, 333*n*
Méchain, Pierre François André, 38, 40*n*
Meek, Ronald, 7*n*
Mégnié, Pierre Bernard, 274
Meidinger, Karl Freiherr von, 304, 308
Meinel, Christoph, 14*n*
Meldrum, Andrew N., 160, 167–168, 171*n*, 176–177, 247, 249, 259*n*
Melhado, Evan M., 95*n*, 103*n*, 247*n*
Mercati, Michele, 93*n*, 95–97
Merzari, G.B., 314
Metzger, Hélène, 77, 99*n*, 103*n*, 110*n*–111, 124*n*, 201*n*, 247, 263*n*, 267, 332
Meurer, Wolfgang, 83*n*
Meusnier de la Place, Jean Baptiste Marie Charles, 183*n*, 185, 274
Meyerson, Émile, 118*n*, 276*n*–277*n*, 327
Michaëlis, Rudolf, 89*n*
Micheli, Gianni, 7*n*, 31*n*
Mieli, Aldo, 247
Mill, John Stuart, 245, 327
Millin, Aubin Louis, 59–60, 288
Molière, Jean Baptiste Pouquelin de, 6, 33
Monboddo, James Burnett, 30, 189
Monge, Gaspard, 38–39*n*, 183*n*, 236, 244, 256
Monnet, Antoine Grimoald, 59, 287
Mons, Jean Baptiste van, 282, 307, 313
Montucla, Jean Etienne, 13*n*
Moravia, Sergio, 65*n*, 67*n*, 70*n*
Moscati, Pietro, 256, 313
Multhauf, Robert P., 75*n*
Mumford, Lewis, 78*n*
Musschenbroek, Petrus van, 164–165

Naigeon, Jean, 34, 189
Napoleon I, 314, 324
Naville, Pierre, 34*n*
Nazari, G.B., 336–337
Newton, Isaac, 4, 6, 8, 18, 29, 32, 141–144, 151, 165, 198*n*, 281*n*
Nicholson, William, 297–298, 359–360*n*, 366
Nietzki, Adam, 68
Nollet, Jean Antoine, 165–166, 218, 256
Neumann, Caspar, 300, 307

Oersted, J.C., 321–322
Olivier, 59

Olschki, Leo, 46*n*, 74*n*
Olsson, Hugo, 90*n*, 137*n*, 317*n*
Opoix, Christophe, 287
Ostwald, Wilhelm, 247

Pagel, Walter, 107*n*, 110*n*
Pandolfus Anglus, 86*n*
Paracelsus, Theophrastus Philippus Aureolus Bombastus von Hohenheim, 16, 18, 75, 77, 79, 87, 106–108, 113, 126
Parmentier, Antoine Augustin, 59, 282
Partington, James Riddik, 90*n*, 125*n*, 171*n*–172*n*, 185*n*, 259*n*, 263*n*, 307*n*, 333*n*, 337*n*
Patrizi, Francesco, 107
Pearson, George, 298
Peart, Edward, 296–297
Pelletier, Bertrand, 59, 316
Perrin, C.E., 168*n*, 181*n*, 183*n*, 247*n*, 254*n*, 282*n*, 311*n*
Philalethes, 345
Picardet, Claudine, 148*n*
Pictet, Marc Auguste, 254
Pinel, Philippe, 68–69, 71, 288
Pini, Ermenegildo, 314
Pliny the Elder, 79, 83–84, 86, 240
Poppe, Johann Heinrich Moritz von, 11*n*
Poppius, Hamerius, 128*n*
Pott, Johann Heinrich, 99, 101, 300
Poyet, Bernard, 63–64
Prescher, Hans, 79*n*, 89*n*
Priestley, Joseph, 23, 143, 167, 171–178, 181, 193, 202, 207–208, 231, 238–239, 242, 252–253, 257, 263, 278, 310, 362
Prieur de la Côte d'Or, Claude Antoine, 41, 288*n*
Proust, Louis, 316

Rabelais, François, 61
Racine, Jean, 6
Ramsay, William, 171*n*–172*n*
Rappaport, Rhoda, 130*n*, 205*n*
Rashed, R., 58*n*
Read, John, 330*n*, 343*n*
Reboul, Henri Paul, 250
Redgrove, H. Stanley, 330*n*
Regnault, Noel, 10*n*
Remler, Johann Christian, 304
Rey, Jean, 21–22
Richter, Jeremias Benjamin, 307–308
Rider, Robin, 29*n*, 38*n*, 324*n*
Rinman, Sven, 92–93*n*, 318, 321
Robespierre, Maximilien, 43
Robinet, André, 196*n*
Roger, Jacques, 31*n*
Romé de l'Isle, Jean Baptiste de, 59, 103, 105
Rössel, Balthasar, 92
Rossi, Paolo, 46–47*n*, 75*n*, 99*n*, 197*n*
Rothe, Gottfried, 300
Rouelle, Guillaume François, 18, 130–132, 136, 138, 145, 149*n*, 151, 161–163, 166, 212, 286, 357, 359, 367
Rousseau, Jean Jacques, 27, 30, 50, 131, 188–189
Rousseau, Nicolas, 189*n*
Rozier, François, 222
Ruland, Martin, 109–110*n*, 124*n*, 352

Sagar, Jean Baptiste Michael, 68
Sage, Balthazar Georges, 59, 187, 214–215, 218, 222–223, 287–288, 313
Saltonsall, 312
Santi, Giorgio, 254, 314
Santinello, Giovanni, 11*n*
Sarton, George, 75
Saussure, Horace Bénédicte de, 237
Sauvages, François Boissier de, 63, 68
Savérien, Alexandre, 13*n*
Scheele, Carl Wilhelm, 181, 207, 211, 239, 278, 307
Scheffer, Henrik Theophilus, 139, 141
Scherer, Johann Andreas, 304
Schindler, C.C., 90

Schlüter, Christoph Andreas, 92–93
Schofield, Robert E., 172*n*, 175*n*
Schmitt, Charles B., 78
Schneider, Hans Georg, 307*n*–308*n*
Schwab, R.N., 5*n*
Scopoli, Giovanni Antonio, 234
Seabra Telles, Vicente de, 316–317
Secretain, Claude, 130*n*
Seguin, Armand, 59, 282
Selle, Christian Gottlieb, 68
Senac, Jean Baptiste, 15–16
Senebier, Jean, 233*n*
Sévrin, L.J., 288
Sgard, Jean, 188*n*–189*n*
Shaw, Peter, 127
Sherwood Taylor, F., 333*n*
Shklar, J., 6*n*
Siegfried, Robert, 160*n*, 166*n*, 168*n*
Slaughter, M.M., 46*n*
Sloan, Philip R., 50*n*
Smeaton, William A., 102*n*, 134*n*, 137, 147*n*, 149*n*, 152*n*, 154*n*, 185*n*, 360*n*
Smith, Adam, 189
Smith, Cyril S., 81*n*
Smith, Edgar F., 251*n*
Smith, James Edward, 59
Smith, Thomas P., 254
Sommerhoff, Johann Christoph, 124*n*, 352–353
Spallanzani, Lazzaro, 256, 312–313, 315
Sparrman, Anders, 319–321
Spinoza, Baruch, 36, 117–118, 135*n*
Sprengel, Kurt Polykarp Joachim, 13*n*
Stafleu, Frans A., 50*n*, 52*n*, 55, 57*n*
Stahl, Georg Ernst, 13, 16, 18, 22, 124–131, 133, 150, 163–165, 167–169, 171, 181–182, 207, 209, 214, 217–218, 234, 238–239, 241, 252, 273, 292, 299–301, 307, 314
Staum, Martin S., 70*n*
Stearn, William T., 50*n*
Stephanus of Alexandria, 335
Stewart, Dugald, 326
Stillman, John Maxon, 129*n*
St. John, James, 187*n*, 228, 289, 300, 302, 365*n*
Stratico, Antonio, 314
Strube, Irene, 124*n*
Swift, Jonathan, 299*n*

Tachenius, Otto, 16
Talleyrand Périgord, Charles Maurice de, 49
Telesio, Bernardino, 107
Tenon, Jacques René, 62–63
Testi, G., 309*n*
Thenard, Louis Jacques, 268, 288, 323
Theophrastus, 78, 84, 86*n*
Thorndike, Lynn, 75, 81
Thouin, Jean André, 59
Thurneisser, Leonhart, 337–339
Tillet, Mathieu, 38, 40*n*
Titley, A.F., 333*n*
Todéricu, Doru, 133*n*
Tournefort, Joseph Pitton de, 51
Trommsdorff, Johann Bartholomäus, 307–308
Trudaine de Montigny, Jean Charles Philibert, 167
Tugnoli Pattaro, Sandra, 124*n*, 129*n*
Turgot, Anne Robert Jacques, 2, 7–8, 15, 131, 240
Tychsen, N., 319

Vandermonde, Alexandre Théophile, 183*n*, 236
Vartanian, Aram, 31*n*–32*n*
Vauquelin, Louis Nicolas, 59, 283, 313
Venel, Gabriel François, 17, 19, 130, 135, 145, 290
Vicq d'Azyr, Félix, 65–68, 71, 256, 288
Volta, Alessandro, 256, 309, 312–313, 315
Voltaire, François Marie Arouet de, 30, 32

Wächter, Eberhard, 90*n*
Wagenbreth, Otfried, 90*n*
Waite, Arthur Edward, 345*n*
Walden, Paul, 332*n*
Wall, Martin, 329
Wallerius, Johan Gottschalk, 90, 96, 104, 318, 352
Waltersdorff, Johann Lucas, 98
Watt, James, 232
Weidler, Johann F., 10*n*
Werner, Abraham Gottlob, 104
Westrumb, Johann Friedrich, 286, 306, 308, 312
Weyer, Jost, 14*n*, 20*n*
Whewell, William, 50*n*, 323, 326–327
White, Robert, 292
Wiegleb, Johann Christian, 234, 307
Wilkins, John, 46, 291
Wolff, Friedrich, 244

Yates, F.A., 75*n*

Zeisig, Johann Caspar, 93*n*–94
Zosimos, 86*n*, 335
Züecker, Johann Friedrich, 92*n*
Zuretti, C.O., 333*n*

Alphabetum Honorij Thebani, quo magitam scribebat, autore Petro

a b c d e f g h i k l m n o p q r

Alphabetum Hichi Francorum natis prisci vath

a b c d e f g h i k l m m n o p ph p

Alphabetum secretum Iaimelis Megalopij regis sapien

b c d e f g h i k l m n o p q r

Alphabetum Nortmannorum, autore D. Beda Presbytero

a b c d e f g h i k l m n o p q r

Alphabetum secretum Iohannis Tritenhemij Abbatis Span

a b c d e f g h i k l m n o p q r s

Alphabetum Vtopiense, teste doctiss. viro Thoma Mauro D. Erasmi Ro

a b c d e f g h i k l m n o p q

Alphabetum Caroli Magni. Teste Alcunio, Gallo, Philosoph.

a b c d e f g h i k l m n o p q r

Alphabetum exploratum Pharamundi regis Francorum pe

a b c d e f g h i k l m n o p q r s

Alphabetum facillimum, secundum anonymum, Trittenhemi

a b c d e f g h i k l m n o p q r

Alphabetum cuiusdam summi Alchimistæ, autore Sphanh

a b c d e f g h i k l m n o p q r

Alphabetum Dorari q̃ Franci vsi sunt, Hunibaldo Historiogr

a b g d e ; f h i k l m m fin. n n fin. o p ph

Alphabetum Vastualdi, qui acta Regum, Ducum & principum per annos Sept hisce notis primo sermone scripsit autore Hunibaldo.

a b g d e z e th i k l m n x o p r

Alphabetum D. Otfridi Monachi VVissenburgensis. Rabani Fulde

a b c d e f g h i k l m n o p q r s